ALL · IN · ONE

CDPSE™

Certified Data Privacy Solution Engineer

EXAM GUIDE

ABOUT THE AUTHOR

Peter H. Gregory, CDPSE, CIPM, CISA, CISM, CRISC, CISSP, DRCE, CCSK, is a 30-year career technologist and a security leader at a telecommunications company. He has been developing and managing information security management programs since 2002 and has been leading the development and testing of secure IT environments since 1990. Peter has also spent many years as a software engineer and architect, systems engineer, network engineer, and security engineer. He has written many articles, white papers, user manuals, processes, and procedures throughout his career, and he has conducted numerous lectures, training classes, seminars, and university courses.

Peter is the author of more than 40 books about information security and technology, including *Solaris Security, CISM Certified Information Security Manager All-In-One Exam Guide,* and *CISA Certified Information Systems Auditor All-In-One Exam Guide*. He has spoken at numerous industry conferences, including RSA, Interop, (ISC)² Congress, ISACA CACS, SecureWorld Expo, West Coast Security Forum, IP3, Source, Society for Information Management, OptivCon, the Washington Technology Industry Association, and InfraGard.

Peter serves on advisory boards for cybersecurity education programs at the University of Washington and the University of South Florida. He was the lead instructor for nine years in the University of Washington certificate program in cybersecurity, a former board member of the Washington State chapter of InfraGard, and a founding member of the Pacific CISO Forum. Peter is a 2008 graduate of the FBI Citizens' Academy and a member of the FBI Citizens' Academy Alumni Association.

Peter resides with his family in Washington state. For more information about Peter, visit his web site at www.peterhgregory.com.

About the Technical Editor

John Clark, CISSP, CISA, CISM, CIPP/E, CIPT, FIP, is an information security executive advisor to CISOs, CIOs, boardrooms, and business executives. John has contributed to many articles, blogs, and presentations addressing privacy program management and has spoken on the topic at industry conferences. With more than 20 years of experience in information security and privacy, he has developed a passion for working with clients to develop sustainable business-aligned information security and privacy management programs that can be applied to emerging regulations with minimal change. In addition to having multiple industry certifications, John has a bachelor's degree in management information systems and an MBA from the University of Houston.

CDPSE™

Certified Data Privacy Solution Engineer

EXAM GUIDE

Peter H. Gregory

New York   Chicago   San Francisco
Athens   London   Madrid   Mexico City
Milan   New Delhi   Singapore   Sydney   Toronto

CDPSE™ Certified Data Privacy Solution Engineer All-in-One Exam Guide

1 2 3 4 5 6 7 8 9   LCR   24 23 22 21

Library of Congress Control Number: 2020945526

ISBN   978-1-260-47482-4
MHID   1-260-47482-8

Sponsoring Editors	**Technical Editor**	**Production Supervisor**
Emily Walters	John Clark	Thomas Somers
Wendy Rinaldi	**Copy Editor**	**Composition**
Editorial Supervisor	Lisa Theobald	KnowledgeWorks Global Ltd.
Janet Walden	**Proofreader**	**Illustration**
Project Managers	Rick Camp	KnowledgeWorks Global Ltd.
Sarika Gupta and Neelu Sahu,	**Indexer**	**Art Director, Cover**
KnowledgeWorks Global Ltd.	Ted Laux	Jeff Weeks

To current and aspiring privacy professionals everywhere who own the mission of protecting personal information about customers, employees, and constituents.

CONTENTS AT A GLANCE

CONTENTS

Figure Credits

Figure 1-2 Courtesy Xhienne: SWOT pt.svg, CC BY-SA 2.5, https://commons.wikimedia
.org/w/index.php?curid=2838770.

Figure 1-3 Courtesy *Hi-Tech Security Solutions* magazine.

Figure 3-2 Source: US National Institute for Standards and Technology (NIST).

Figure 5-1 Courtesy of Oxford University Press, Inc. From Christopher Alexander, et al.,
The Oregon Experiment, 1975, p. 44. Used by Permission of Oxford University Press, Inc.

ACKNOWLEDGMENTS

I am immensely grateful to Wendy Rinaldi for affirming the need to have this book published on a tight timeline. My readers, including current and future security managers, deserve nothing less.

Heartfelt thanks to Emily Walters for proficiently managing this project, facilitating rapid turnaround, and equipping me with the information and guidance I needed to produce the manuscript.

I want to thank my former consulting colleague, John Clark, who took on tech reviewing the manuscript. John, a Fellow of Information Privacy and a member of the International Association of Privacy Professionals, carefully and thoughtfully scrutinized the entire draft manuscript and made scores of useful suggestions that have improved the book's quality and value for readers.

Next, I want to thank my former consulting colleague, Greg Tyler, with whom I worked in a consulting role in data protection projects. His insight has been invaluable to our clients and to me.

Many thanks to Janet Walden, Sarika Gupta, and Neelu Sahu for managing the editorial and production ends of the project and to Lisa Theobald for copyediting the book and further improving readability. I appreciate KnowledgeWorks Global Ltd. for expertly rendering my sketches into beautifully clear line art and laying out the pages. Like stage performers, they make hard work look easy.

Many thanks to my literary agent, Carole Jelen, for diligent assistance during this and other projects. Sincere thanks to Rebecca Steele, my business manager and publicist, for her long-term vision and for keeping me on track.

Virtually all of the work producing this book was completed during the COVID-19 pandemic. In addition to life's normal pressures and challenges, everyone involved in this project stayed on task and completed their typically high-quality work on schedule. This effort was likely quite difficult for some of you. I admire your drive and your dedication to serve our readers with nothing but the best. Privacy professionals around the world depend upon it.

Despite having written more than 40 books, I have difficulty putting into words my gratitude for my wife, Rebekah, for tolerating my frequent absences (in the home office) while I developed the manuscript. This project could not have been completed without her loyal and unfailing support and encouragement.

INTRODUCTION

The information revolution has transformed businesses, governments, and people in profound ways. Virtually all business and government operations are now digital, resulting in everyone's personal details being stored in information systems.

Two issues have arisen out of this transformation: the challenge to safeguard personal information from criminal organizations, and the challenge to ensure that personal information is used only for clearly stated purposes. Difficulties in meeting these challenges have helped create and emphasize the importance of the cybersecurity and information privacy professions. Numerous security and privacy laws, regulations, and standards have been enacted and created, imposing a patchwork of new requirements on organizations and governments to enact specific practices to protect and control the use of our personal information.

These developments continue to drive demand for information privacy, information security professionals, and leaders in both privacy and security. These highly sought-after professionals play a crucial role in developing better information privacy and security programs that result in reduced risk and improved confidence.

The Certified Data Privacy Solutions Engineer (CDPSE™) certification, established by ISACA in 2020, will light the path for tens of thousands of privacy and security professionals who need to demonstrate competence in the privacy field. ISACA, the creator of the Certified Information Systems Auditor (CISA, established in 1978), the Certified Information Security Manager (CISM, established in 2002), and other certifications, is one of the world's leading security, privacy, and IT management and professional development organizations.

Purpose of This Book

Let's get the obvious out of the way: this is a comprehensive study guide for the privacy professional who needs a reliable reference for individual or group-led study for the Certified Data Privacy Solutions Engineer (CDPSE) certification. The content in this book contains the information that CDPSE candidates are required to know. This book is one source of information to help you prepare for the CDPSE exam but should not be thought of as the ultimate collection of *all* the knowledge and experience that ISACA expects qualified CDPSE candidates to possess. No one publication covers all of this information.

The CDPSE certification may be new, but the privacy profession is relatively mature, although changing rapidly with the onset of watershed privacy laws, including GDPR, CCPA, and CPRA. Further, ISACA has a long history of developing and managing the highest quality professional certifications in the information-processing industry, beginning with creating the CISA (Certification Information Systems Auditor) in 1978.

I have participated in item-writing workshops for ISACA for other certifications, and I can attest to the organization's unwavering attention to accuracy and quality. Hence, it is certain that the CDPSE certification will succeed.

This book also serves as a reference for aspiring and practicing privacy professionals and leaders. The content required to pass the CDPSE exam is the same content that practicing privacy professionals need to be familiar with in their day-to-day work. This book is an ideal CDPSE exam study guide as well as a desk reference for those who have already earned their CDPSE certification.

The pace of change in the privacy and information security industries and professions is high. Rather than contain every detail and nuance of every law, practice, standard, and technique in privacy and security, this book shows the reader how to stay current in the profession. Indeed, the pace of change is one of many reasons that ISACA and other associations require continuous learning to retain one's certifications. It is important to understand key facts and practices in privacy and how to stay current as they continue to change.

This book is also invaluable for privacy professionals who are not in a leadership position. You will gain considerable insight into today's privacy challenges. This book is also useful for IT, security, and business management professionals who work with privacy professionals and need a better understanding of what they are doing and why.

Finally, this book is an excellent guide for anyone exploring a career in privacy. The study chapters explain all the relevant technologies, techniques, and processes used to manage a modern privacy program. This is useful if you are wondering what the privacy profession is all about.

How This Book Is Organized

This book is logically divided into four major sections:

- **Introduction**   The front matter (this chapter) provides an overview of the CDPSE certification and the privacy profession.
- **CDPSE study material**   Chapters 1 through 8 contain everything a studying CDPSE candidate is responsible for. This same material is a handy desk reference for aspiring and practicing privacy professionals.
- **Glossary**   There are nearly 550 terms used in the privacy profession.
- **Practice exams**   The Appendix explains the online CDPSE practice exam and TotalTester software accompanying this book.

Information privacy is a big topic, and it depends heavily upon sound information security practices. Many security and audit topics are summarized in this book, and there are numerous references to two other books that offer considerable depth in information security and information systems audit:

- *CISM Certified Information Security Manager All-In-One Exam Guide*
- *CISA Certified Information Systems Auditor All-In-One Exam Guide*

Earning and Maintaining the CDPSE Certification

In this section, I'm going to talk about

- What it means to be a Certified Data Privacy Solutions Engineer (CDPSE) professional
- ISACA, its code of ethics, and its standards
- The certification process
- How to apply for the exam
- How to maintain your certification
- How to get the most from your CDPSE journey

Congratulations on choosing to become a Certified Data Privacy Solutions Engineer! Whether you have worked for several years in the field of privacy, or you have recently been introduced to the world of privacy and information security, don't underestimate the hard work and dedication required to obtain and maintain CDPSE certification. Although ambition and motivation are required, the rewards can far exceed the effort.

You may not have imagined you would find yourself working in the privacy world or looking to obtain a privacy certification. Perhaps the explosion of privacy laws led to your introduction to this field. Or possibly you have noticed that privacy-related career options are increasing exponentially, and you have decided to get ahead of the curve.

By selecting the CDPSE certification, you're hitching your wagon to the ISACA star. Founded in 1967, ISACA has more than 145,000 members and more than 100,000 professionals who have earned one or more of its certifications: CISA (established in 1978), CISM (established in 2002), CGEIT (established in 2007), and CRISC (established in 2010). These certifications have won industry awards too numerous to count. CDPSE will soon earn such awards. It's hard to find a professional certification organization with so many accolades. Welcome to the journey and the amazing opportunities that await you.

I have put together this information to help you further understand the commitment you'll need to prepare for the exam and to maintain your certification. It is my wish to see you pass the exam with flying colors, and in this book, I provide you with the information and resources you need to maintain your certification and to proudly represent yourself and the professional world of privacy with your new credentials.

If you're new to ISACA, I recommend you tour the organization's web site and become familiar with the available guides and resources. Also, if you're near one of the 200-plus local ISACA chapters in more than 80 countries, consider taking part in the activities and even reaching out to the chapter board for information on local meetings, training days, conferences, and study sessions. You may meet other privacy professionals who can give you additional insight into the CDPSE certification and the privacy profession.

CDPSE certification primarily focuses on privacy governance, technical architecture, and data privacy management operations. It certifies the individual's knowledge of establishing information privacy strategies, building and managing a privacy program, and preparing for and responding to privacy incidents, and their knowledge of information security. Organizations seek out qualified personnel for assistance with developing and maintaining strong and effective privacy programs. A CDPSE-certified individual is a great candidate for this.

Benefits of CDPSE Certification

Obtaining the CDPSE certification offers several significant benefits:

- **Expands knowledge and skills, and builds confidence** Developing knowledge and skills in privacy and data protection, building and managing a privacy program, and responding to privacy incidents can prepare you for advancement or expand your scope of responsibilities. The personal and professional achievement can boost confidence that encourages you to move forward and seek new career opportunities.

- **Increases marketability and career options** Because of various legal and regulatory requirements, such as the Health Insurance Portability and Accountability Act (HIPAA), Gramm-Leach-Bliley Act (GLBA), the European General Data Protection Regulation (GDPR), the California Consumer Privacy Act (CCPA), and the California Privacy Rights Act (CPRA), demand is growing for individuals with experience in developing and running privacy programs. Besides, obtaining your CDPSE certification demonstrates to current and potential employers your willingness and commitment to improve your knowledge and skills in privacy. Having a CDPSE certification can provide a competitive advantage and open up many opportunities in various industries and countries.

- **Meets employment requirements** Many government agencies and organizations are requiring certifications for positions involving privacy and information security. Although the CDPSE certification is still somewhat new, it's only a matter of time before government agencies and the privacy industry require a leading privacy certification for their privacy professionals.

- **Builds customer confidence and international credibility** Prospective customers needing privacy work will have faith that the quality of the strategies and execution are in line with internationally recognized practices and standards.

Regardless of your current position, demonstrating knowledge and experience in the areas of privacy can expand your career options. The certification does not limit you to privacy or privacy management; it can provide additional value and insight to those currently holding or seeking the following positions:

- Executives such as chief privacy officers (CPOs), data protection officers (DPOs), chief operating officers (COOs), chief financial officers (CFOs), chief compliance officers (CCOs), and chief information officers (CIOs)
- Records management executives and practitioners
- Marketing management executives and practitioners

- IT management executives such as chief information officers (CIOs), chief technology officers (CTOs), directors, managers, and staff
- Chief audit executives, audit partners, and audit directors
- Compliance executives and management
- Security and audit consultants

Finally, because privacy and cybersecurity are so closely related, many cybersecurity leaders and professionals see their span of responsibilities expanding to include privacy. Soon, cybersecurity professionals lacking privacy certifications and experience may find themselves disadvantaged in their organizations and in the employment market.

Becoming a CDPSE Professional

To become a CDPSE professional, you are required to pay the exam fee, pass the exam, prove that you have the required experience and education, and agree to uphold ethics and standards. To keep your CDPSE certification, you are required to take at least 20 continuing education hours each year (120 hours in three years) and pay annual maintenance fees. This life cycle is depicted in Figure 1.

 EXAM TIP When preparing for any certification exam, you should routinely refer to the certifying body's web site for the latest information about objectives and exam requirements. Visit the ISACA web site at www.isaca .org/cdpse for up-to-date information on the CDPSE certification and exam requirements.

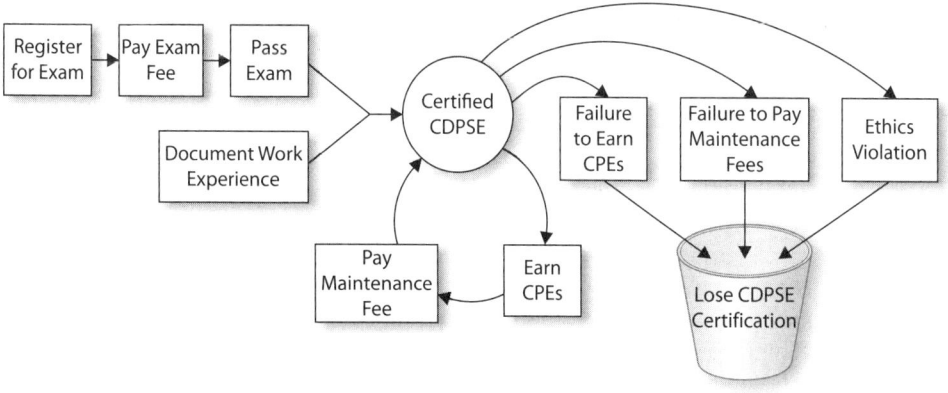

Figure 1 The CDPSE certification life cycle

The following list outlines the primary requirements for becoming certified:

- **Experience**   A CDPSE candidate must submit verifiable evidence of at least five years of experience, with a minimum of three years of professional work experience in CDPSE job practice areas. Experience must be verified and must be gained within the ten-year period preceding the application date for certification or within five years from the date of passing the exam. Substitution and waiver options for up to three years of experience are available.

- **Ethics**   Candidates must commit to adhere to ISACA's Code of Professional Ethics, which guides the personal and professional conduct of those certified.

- **Exam**   Candidates must receive a passing score on the CDPSE exam. A passing score is valid for up to five years, after which the score is void. This means that a CDPSE candidate who passes the exam has a maximum of five years to apply for CDPSE certification; candidates who pass the exam but fail to act after five years will have to retake the exam if they want to become CDPSE certified.

- **Education**   Those certified must adhere to the CDPSE Continuing Professional Education Policy, which requires a minimum of 20 continuing professional education (CPE) hours each year, with a total requirement of 120 CPEs over the certification period (three years).

- **Application**   After successfully passing the exam, meeting the experience requirements, and reading through the Code of Professional Ethics and Standards, a candidate is ready to apply for certification. An application must be received within five years of passing the exam.

Experience Requirements

To qualify for CDPSE certification, you must have completed the equivalent of five years of total work experience. These five years can take many forms, with some substitutions available. Additional details on the minimum certification requirements, substitution options, and various examples are discussed next.

NOTE   Although not recommended, a CDPSE candidate can take the exam before completing any work experience directly related to privacy management. As long as the candidate passes the exam and the work experience requirements are fulfilled within five years of the exam date, and within ten years from the application for certification, the candidate is eligible for certification.

Direct Work Experience

You are required to have a minimum of three years of work experience in privacy and security strategy and management. This is equivalent to 6000 actual work hours, which must be related to at least two of the three CDPSE job practice areas:

- **Privacy governance**   Establish and/or maintain a privacy governance and risk management framework and supporting processes to ensure that the privacy strategy is aligned with applicable laws, organizational goals, and objectives.
- **Privacy architecture**   Ensure that technology architecture, infrastructure, operations, and controls provide adequate safeguards for the protection and proper use of personal information.
- **Data cycle**   Develop and maintain an information inventory and processes to ensure data quality, data limitation, data minimization, and protection throughout the data's documented life cycle.

All work experience must be completed within the ten-year period before completing the certification application or within five years from the date of initially passing the CDPSE exam. You will need to complete a separate Verification of Work Experience form for each segment of experience.

Substitution of Experience

Two years of direct work experience can be substituted with the following to meet the five-year experience requirement:

- Certified Information Systems Auditor (CISA) in good standing
- Certified Information Security Manager (CISM) in good standing
- Certified in Governance of Enterprise IT (CGEIT) in good standing
- Certified in Risk and Information Systems Control (CRISC) in good standing
- CSX Cybersecurity Practitioner (CSX-P) in good standing
- Fellow of Information Privacy (FIP) in good standing (this is a certification from iapp.org)

 **TIP**   I recommend you also read the CDPSE certification qualifications on the ISACA web site. From time to time, ISACA does change the qualification rules, and I want you to have the most up-to-date information available.

ISACA Code of Professional Ethics

Becoming a CDPSE professional means you agree to adhere to the ISACA Code of Professional Ethics, a formal document outlining those things you will do to ensure the utmost integrity and to support and represent the organization and certification to the best of your abilities.

The ISACA code of ethics requires ISACA members and certification holders to do the following (paraphrased from the code of ethics):

- Support and promote appropriate practices for the effective management of information systems.
- Work with objectivity, due diligence, and professional care according to appropriate standards.

- Perform duties legally and with high standards of professional conduct. Do not discredit the profession.

- Ensure the privacy and confidentiality of information according to applicable laws, standards, and practices. Do not use sensitive information for personal gain.

- Work with an attitude of continuous learning in the lifelong development of skills and knowledge. Promote the same behavior in others in the profession.

- Ensure truth and accuracy in reporting. Do not use facts to mislead others.

Failure to comply with this Code of Professional Ethics can result in an investigation into a member's or certification holder's conduct and, ultimately, in disciplinary measures.

You can find the full text and terms of enforcement of the ISACA Code of Ethics at www.isaca.org/credentialing/code-of-professional-ethics.

The Certification Exam

ISACA offers certification throughout the year in several examination windows. I highly recommend you plan ahead and register early.

NOTE As is the case with professional certifications and examinations, the terms, conditions, locations, and rules for certification exams are likely to change from time to time. Readers should thoroughly review the exam logistics described at www.isaca.org/cdpse for the most current information.

Once you have registered for the exam, you will receive one or more e-mail messages that describe the steps you must undergo to take the CDPSE exam. You may be directed to select a location to take your exam, or you may be directed to meet requirements to take the exam remotely. As you decide about test locations, dates, and other conditions, ISACA or a third-party exam service will send you confirmations.

CAUTION It is essential that you thoroughly understand the rules regarding the exam. Failure to abide by these rules may result in your disqualification for the CDPSE certification.

Whether you take the exam remotely or at a test center, you will be supervised by a proctor. You should expect to be monitored by video surveillance to ensure that no one can cheat on the exam.

Each registrant has a time limit to take the multiple-choice question exam. There are 100 or more questions on the exam representing the three job practice areas. The exam is computerized. Each question has four answer choices; test-takers can select only one best answer. You can skip questions and return to them later, and you can also flag questions that you want to review later if time permits. While you are taking your exam, the time remaining will appear on the screen.

Domain	CDPSE Job Practice Area	Percentage of Exam
1	Privacy Governance	34%
2	Privacy Architecture	36%
3	Data Cycle	30%

Table 1 CDPSE Exam Practice Areas

When you have completed the exam, you are directed to close the exam. At that time, the exam may display your preliminary pass or fail status, with a reminder that your score and passing status is subject to review.

Exam questions are derived from a job practice analysis conducted by ISACA. The areas selected represent those tasks performed in a CDPSE's day-to-day activities and represent the background knowledge required to develop and manage an information privacy program. You can find more detailed descriptions of the task and knowledge statements at www.isaca.org/CDPSE.

The CDPSE exam is quite broad in its scope. The exam covers three job practice areas, as shown in Table 1.

Independent committees have been developed to determine the best questions, review exam results, and statistically analyze the results for continuous improvement. Should you come across a horrifically difficult or strange question, do not panic. This question may have been written for another purpose: a few questions on the exam are included for research and analysis purposes and will not count against your score. The exam contains no indications in this regard, so you should consider each question as one that contributes to your final score.

Exam Preparation

The CDPSE certification requires a great deal of knowledge and experience from the CDPSE candidate. You need to map out a long-term study strategy to pass the exam. The following sections offer some tips and are intended to help guide you through and beyond exam day.

Before the Exam

Consider the following list of tips on tasks and resources for exam preparation. They are listed in sequential order.

- *Read the exam candidate guide.* For information on the certification exam and requirements for the current year, find the ISACA Exam Candidate Guide. Go to www.isaca.org/cdpse and find the link.

- *Register for the exam.* If you are able, register early for any cost savings and to solidify your commitment to moving forward with this professional achievement.

- *Schedule your exam.* Find a location (where applicable), date and time, and commit.

- *Become familiar with the CDPSE job practice areas.* The job practice areas serve as the basis for the exam and requirements. Ensure that your study materials align with the current list at www.isaca.org/cdpse.

- *Know your best learning methods.* Everyone has preferred learning styles, whether it's self-study, a study group, an instructor-led course, or a boot camp. Try to set up a study program that leverages your strengths.

- *Self-assess by taking practice exams.* Run through practice exam questions available online (see the Appendix for more information). ISACA may offer a free 50-question CDPSE self-assessment like they do for their other certifications.

- *Study iteratively.* Depending on how much work experience in privacy you have already, I suggest you plan your study program to take at least two months but as long as six months. During this time, periodically take practice exams and note your areas of strength and weakness. Once you have identified your weak areas, focus on those areas weekly by rereading the related sections in this book and retaking practice exams, and note your progress.

- *Avoid cramming.* We've all seen the books on the shelves with titles that involve last-minute cramming. Just one look on the Internet reveals various web sites that cater to teaching individuals how to cram for exams most effectively. Research sites claim that exam cramming can lead to colds and flu, sleep disruptions, overeating, and digestive problems. One thing is certain: many people find that good, steady study habits result in less stress and greater clarity and focus during the exam. Because of the complexity of this exam, I highly recommend the long-term, steady-study option. Study the job practice areas thoroughly. There are many study options. If time permits, investigate the many resources available to you.

- *Find a study group.* Many ISACA chapters and other organizations have formed specific study groups or offer less expensive exam review courses. Contact your local chapter to see whether these options are available to you. Also, be sure to keep your eye on the ISACA web site. And use your local network to find out whether there are other local study groups and other helpful resources.

- *Check your confirmation letter.* Recheck your confirmation letter. Do not write on it or lose it. Put it in a safe place, and take note of what date, time, and place you are registered to take the exam. Note this on your calendar. If you are taking the exam at a testing center, confirm that the location is the one you selected and is located reasonably close by. Understand all specific requirements and plan ahead.

- *Check logistics.* If you are taking the exam at a test center, check the candidate guide and your confirmation letter for the exact time required to report to the test site. Check the site a few days before the exam—become familiar with the location and tricks to getting there. If you are taking public transportation, be sure you are looking at the schedule for the day of the exam: if your CDPSE exam is scheduled on a Saturday, public transportation schedules may differ from weekday schedules. If you are driving, know the route and where to park your vehicle. If you are taking the exam online, check the candidate guide and your confirmation letter for your exam's exact time and to ensure that you have the required equipment, software, and materials available.

- *Pack what you need.* If you are taking the exam at a test center, place your confirmation letter and a photo ID in a safe place, ready to go. Your ID must be a current, government-issued photo ID that matches the name on the confirmation letter and must not be handwritten. Examples of acceptable forms of ID are passports, driver's licenses, state IDs, green cards, and national IDs. Make sure you leave food, drinks, laptops, cell phones, and other electronic devices behind, as they are not permitted at the test site.

- *Decide on a mode of notification.* When you complete the exam, you'll receive a preliminary pass/fail result (see next page). To receive your actual test results, you can decide to have them sent via postal mail or e-mail. If you are fully paid (zero balance on the exam fee) and have consented to the e-mail notification, you should receive a one-time e-mail approximately eight weeks from the exam date with the results.

- *Get some sleep.* Make sure you get a good night's sleep before the exam. Research suggests that you avoid caffeine at least four hours before bedtime, keep a notepad and pen next to the bed to capture late-night thoughts that might keep you awake, eliminate as much noise and light as possible, and keep your room a comfortable temperature for sleeping. In the morning, rise early so as not to rush and subject yourself to additional stress.

Day of the Exam

On the day of the exam, follow these tips:

- *Dress comfortably.* Certification exams are difficult and require long periods of intense concentration. It is important, therefore, to ensure that you will be as comfortable as possible physically. Avoid tight-fitting clothes, and dress in layers to stay comfortable throughout the exam.

- *Arrive early.* If you are taking the exam at a test center, check the Bulletin of Information and your confirmation letter for the exact time you are required to report to the test site. The confirmation letter or the candidate guide explains that you must be at the test site *no later* than approximately 30 minutes *before* testing time. The examiner will begin reading the exam instructions at this time, and any latecomers will be disqualified from taking the test and will *not* receive a refund of fees.

- *Observe test center rules.* There may be rules about taking breaks. The examiner will discuss this along with exam instructions. If you need something at any time during the exam and are unsure as to the rules, be sure to ask first.

- *Answer all exam questions.* Read questions carefully, but do not try to overanalyze. Remember to select the *best* solution. There may be several reasonable answers, but one is *better* than the others. If you aren't sure about an answer, you can mark the question and return to it later. After going through all the questions, you can return to the marked questions (and any others) to read them and consider them more carefully. Above all, don't try to overanalyze questions and do trust your instincts.

Do not try to rush through the exam; you'll have plenty of time to take as much as a few minutes for each question. But at the same time, do watch the clock so that you don't find yourself going so slowly that you won't be able to answer every question thoughtfully.

- *Note your exam result.* When you have completed the exam, you should see your preliminary pass/fail result. Your results may not be in large, blinking text; you may need to read the fine print to see your preliminary results. If you passed, congratulations! If you did not pass, do observe any remarks about your status; you will be able to retake the exam—there is information about this on the ISACA web site.

If You Did Not Pass

If you did not pass your exam on the first attempt, don't lose heart. Instead, remember that failure is a stepping stone to success. Thoughtfully take stock and determine your improvement areas. Go back to this book's practice exams and be honest with yourself regarding those areas where you need to learn more. Reread the chapters or sections where you require additional study. If you participated in a study group or training, contact your study group coach or class instructor for advice on studying the topics you need to master. Take at least several weeks to study those topics, refresh yourself on other topics, and then give it another go. Success is granted to those who are persistent and determined.

After the Exam

A few weeks after you take the exam, you will receive your exam results by e-mail or postal mail. Each job practice area score may be noted in addition to the overall final score. All scores are scaled. Should you receive a passing score, you will also receive the application for certification or guidance on where to find the application.

Those unsuccessful in passing will also be notified. These individuals will want to closely examine the job practice area scores to determine areas for further study. They may retake the exam as many times as needed on future exam dates, as long as they have registered and paid the applicable fees. Regardless of pass or fail, exam results will not be disclosed via telephone, fax, or e-mail (except for the consented e-mail notification).

 **NOTE** You are not permitted to display the CDPSE moniker until you have completed certification. Passage of the exam is *not* sufficient to use the CDPSE anywhere, including e-mail, resumes, correspondence, or social media.

Applying for CDPSE Certification

To apply for certification, you must submit evidence of a passing score and related work experience. Remember that you have five years to use this score on a CDPSE certification application once you receive a passing score. After this time, you will need to retake the exam. Also, all work experience submitted must have been within ten years of your new certification application.

NOTE Readers are advised to confirm this information at www.isaca.org/cdpse.

To complete the application process, you need to submit the following information:

- **CDPSE application** Note the exam ID number in your exam results letter, list the privacy experience and any experience substitutions, and identify which CDPSE job practice area (or areas) your experience pertains to.
- **Verification of Work Experience forms** These must be filled out and signed by your immediate supervisor or a person of higher rank in the organization to verify your work experience noted on the application. You must fill out a complete set of Verification of Work Experience forms for each separate employer.
- **Certification confirmation** If you are using a certification waiver, you must submit suitable evidence of your certification status.

As with the exam, after you've successfully mailed the application, you must wait approximately eight weeks for processing. If your application is approved, you will receive an e-mail notification, followed by a package in the mail containing your letter of certification, certificate, and a copy of the Continuing Professional Education Policy. You can then proudly display your certificate and use the "CDPSE" designation on your résumé, e-mail and social media profiles, and business cards.

NOTE You are permitted to use the CDPSE moniker *only* after receiving your certification letter from ISACA.

Retaining Your CDPSE Certification

There is more to becoming a CDPSE professional than merely passing an exam, submitting an application, and receiving a paper certificate. Becoming a CDPSE professional is not a destination; instead, it should be considered a lifestyle. Those with CDPSE certification agree to abide by the code of ethics, meet ongoing education requirements, and pay annual certification maintenance fees. Let's take a closer look at the education requirements and explain the fees involved in retaining certification.

Continuing Education

The goal of continuing professional education requirements is to ensure that individuals maintain CDPSE-related knowledge to develop and manage better privacy and security management programs. To maintain CDPSE certification, individuals must obtain 120 continuing education hours within three years, with a minimum requirement of 20 hours per year. Each CPE hour is to account for 50 minutes of active participation in educational activities.

What Counts as a Valid CPE Credit?

For training and activities to be utilized for CPEs, they must involve technical or managerial training directly applicable to information privacy, information security, and information privacy and security management. The following list of activities has been approved by the CDPSE certification committee and can count toward your CPE requirements:

- ISACA professional education activities and meetings
- ISACA members can take *Information Systems Control Journal* CPE Quizzes online or participate in monthly webcasts; for each webcast, CPEs are rewarded after you pass a quiz
- Non-ISACA professional education activities and meetings
- Self-study courses
- Vendor sales or marketing presentations (ten-hour annual limit)
- Teaching, lecturing, or presenting on subjects related to job practice areas
- Publication of articles and books related to the profession
- ISACA certification exam question development and review
- Passing related professional examinations
- Participation in ISACA boards or committees (20-hour annual limit per ISACA certification)
- Contributions to the information security management profession (ten-hour annual limit)
- Mentoring (ten-hour annual limit)

**NOTE**   For more information on what is accepted as a valid CPE credit, see the Continuing Professional Education Policy at www.isaca.org/credentialing/how-to-earn-cpe.

Tracking and Submitting CPEs

Not only are you required to submit a CPE tracking form for the annual renewal process, but you should also keep detailed records for each activity. Records associated with each activity should include the following:

- Name of attendee
- Name of sponsoring organization
- Activity title
- Activity description
- Activity date
- Number of CPE hours awarded

It is in your best interest to track all CPE information in a single file or worksheet. ISACA has developed a tracking form for your use, which can be found online in the Continuing Professional Education Policy. Consider keeping all related records such as receipts, brochures, and certificates in the same place. Documentation should be retained throughout the three-year certification period and for at least one additional year afterward. This is especially important because you may someday be audited. If this happens, you would be required to submit all paperwork as proof of your continuous learning. So why not be prepared?

For new CDPSEs, the annual and three-year certification period begins January 1 of the year following certification. You are not required to report CPE hours for the first partial year after your certification; however, the hours earned from the time of certification to December 31 can be utilized in the first certification reporting period the following year. Therefore, should you get certified in January, you will have until the following January to accumulate CPEs. You will not have to report them until you report the totals for the following year, which will be in October or November. This is known as the *renewal period*. During this time, you will receive an e-mail directing you to the web site to enter CPEs earned over the course of the year. Alternatively, the renewal will be mailed to you, and then CPEs can be recorded on the hard-copy invoice and sent with your maintenance fee payment. CPEs and maintenance fees must be received by January 15 to retain certification.

Notification of compliance from the certification department is sent after all the information has been received and processed. Should ISACA have any questions about the information you have submitted, it will contact you directly.

Sample CPE Submission

Table 2 shows an example of a CPE submission consisting of several activities and events attended throughout the year.

Name <u>Kamala Khan</u>

Certification Number <u>67895787</u>

Certification Period <u>1/1/2021 to 12/31/2021</u>

Activity Title/Sponsor	Activity Description	Date	CPE Hours	Support Docs Included?
ISACA presentation/lunch	CCPA compliance	2/12/2021	1 CPE	Yes (receipt)
ISACA presentation/lunch	Security in SDLC	3/12/2021	1 CPE	Yes (receipt)
Regional conference, RIMS	Compliance, risk	1/15–17/2021	6 CPEs	Yes (CPE receipt)
Brightfly webinar	Governance, risk, and compliance	2/16/2021	3 CPEs	Yes (confirmation e-mail)
ISACA board meeting	Chapter board meeting	4/9/2021	2 CPEs	Yes (meeting minutes)
Presented at ISSA meeting	Privacy management presentation	6/21/2021	1 CPE	Yes (meeting notice)
Published an article in XYZ	Journal article on GDPR	4/12/2021	4 CPEs	Yes (article)
Vendor presentation	Learned about GRC tool capability	5/12/2021	2 CPEs	Yes
Employer-offered training	Change management course	3/26/2021	7 CPEs	Yes (certificate of course completion)

Table 2 Sample CPE Submission

Certification Maintenance Fees

To remain CDPSE certified, you must pay maintenance fees each year. These fees are in addition to ISACA membership and local chapter dues (neither is required to maintain your CDPSE certification).

 **NOTE**   Because you might not receive an e-mail reminder, I recommend that you create calendar entries or other suitable ways to remind you to record your CPEs and pay your certification maintenance fees each year.

Revocation of Certification

A CDPSE-certified individual may have his or her certification revoked for the following reasons:

- Failure to complete the minimum number of CPEs during the period
- Failure to document and provide evidence of CPEs in an audit
- Failure to submit payment for maintenance fees
- Failure to comply with the Code of Professional Ethics, which can result in investigation and ultimately lead to revocation of certification

If you have received a revocation notice, you will need to contact the ISACA Certification Department at certification@isaca.org for more information.

Summary

Becoming and being a CDPSE professional is a lifestyle, not just a one-time event. It takes motivation, skill, good judgment, persistence, and proficiency to be a strong and effective contributor in the world of privacy. The CDPSE certification was designed to help you navigate the privacy world with greater ease and confidence.

In the following chapters, each CDPSE job practice area will be discussed in detail and additional reference material will be presented. Not only is this information useful for studying before the exam, but it is also meant to serve as a resource throughout your career as a privacy professional.

PART I

Privacy Governance

Governance

In this chapter, you will learn about

- Business alignment
- Privacy and security governance activities
- Privacy and security strategy development
- Resources needed to develop and execute a privacy and security strategy
- Obstacles to strategy development and execution
- Privacy and security metrics

This chapter covers Certified Data Privacy Solutions Engineer (CDPSE) job practice 1, "Privacy Governance," part A, "Governance." The entire Privacy Governance domain represents 34 percent of the CDPSE examination.

Properly implemented, *governance is a process* whereby senior management exerts strategic control over business functions through policies, objectives, delegation of authority, and monitoring. Governance is management's continuous oversight of an organization's business processes to ensure that they effectively meet the organization's business vision and objectives.

Organizations often establish governance through a committee or formalized position that is responsible for setting long-term business strategy, and by making changes to ensure that business processes continue to support business strategy and the organization's overall needs. Effective governance is enabled through the development and enforcement of documented policies, standards, requirements, and various reporting metrics.

Introduction to Privacy Governance

Privacy governance is a set of established activities that typically focuses on several fundamental principles and outcomes designed to enable management to have a clear understanding of the state of the organization's privacy program, its current risks, its direct activities, and its alignment to the organization's business objectives and practices. A goal of the privacy program is enabling the fulfillment of the privacy strategy, which itself will continue to align with the business, business objectives, and developing regulations. The processes supporting these principles and outcomes include privacy policy, data governance, compliance, risk management, and cybersecurity. Whether the organization has a board of directors, council members, commissioners, or some other top-level governing body, governance begins with establishing top-level strategic objectives that are translated into actions and roles and responsibilities through policies, processes, procedures, and other activities downward through each level in the organization.

Privacy is a business issue, and organizations that are not yet properly managing or adequately protecting personal information have a business problem. The reason for this is almost always a lack of understanding and commitment by boards of directors and senior executives. For many, privacy is viewed as a security issue that focuses on data protection problems at the tactical level or data usage problems, and it's not about usage at all. The challenge is that, because of a lack of awareness or experience in privacy, organizations still struggle with how to organize, manage, and communicate about privacy successfully at the executive leadership and boardroom levels.

To manage privacy successfully, organizations need to understand that privacy is also a people issue. When people at each level in the organization—from board members to individual contributors—understand the importance of privacy and security within their own roles and responsibilities, an organization will be in a position of reduced risk. This reduction in risk or identification of potential privacy or security events results in fewer incidents with less impact on the organization's ongoing reputation and operations.

NOTE Because modern privacy practices are heavily influenced by privacy laws such as the European Union General Data Protection Regulation (EU GDPR) and the California Consumer Privacy Act (CCPA), organizations should rely upon qualified legal counsel as a part of the overall governance process. Including legal counsel helps to ensure that the organization's privacy policies and practices comply with these and other laws.

Think of privacy as having two main components: proper data management and usage, and data protection—commonly referred to as cybersecurity, data security, or information security. A privacy program cannot succeed without effective cybersecurity. Further, cybersecurity cannot succeed without a solid foundation in IT and IT operations. IT is the enabler and force multiplier that facilitates business processes that fulfill organization objectives. Without effective IT governance, privacy and information security governance will not reach their full potential. The result may be that the IT bus will travel safely but to the wrong destination. Figure 1-1 shows how the business vision, strategy, and objectives of privacy and information security governance flow downward in an organization through its privacy and IT security strategies, policies, standards, and processes.

NOTE Although the CDPSE certification is not directly tied to IT governance, this implicit dependence of privacy and security governance on IT governance cannot be understated. IT and security professionals specializing in IT governance itself may be interested in ISACA's Certified Information Security Manager (CISM) and Certified in the Governance of Enterprise IT (CGEIT) certifications, which specialize in these domains.

PART I

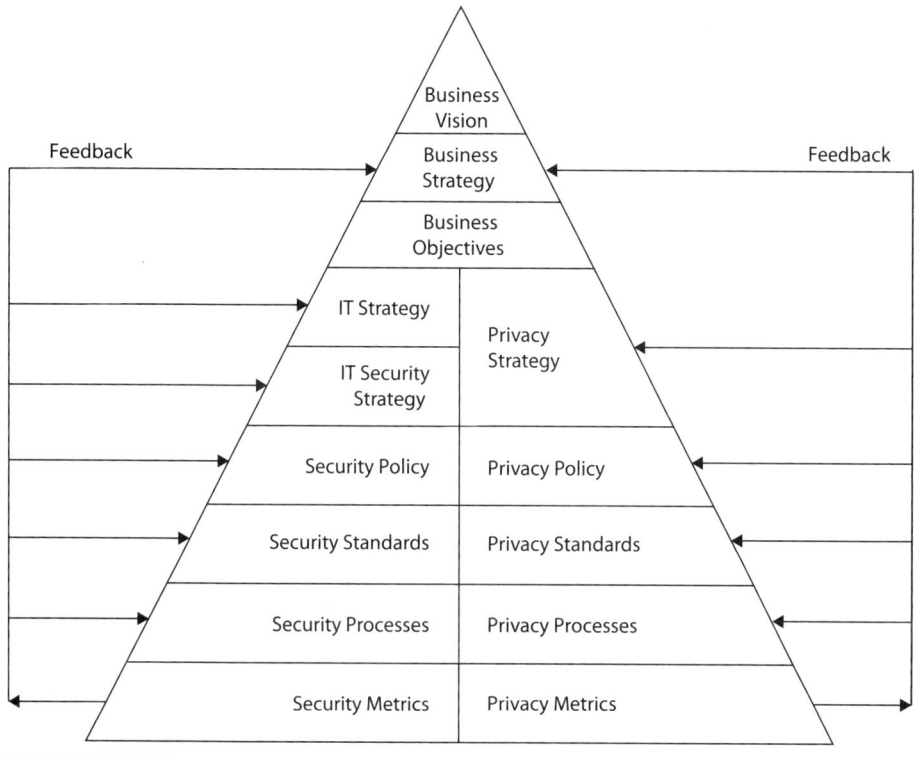

Figure 1-1 Business vision flows downward in an organization.

While IT governance, information security governance, and privacy governance may be separate activities, in many organizations, these activities will closely resemble or rely upon one another. Many issues will span IT, security, and privacy governance bodies, and many individuals will participate actively in all three areas. Some organizations may integrate IT, information security, and privacy governance into a single set of participants, activities, and business records. The most important thing is that organizations figure out how to establish governance programs that are effective for achieving formally established business outcomes.

Privacy governance will enable alignment of the organization's privacy program with customer or constituent expectations, applicable regulations, identified risks, and business needs. An objective of privacy governance is to provide assurance of the proper protection and use of personal information from a strategic perspective to ensure that required privacy practices align with the business practices. Here are some of the artifacts and activities that flow out of sound privacy governance:

- **Objectives**   The desired capabilities or end states, ideally expressed in achievable, measurable terms.

- **Established legal basis**   The manner in which the organization may lawfully collect and process personal information about data subjects.

- **Consent**   The mechanisms through which the organization directly or indirectly obtains permission from data subjects to collect and process their personal information.

- **Strategy**   The plan to achieve one or more objectives.

- **Policy**   The mission, objectives, and goals of the overall organization that align with constituent expectations and applicable laws.

- **Priorities**   The main concerns of the privacy program, which should flow directly from the organization's mission, objectives, and goals. Whatever is most important to the organization as a whole should be relevant to privacy and information security.

- **Standards**   The technologies, protocols, and practices used by IT that should reflect the organization's needs. On their own, standards help to drive a consistent approach to solving business challenges; the choice of standards should facilitate solutions that meet the organization's needs in a cost-effective and secure manner.

- **Processes**   The formalized descriptions of repeated business activities that include instructions to applicable personnel. Processes include one or more procedures, as well as definitions of business records and other facts that help workers understand how things are supposed to be done.

- **Controls**   The formal descriptions of critical activities performed to ensure desired outcomes.

- **Program and project management**   The ways in which the organization's privacy, security, and IT programs and projects are organized and performed, which should be in a consistent manner that reflects business priorities and supports the business.

- **Metrics/reporting**   The formal measurement of processes and controls that management can understand and measure.

- **Review/audit**   The formal evaluation of processes and controls to determine their effectiveness.

To the greatest extent possible, privacy governance in an organization should be practiced in the same way that the organization performs cybersecurity, IT, and overall corporate governance. Privacy governance should mimic organizational and/or security and IT governance processes, or privacy governance may be integrated into corporate, cybersecurity, or IT governance processes.

Though privacy governance contains the elements just described, strategic planning is also a key component of governance. Strategy development is discussed in the next section.

> ## Privacy and Security: Together or Separate?
> Should privacy and security be managed separately or together? Although there's no right or wrong answer, know this: privacy cannot succeed without information security. The objectives of a privacy program are the protection and proper handling of personal information. The protection part is done by information security, and the proper handling part is solely the domain of privacy.
>
> Privacy needs information security to be successful. Security is a prerequisite to privacy, but privacy adds more: the proper *handling* of information and its *protection*.
>
> This is why privacy and security are discussed hand-in-hand throughout this chapter and in most of this book. To discuss privacy alone, without security, tells only half of the entire story that needs to be told.

Privacy Governance Influencers

An organization's privacy program must focus on several internal and external events and activities. Privacy professionals realize that some of these factors can be influenced to some degree, while others are entirely out of the privacy professional's sphere of influence. We must be informed and able to react to these influencers.

The Nature of Personal Data and Information

Much of the information contained in information systems is about people. In both government and business organizations, information systems keep track of property owners, taxpayers, voters, patients, clients, customers, and potential customers. Often, the information retained about people is sensitive in nature, or even secretive, and all parties have a vested interest in the adequate protection of that information.

In most situations, transactions between individuals and businesses, governments, and healthcare organizations are considered confidential, not to be disclosed, and to be used for official business purposes only. When this information migrated from paper to information systems, and with advancements in information technology, organizations developed numerous techniques by which more value could be obtained from the information about their citizens, patients, customers, and constituents. Abuses of these practices have given rise to privacy laws intended to curb such activities.

Privacy laws are discussed in detail in this section. Note that there are many variances among these laws in the following areas:

- **Definitions of personal information**   Privacy laws provide sometimes specific, sometimes vague definitions of which type of data is considered sensitive and which is not. Most laws consider the aggregation of someone's name, together with other items, such as financial account numbers, medical records, political affiliation, and more, as personal information that is to be safeguarded and used within stated guidelines.

- **Data subject rights**   Privacy laws define a number of rights that vary somewhat from one regulation to another. These rights cover transparency and limitations of use, adequate protection, personal data correction, and data removal.

- **Protection of personal information**   Laws require organizations to take measures to ensure the adequate protection of personal information so that it cannot be accessed, altered, stolen, or destroyed by unauthorized parties.

- **Use of personal information**   Laws require transparency regarding the uses of personal information so that persons can be aware of these uses.

- **Notification of breach**   Laws require organizations to disclose to affected individuals any instances in which their personal information was improperly accessed, used, or compromised.

- **Jurisdiction**   Many privacy laws today are *extraterritorial*, meaning that they intend to regulate the activities of organizations located outside of political boundaries.

Data privacy and data protection laws are being enacted at a relatively fast pace, as a reflection of vast expansions of the collection and use of personal information, abuses and breaches by the organizations collecting and using personal data, and still-developing social norms regarding the definitions and expectations of privacy.

The Imperfect Lexicon of Privacy

As in every profession, privacy and information security professions include some special vocabularies. In the privacy profession are the terms *personal information* and *data subject request*. Is there really a distinction and a valid reason why "personal information" uses the term *information* while "data subject request" uses the term *data*?

Looking at dictionary definitions, *data* and *information* have similar definitions. However, if we go deeper and more specific, we find definitions (in this case, from www.diffen.com) along these lines: "Data are simply facts or figures—bits of information, but not information itself. When data are processed, interpreted, organized, structured, or presented so as to make them meaningful or useful, they are called information. Information provides context for data."

Perhaps this is a clue. The remainder of this exercise is left to the reader.

Privacy Laws

Laws passed by governments at national, state, and provincial levels are imposing requirements on organizations that store or process personal information about natural persons. These laws have been enacted to respond to citizens' outcry at the abuse of their personal information and violations of their desire for privacy as the use of technology and the

Internet has become part of our daily lives. Activities such as telemarketing and tracking people's locations and habits have been the focus of growing concerns by citizens and privacy advocates. Advances in the capabilities of information systems and data mining of large databases containing personal information have enabled practices that many private citizens find inappropriate and even intrusive. Abuses of such capabilities would allow for the creation of a new and very invasive form of a surveillance police state, which is not difficult to imagine, as some governments in the world are already there. Legislators in many countries have sought to counterbalance these capabilities by defining the rights of natural persons concerning the data that is collected about them and used in various ways.

The most influential privacy laws include the EU GDPR, the US Health Information Portability and Accountability Act (HIPAA), the US Fair Credit Reporting Act (FCRA), the US Electronic Communications Privacy Act (ECPA), the Canadian Personal Information Protection and Electronic Documents Act (PIPEDA), the CCPA, and the China Cybersecurity Law (CCSL). These and other laws are discussed in the remainder of this section. This is not intended to be an exhaustive or authoritative list, but instead a sampling of better-known privacy laws.

 **CAUTION** Because privacy and data-protection laws are being rapidly enacted and changed, organizations should devise a way to remain fully aware of new and changing laws.

EU General Data Protection Regulation   One of the most notable modern privacy laws, the GDPR enacts sweeping requirements upon organizations within and beyond the European Union that store, process, or transmit personal data about EU citizens and residents. The GDPR was passed in April 2016 and became effective in May 2018. The main privileges enacted by the GDPR include the following:

- **Rights of data subjects**   Any organization collecting information about EU residents is required to operate with transparency in collecting and using their personal information. Chapter III of the GDPR defines eight data subject rights that have become foundational for other privacy regulations around the world:

  - *Right to access personal data.* Data subjects can access the data collected on them.

  - *Right to rectification.* Data subjects can request a modification of their data to correct errors and the updating of incomplete information.

  - *Right to erasure.* Also referred to as the right to be forgotten, data subjects can request that their personal data be erased from an entity's processing activities.

  - *Right to restrict processing.* In certain circumstances, data subjects can request that processing of their personal data be stopped.

  - *Right to be notified.* Data subjects must be notified about what information is being collected at or before collection, and the use of that information.

- *Right to data portability.* Data subjects can request that their personal data be provided in a commonly used, machine-readable format.
- *Right to object.* Data subjects can object to processing of their data when a controller attempts to argue legitimate grounds for the processing that override the interests, rights, and freedoms of the data subject or for legal purposes.
- *Right to reject automated individual decision-making.* Data subjects can refuse the automated processing of their personal data to make decisions about them.

- **Definitions of data controller and data processor**   The GDPR defines a *data controller* as an organization that directs the use of personal data. A *data processor* is an organization that processes personal data as directed by a data controller. Controllers and processors are required to maintain records regarding the processing of personal information.

- **Data protection and privacy by design and by default**   Organizations are required to design, operate, and maintain their business processes and information systems with privacy and security as a part of their default design.

- **Cybersecurity**   Organizations that process personal information are required to enact cybersecurity capabilities to protect that data.

- **Breach notification**   Organizations are required to notify supervisory authorities and affected data subjects in the event of a privacy or security breach of personal data.

- **Data protection impact assessment (DPIA)**   The DPIA is a formal process to identify and minimize the data protection risks of a data-processing activity. Organizations are required to perform DPIAs whenever they are implementing new, or making significant changes to, business processes or information systems.

- **Data protection officer (DPO)**   Organizations are required to appoint a DPO if they process personal data on a large scale. The DPO is expected to have expert knowledge of data protection law and practices.

- **Certification**   The GDPR permits the creation of certification authorities and voluntary certifications of organizations that process personal data.

NOTE   At the time of this writing, no GDPR certification has been approved by the national supervisory authorities or the European Data Protection Board.

- **Cross-border data transfers**   The GDPR contains rules regarding the transfer of personal data out of the European Union.

- **Binding corporate rules**   The GDPR accommodates the use of binding corporate rules that multinational organizations may enact to ensure the protection and appropriate use of personal data when transferred outside the European Union to their overseas systems.

- **Supervisory authority** Each EU member state establishes a supervisory authority that is responsible for monitoring GDPR compliance.
- **Penalties** The GDPR provides supervisory authorities the power to enact fines against organizations that violate terms of the GDPR. Administrative fines can be as high as €20 million or 4 percent of global turnover, whichever is greater.

The GDPR claims to have extraterritorial jurisdiction to companies not based in the European Union that provide services to EU citizens in EU member states. This claim of jurisdiction beyond its own borders has yet to be seriously tested in courts, so it remains to be seen how such an assertion will be practiced. EU data protection authorities fined Marriott £99 million for a 2018 breach and Google €50 million for alleged abuses of privacy settings. Subsequent appeals of these and other fines against non-EU companies should begin to reveal the law's true reach over several years.

US Health Information Portability and Accountability Act Enacted in 1996, HIPAA includes two rules that are concerned with the protection of protected health information (PHI) and electronic protected health information (ePHI). Covered entities are organizations that store or process medical information and are subject to one or both of these rules:

- **Security Rule** This part of HIPAA requires that organizations enact several administrative, physical, and technical safeguards to protect ePHI. Many of these controls are compulsory, while others are "addressable" (an organization can rationalize their disuse).
- **Privacy Rule** Effective in 2003, this part of HIPAA requires that organizations protected PHI, mainly in hard copy form.

Note that additional provisions of HIPAA are not related to security or privacy.

HIPAA defines various civil and criminal penalties for organizations that violate the law. Covered entities are required to enact business associate agreements (BAAs) with all third parties that store or process ePHI on behalf of covered entities.

Health Information Technology for Economic and Clinical Health Act Enacted in 2009, HITECH extends the Security Rule and Privacy Rule in HIPAA by expanding security breach notification requirements and expanding the disclosures of the use of a patient's PHI.

US Fair Credit Reporting Act Enacted in 1970, the FCRA provides visibility, remedies, and assurances through civil liability that the information contained in consumers' credit history is accurate. Since credit reports are used by banks and other financial institutions (as well as employers in many states for making hire/no-hire decisions), FCRA provides a means for consumers to obtain copies of their credit reports and procedures for making corrections to erroneous information contained in those credit reports.

FCRA is an early example of a privacy law that provides persons with the ability to know what personal information is being stored, how it is used, and what methods can be used for obtaining copies and requesting corrections.

If consumers' rights are violated, they may recover damages, attorney fees, and court costs, and punitive damages may be awarded if actions against them were willful.

Canadian Personal Information Protection and Electronic Documents Act   PIPEDA went into effect in 2000 and seeks to ensure consumer data privacy in the context of e-commerce. In part, PIPEDA was enacted to provide assurances to European countries and consumers that their personal information present in Canadian companies' information systems would be safe and free from abuses.

PIPEDA also gives Canadians the right to know why organizations collect, use, or disclose their personal information, and the assurance that their information will not be used for any other purpose. They can further know who within the organizations are responsible for protecting their information. They can contact the organizations to ensure that their personal data is accurate and can lodge complaints if they believe their privacy rights have been violated. Canadian companies must obtain consent for the collection of personal information, and they cannot refuse to provide service to a Canadian citizen if the citizen refuses to provide such consent.

California Consumer Protection Act   Made effective on January 1, 2020, CCPA is a state law designed to improve California residents' privacy rights. Provisions of CCPA give California residents certain rights, including the following:

- Knowledge of what personal information is being collected
- Notification of whether such personal information is subsequently transferred or disclosed to another party
- The ability to prohibit an organization from transferring or selling personal information
- The ability to examine the personal information held by organizations, with the right to request that the information be corrected or removed
- The freedom from discrimination should individuals choose to exercise their privacy rights

Californians can sue for damages when unencrypted and unredacted personal information is subject to unauthorized access and exfiltration, theft, or disclosure as a result of a business's violation of the duty to implement and maintain reasonable security procedures.

Like the EU GDPR, the CCPA claims jurisdiction over companies not located in California if they collect personal information about California residents. And like the GDPR, this provision has yet to be tested in court.

 **CAUTION**   The CCPA is expected to change significantly, and enforcement actions have yet to begin as of the writing of this book. Organizations subject to CCPA should watch this law, its enforcement, and resulting case law carefully.

China Cyber Security Law Enacted in 2016 and effective in 2017, the CCSL consists of three main parts:

- **Data protection** Organizations holding personal information about Chinese citizens must take measures to protect that information.
- **Data localization** Organizations collecting information about Chinese citizens must keep such data within the country of China. Organizations that want to transfer data out of China must undergo a data assessment by the Chinese government.
- **Cybersecurity** Organizations are required to enact controls to prevent malware, intrusions, and other attacks. The law includes mandatory standards, assessments, and certifications for network devices.

CCSL's data protection principles resemble those of the GDPR. Both laws require that organizations inform citizens if organizations hold their personal data, explain the use of their personal data, identify collection methods, and obtain consent for continued use.

Brazilian General Data Protection Law Enacted in 2019 and effective in August 2020, the LGPD is similar to the EU GDPR. The LGPD establishes a National Data Protection Authority that is designated as the federal agency responsible for overseeing the data protection regulation.

The LGPD establishes a number of individual rights over personal data, many of which are similar to what is provided by the GDPR. However, some additional rights are provided by the LGPD, including access to information about entities with which an organization has shared the individual's personal data.

Like the GDPR and the CCPA, the LGPD claims jurisdiction over companies not located in Brazil if one of the following criteria are met:

- The processing operation is carried out in Brazil.
- The purpose of the processing activity is to offer or provide goods or services to individuals located in Brazil.
- The personal data was collected in Brazil.

Similar to the GDPR and the CCPA, this provision has yet to be tested in court.

Privacy Laws Enforced by the FTC In the United States, the Federal Trade Commission (FTC), the agency that monitors the rights of consumers, has brought action legal against scores of companies in alleged violation of consumer protection laws, including violations of posted privacy policies, deceptive practices with regards to US-EU Safe Harbor registration, breaches of personal information, improper collection of personal data without consent, and failures to protect personal information.

The FTC enforces more than 70 laws, including the following:

- Federal Trade Commission Act, which protects consumers from unfair and deceptive practices

- Children's Online Privacy Protection Act (COPPA), which protects children's online privacy, particularly those under age 13

- Do-Not-Call Implementation Act, which provides for consumers to opt out of all telemarketing calls

- Controlling the Assault of Non-Solicited Pornography and Marketing Act (CAN-SPAM), which prevents misleading advertising and requires consumer opt out

- Gramm-Leach-Bliley Act, which protects and secures the privacy of consumer personal information by financial institutions

- HITECH, which extends the scope of HIPAA

- Identity Theft Assumption and Deterrence Act, which provides a central clearinghouse of identity theft complaints

- Fair Credit Reporting Act, which protects personal information collected by consumer credit bureaus, medical information companies, and tenant screening services

- Clayton Act, which prevents illegal contracts, mergers, and acquisitions

Other Legal Obligations

In addition to abiding by applicable laws and regulations, organizations may negotiate additional terms and conditions regarding the protection of personal information. Such obligations may be related to specific services rendered by third-party service providers that store or process personal information on behalf of other organizations. For instance, a service provider may comply with all applicable laws, agree to specific protective measures, undergo periodic audits or examinations, or provide specific reporting regarding the storage or use of personal data.

Privacy Practices

Organizations need to identify basic privacy operational practices and the legal grounds under which they operate, or intend to operate. Applicable privacy laws define and constrain these activities.

Consent In the context of data privacy, *consent* is a distinct action taken by a data subject to grant an organization permission to collect and/or process his or her personal information. There are several ways in which consent is given and obtained, including the following:

- **At the time of data collection** When a data subject is providing one or more items of personal information to an organization, the data subject also provides consent for the collection and processing of that information. In an online context, consent often takes the form of a checkbox with words citing agreement with a privacy policy that can be read in its entirety. On a paper form, consent often appears in or near the signature block.

- **Prior to data collection**   When a data subject is establishing a relationship with an organization, a part of the agreement may include the collection of consent for instances of data collection that will take place in the future.
- **Consent obtained through a third party**   In some instances, it is not feasible for an organization to collect consent directly from a data subject. For example, in the case of advertising that is specifically chosen for and delivered to a data subject, the organization displaying the advertising will have collected consent "for other uses" including advertising. The advertiser will have been assured that consent has been obtained from all data subjects to whom the advertiser is displaying ad content.

Regardless of the method used to collect consent, it is obligatory for organizations to record the specific date, time, stipulations, and circumstances in which that consent was collected. It is important for organizations to be quite specific with regard to this collection. For instance, if an organization collects items of personal information for use in a specific context or event, and consent for use of that information is for that context or event only, the organization cannot later use that information for other contexts or events.

NOTE   GDPR and CCPA treat consent somewhat differently. GDPR requires explicit consent prior to the collection and processing of personal information, meaning that data subjects must opt in for any collection and use of their personal data. On the other hand, CCPA permits collection and processing of personal information but requires that data subjects be able to opt out of all such processing.

Legal Basis   Within the context of data privacy and privacy regulations, organizations must identify specifically the legal basis under which they are collecting and/or processing personal information. Simply put, it must be lawful for the organization to collect and use data subjects' personal information, and the organization must be able to cite specifically how it is lawful.

In Article 6.1, the GDPR provides five possible avenues of legal basis (quoting directly from GDPR):

- *processing is necessary for the performance of a contract to which the data subject is party or in order to take steps at the request of the data subject prior to entering into a contract;*
- *processing is necessary for compliance with a legal obligation to which the controller is subject;*
- *processing is necessary in order to protect the vital interests of the data subject or of another natural person;*
- *processing is necessary for the performance of a task carried out in the public interest or in the exercise of official authority vested in the controller;*

- *processing is necessary for the purposes of the legitimate interests pursued by the controller or by a third party, except where such interests are overridden by the interests or fundamental rights and freedoms of the data subject which require protection.* [This does not apply to public authorities in the performance of their duties.]

EU member states are permitted to include additional provisions.

The CCPA does not treat this in the same way as the GDPR. Instead, entities subject to CCPA must follow the law in general. Certain use cases, such as healthcare and employment, with specific provisions are cited.

Legitimate Interest   The term *legitimate interest* is coined in the GDPR as one of five bases for legal collection and processing of personal information. One might consider legitimate interest a loophole, but it is the author's opinion that this provision was included so that GDPR did not need to enumerate every possible use of personal information (which would soon be out of date, requiring frequent updates to the law).

Legitimate interest gives organizations a basis for collecting and processing personal information if they benefit from doing so. But it doesn't end there: the real issue of legitimate interest lies in the fact that organizations must balance their interests with that of the data subject.

Online advertising is an interesting case to consider. Advertisers have an interest in serving advertising content to readers and viewers. To stay in business and earn advertising revenue (which is the business model for a considerable number of web sites), advertisers claim that readers and viewers benefit (their interest) in advertising that is aligned with topics of interest to them (the readers' and viewers' interest). Thus, there is an argument that online advertising benefits the advertiser (and web site operator) as well as the viewer (the data subject).

   CAUTION   Organizations are cautioned to not lean upon legitimate interest too readily. Some long-established industry and business practices may be forever altered as a result of GDPR, CCPA, and other privacy laws.

Data Governance and Data Management

The business processes and information system capabilities in an organization are at the heart of an organization's privacy program. Data governance and data management are discussed here as *influencers* of an organization's privacy program. Also, quite clearly, an organization's privacy program will influence its data governance and data management. Such is the two-way street of the needs of an organization's privacy program and the needs of the organization itself to conduct its business operations.

The practices of data governance and data management are explored fully throughout this book, particularly in Chapters 7 and 8.

Cybersecurity Practices

An organization's cybersecurity practices and capabilities are a part of the core of an organization's privacy program. Like data governance and data management, cybersecurity is an *influencer* of an organization's privacy program. Also, like data governance and management, a privacy program will often influence the organization's cybersecurity policies, processes, and tools to implement all of the safeguards necessary to protect personal information adequately.

Cybersecurity practices and controls are discussed throughout this book, particularly in Chapters 5, 6, and 7.

Information Systems and Their Capabilities

In the age of digital transformation, it is said that organizations can do only what their information systems are capable of enabling. This includes both how information systems are expected to be used and how they are actually used to support business processes. The capabilities of an organization's IT systems greatly influence an organization's privacy program. Like cybersecurity, however, privacy and information systems are a two-way street, inasmuch as a privacy program needs to drive more capabilities into information systems and supporting processes so that the organization attains and maintains its compliance with applicable privacy laws.

Reasons for Privacy Governance

Whether you attribute the emergence of sweeping data privacy laws to citizen backlash or merely a coming of age, organizations everywhere are becoming aware of the fact that people's privacy rights matter, and that ignoring these rights can land an organization in hot water. For the most part, organizations are being forced to change their practices, their information systems, and sometimes even their business models to align with the new reality: organizations must be transparent about how they obtain, collect, process, and pass on personal information.

Governance is management's sharpest tool for getting things done. With regard to privacy laws such as the GDPR and CCPA, organizations have put governance structures in place to oversee the transformation in their business processes and information systems from practices of opaqueness to practices of transparency. In many cases, this transformation meant an about-face on internal practices. Indeed, this has prompted numerous (dare I say the majority of) organizations to "discover" how they are using personal information internally, as though the proverbial foxes have been in charge of the henhouse.

Simply put, privacy governance is all about keeping organizations out of trouble with regulators, outraged citizens, and the courts. Many organizations have had no desire to change their business models, and many complain that it will hurt them financially. Just as the do-not-call lists have curbed the use of unsolicited "robo-calls" in the United States, privacy laws will forever alter business models that include mining and monetizing personal data behind the dark curtains of organizations' marketing machines.

While the foregoing portrays the darker side of some organizations, many others were already "doing the right thing" with regard to transparency in managing personal data. For them, the new journey to privacy compliance has been less impactful.

Because privacy governance is driven by emerging privacy laws, many organizations have legal counsel in their governance structure as experts on the law and its interpretation. As even more privacy laws are enacted, and as case law begins to emerge, organizations will be watching privacy laws develop and will adjust processes and systems accordingly.

The flexibility and capabilities of information systems make it all too easy for organizations to exceed implicitly or explicitly stated purposes for the collection and use of personal information, leading to potential abuses and overreach. As a privacy professional, you must understand the workings of the business with regard to the data about natural persons, how that data is collected, how that data is used, and how it is protected. These four things should be considered when you're building out a privacy governance structure, but the type of information used by the organization will drive the priority that is given to managing data usage and protection. Privacy governance, then, is needed to ensure that the organization's data management and data protection activities do not lead to incidents that can bring harm to affected persons or the organization itself.

Privacy and Security Governance Activities and Results

Within effective privacy and security governance programs in place, an organization's senior management team will see to it that information systems necessary to support business operations are adequately protected and that data about natural persons is properly collected, managed, and used. These are some of the activities required to protect personal data and information:

- **Risk management**   Management will ensure that risk assessments are performed to identify risks in business processes and information systems. Follow-up actions will be carried out to reduce the risk of system failure and compromise.

- **Process improvement**   Management will ensure that key changes are made to business processes that will result in compliance with privacy laws and improvements in cybersecurity.

- **Event identification**   Management will put technologies and processes in place to ensure that privacy and security events and incidents will be recognized and acted upon as quickly as possible.

- **Incident response**   Management will implement incident response procedures that will help to avoid incidents, reduce the impact and probability of incidents, and improve response to incidents so that their impact on the organization is minimized.

- **Improved compliance**   Management will be sure to identify all applicable laws, regulations, standards, and other legal obligations and carry out activities to confirm that the organization is able to attain and maintain compliance.

- **Metrics**   Management will establish processes to measure key privacy and security events such as subject requests, complaints, incidents, policy changes, violations, audits, and training.

- **Resource management**   Management will monitor the allocation of manpower, budget, and other resources to meet privacy and security objectives.

- **Improved IT governance**   Management will implement an effective privacy governance program that will result in better strategic decisions that keep risks at an acceptably low level.

These and other governance activities are carried out through scripted interactions among key business, privacy, security, and IT executives at regular intervals. Meetings will include a discussion of the impact of regulatory changes, alignment with business objectives, the effectiveness of measurements, recent incidents, recent audits, and risk assessments. Other discussions may include such things as changes to the business, recent business results, and any anticipated business events such as mergers or acquisitions.

There are two key results of an effective privacy governance program:

- **Increased trust**   Customers, suppliers, and partners trust the organization to a greater degree when they see that privacy is managed effectively.
- **Improved reputation**   The business community, including customers, investors, and regulators, will hold the organization in higher regard.

Business Alignment

An organization's information privacy program needs to fit in with the rest of the organization. This means that the program needs to align with the organization's highest level guiding principles, including the following:

- **Mission**   Why does the organization exist? Who does it serve, and through what products and services?
- **Goals and objectives**   What achievements does the organization want to accomplish, and when does it want to achieve them?
- **Strategy**   What are the activities that need to take place so that the organization's goals and objectives can be fulfilled?

To be business aligned, privacy and security professionals should be aware of several characteristics of the organization, including these:

- **Business model and processes**   This includes the organization's data flows (particularly flows of personal information), its use of information systems, and its sources of revenue.
- **Sources and uses of personal information**   At the core of a privacy program, it's vital that all sanctioned and unsanctioned uses of personal information are understood, documented, rationalized, and managed.
- **Culture**   This includes how personnel in the organization work, think, and relate to each other. Of utmost importance is the cultural attitude toward the treatment of personal information.
- **Asset value**   This includes information the organization uses to operate. This often consists of intellectual property such as designs, source code, production costs, and pricing, as well as sensitive information related not only to the organization's personnel but its customers, its information processing infrastructure, and its service functions.

- **Risk tolerance**   Risk tolerance for the organization's privacy and information security programs needs to align with the organization's overall tolerance for risk.

- **Legal obligations**   What external laws and regulations govern what the organization does and how it operates? These laws and regulations include GLBA, GDPR, CCPA, and HIPAA. Also, contractual obligations with other parties often shape the organization's behaviors and practices.

- **Market conditions**   How competitive is the marketplace in which the organization operates? What strengths and weaknesses does the organization have in comparison with its competitors? How does the organization want its privacy and security differentiated from its competitors?

- **Privacy law enforcement**   Are regulators and other authorities actively enforcing privacy laws and regulations, or are those laws "paper tigers" that stand unenforced? Organizations are generally reluctant to devote resources to changing business models, business processes, and information systems to comply with laws that may not be enforced.

Goals and Objectives

An organization's goals and objectives specify the activities that are to take place in support of the organization's overall strategy. Goals and objectives are typically statements in the form of imperatives that describe the development or improvement of business capabilities. For instance, goals and objectives may be related to increases in capacity, improvements of quality, or the development of entirely new capabilities. Goals and objectives further the organization's mission, helping it to continue to attract new customers or constituents, increase market share, and increase revenue and/or profitability.

Risk Appetite

Each organization has a particular "appetite" for risk, although few have documented that appetite. ISACA defines *risk appetite* as "the level of risk that an organization is willing to accept while in pursuit of its mission, strategy, and objectives, and before action is needed to treat the risk."

Risk capacity is related to risk appetite. ISACA defines *risk capacity* as "the objective amount of loss that an organization can tolerate without its continued existence being called into question."

Generally, only highly risk-averse organizations such as banks, insurance companies, and public utilities will document and define risk appetite in concrete terms. Other organizations are more tolerant of risk and make individual risk decisions based on gut feeling or qualitative risk analysis. However, because of increased regulation, as well as influence and mandates by customers, many organizations are finding it necessary to document and articulate the risk posture and appetite of the organization. This is an emerging trend in the marketplace but is still relatively new to many organizations.

In a properly functioning risk management program, the chief information security officer (CISO) is rarely the person who makes a risk-treatment decision and is rarely accountable for that decision. Instead, the CISO is a *facilitator* for risk discussions that eventually lead to risk treatment decisions. The only time the CISO would be the accountable party would be when risk treatment decisions directly affect the risk management program itself, such as in the selection of a governance, risk, and compliance (GRC) tool for managing and reporting on risk.

 **NOTE** The data privacy officer (DPO) plays a role in privacy-related risk decisions. Like the CISO, the DPO is a domain expert and guides the business toward decisions that align with applicable laws, internal policies, and the expectations of its affected constituents. Generally, it is business leaders who will make those decisions.

Monitoring Privacy Responsibilities

The practice of monitoring privacy responsibilities helps an organization confirm that personal information is being properly collected, used, protected, and then discarded when no longer needed, and that subject requests are being handled timely and properly. There is no single approach, but several activities provide information to management, including the following:

- **Controls and internal audit**   Developing one or more controls around specific responsibilities increases management's ability to direct key activities. An internal audit of privacy and security controls provides objective analysis on the controls' effectiveness.

- **Metrics and reporting**   Developing metrics for routine activities helps management better understand work output and quality.

- **Work measurement**   This structured activity is used to measure routine tasks carefully to help management better understand the volume of work performed.

- **Performance evaluation**   This traditional qualitative method is used by management to evaluate employee performance.

- **360 feedback**   Soliciting structured feedback from peers, subordinates, and management helps subjects and management better understand characteristics related to specific responsibilities.

- **Position benchmarking**   This technique is used by organizations that want to compare job titles and people holding them with equivalent roles in other organizations. Benchmarking does not involve direct monitoring of responsibilities, but it helps an organization determine whether the appropriate positions are in place and ensure that they are staffed by competent and qualified personnel. This may be useful for organizations that are troubleshooting employee performance.

Privacy Governance Metrics

Metrics are the means through which management can measure key processes and determine whether their strategies are working. Metrics are used in many operational processes, but in this discussion, metrics in the privacy governance context are the emphasis.

In other words, there is a distinction between tactical privacy metrics and those that reveal the state of the overall privacy program.

Privacy metrics are often used to observe technical privacy controls and processes to determine whether they are operating correctly. This helps management better understand the impact of past decisions and can help drive future decisions. Here are some examples of technical metrics:

- Number of personal information records received
- Number of personal information records purged
- Number of personal information records anonymized
- Number of personal information records accessed
- Number of subject data requests received

While useful, these metrics do not address the bigger picture of the effectiveness or alignment of an organization's overall privacy program. They do not answer key questions that boards of directors and executive management often ask, such as the following:

- How much security is enough?
- How should security resources be invested and applied?
- What is the potential impact of a threat event?
- Is our privacy policy aligned with applicable privacy laws?
- Are our privacy practices aligned with customer or constituent expectations?

These and other business-related questions can be addressed through the appropriate metrics, discussed in the remainder of this section.

Privacy strategists sometimes think about metrics in simple categorizations such as these:

- **Key risk indicators (KRIs)**   These metrics are associated with the measurement of risk.
- **Key goal indicators (KGIs)**   These metrics represent the attainment of strategic goals.
- **Key performance indicators (KPIs)**   These metrics are used to show efficiency or effectiveness of privacy- or security-related activities.

Effective Metrics

For metrics to be effective, they need to be measurable. A common way to ensure the quality and effectiveness of a metric is to use the SMART method. A metric that is SMART is:

- Specific
- Measurable
- Attainable

- Relevant
- Timely

 **NOTE**   You can find more information about the development of security metrics in National Institute of Standards and Technology (NIST) Special Publication (SP) 800-55 Revision 1, *Performance Measurement Guide for Information Security*, available at www.nist.gov.

Risk Management

Effective risk management is the culmination of the highest-order activities in information privacy and security programs; these include risk analyses, the use of a risk ledger, formal risk treatment, and adjustments to the suite of privacy and security controls.

While it is difficult to measure the success of a risk management program effectively and objectively, it is possible to take indirect measurements—much like measuring the shadow of a tree to gauge its height. Thus, the best indicators of a successful risk management program would be improving trends in metrics involved with the following:

- Reduction in the number of privacy and security incidents
- Reduction in the impact of privacy and security incidents
- Reduction in the time to remediate privacy and security incidents
- Reduction in the time to remediate vulnerabilities
- Reduction in the number of new unmitigated risks

Regarding the reduction in the number of privacy and security incidents, a privacy and security program improving its maturity from low levels should first expect to see the number of incidents increase. This would be not because of lapses in privacy or security controls but because of the development of—and improvements in—mechanisms used to detect and report privacy and security incidents. If a tree falls in the forest, it will be heard if microphones are installed in key locations. Similarly, as a privacy and security program is improved and matures over time, the number of new risks will, at first, increase and then should later decrease.

Performance Measurement

Metrics on the performance of privacy information security provide measures of timeliness and effectiveness. Generally speaking, performance measurement metrics provide a view of tactical privacy and security processes and activities. As discussed earlier in this section, performance measurements are often the operational metrics that need to be transformed into executive-level metrics for those audiences.

Performance measurement metrics can include any of the following:

- Time to detect privacy and security incidents
- Time to remediate privacy and security incidents
- Time to provision user accounts

- Time to deprovision user accounts
- Time to respond to subject access requests
- Time to discover vulnerabilities
- Time to remediate vulnerabilities

Nearly every operational activity that is privacy- or security-related and measurable is a candidate for performance metrics.

Convergence

Larger organizations with multiple business units, geographic locations, privacy functions, or security functions (often as a result of mergers and acquisitions) may be experiencing issues related to overlapping or underlapping coverage or activities. For instance, an organization that recently acquired another company may have some duplication of effort in the asset management and risk management functions. In another example, local privacy personnel in a large, distributed organization may be performing privacy functions that are also being performed on their behalf from other personnel at headquarters.

Metrics in the category of convergence will be highly individualized, based on specific circumstances in an organization. Categories of metrics may include these:

- Gaps and overlaps in asset coverage
- Gaps and overlaps in data management tools coverage
- Consolidation of licenses for privacy and security tools
- Gaps or overlaps in skills, responsibilities, or coverage

Resource Management

Resource management metrics are similar to value delivery metrics; both convey an efficient use of resources in an organization's privacy program. But because the emphasis here is on program efficiency, these are areas where resource management metrics may be developed:

- Standardization of privacy-related processes—because consistency drives costs down
- Privacy and security involvement in every procurement and acquisition project
- Percentage of personal information records protected by privacy and security controls

Developing Metrics in Layers for Audience Relevance

When embarking on the quest for privacy and security metrics development, a common pitfall is the development of a one-dimensional metrics framework that publishes a set of metrics to all audiences. For instance, a metrics program may publish figures on vulnerabilities discovered, vulnerabilities remediated, privacy incidents, security incidents, and internal audits and their exceptions. Publishing this or a similar set of metrics to various stakeholders will add little value to some audiences and no value to others.

A better approach is the development of operational metrics, which are usually easily discovered and measured. The next step is to transform those operational metrics into different metrics, stated in business terms, for business audiences. In a given organization, a privacy program may employ two, three, or more layers of metrics, usually related to each other, and stated in relevant technical or business terms for each respective audience.

While it may be a good starting point to ask business leaders what kinds of metrics they want to see, in many cases, privacy and security leaders will be asked what metrics *they* want to see. This can be a challenge at times, but by understanding the business, culture, compliance climate, and individuals involved at the stakeholder level, the privacy and security leader can start with a set of metrics that shows success in investments made or use metrics as a call to action by the leadership team.

Privacy Strategy Development

Among business, technology, privacy, and security professionals, there are many different ideas about the meaning of a strategy and the techniques used to develop it, and this can result in general confusion. Although a specific strategy itself may be complex, the concept of a strategy is quite simple. A strategy can be defined as "the plan to achieve an objective."

The effort to build a strategy requires more than saying those six words. Again, however, the idea is not complicated. The concept is this: Understand where you are now and where you want to be. The strategy is the path you must follow to get from where you are (current state) to where you want to be (strategic objective).

The remainder of this section explores strategy development in more detail.

Strategy Objectives

As stated earlier in this section, a strategy is a plan to achieve an objective. The objective (or objectives) is the desired future state for the organization's privacy and security posture and level of risk.

There are, in addition, objectives *of* a strategy:

- **Strategic alignment**   The desired future state, and the strategy to get there, must be in alignment with the organization and *its* strategy and objectives.
- **Effective risk management**   Privacy and security programs must include a risk management policy, processes, and procedures. Without risk management, decisions are made blindly without regard to their consequences or level of risk.

- **Value delivery**   The desired future state of a privacy or security program should include a focus for continual improvement and increased efficiency. No organization has unlimited funds for privacy and security; instead, organizations need to reduce the right risks for the lowest reasonable cost.

- **Resource optimization**   Similar to value delivery, strategic goals should efficiently utilize available resources. Among other things, this means having only the necessary staff and tools required to meet strategic objectives.

- **Performance measurement**   While strategic objectives need to be SMART, the ongoing privacy and privacy-related business operations should themselves be measurable, enabling management to drive continual improvement.

- **Assurance process integration**   Organizations typically operate one or more separate assurance processes in silos that are not integrated. An effective strategy would work to break down these silos and consolidate assurance processes, reducing hidden risks.

All of these should be developed in a way that makes them measurable. This is why these six topics were discussed earlier when discussing governance metrics. These components were made to fit together in this way.

Control Frameworks

While every organization may have unique missions, objectives, business models, tolerance for risk, and so on, organizations need not invent governance frameworks from scratch to manage their privacy and security objectives.

In the context of strategy development, some organizations may already have a suitable control framework in place, while others may not. While it is not always necessary for an organization to select an industry-standard control framework, it is advantageous to do so. Industry-standard control frameworks have been in use in thousands of companies, and they are regularly updated to reflect changing business practices, emerging threats, and new technologies.

Information security is a somewhat more mature profession than privacy. Thus, organizations developing privacy programs and privacy controls are often overlaying privacy controls over an existing information security framework. Doing this is the expected approach, and this idea is reinforced with the recent publication of privacy-specific control frameworks that are extensions of existing information security and risk management frameworks.

It is often considered a mistake to select a control framework because of the presence or absence of a small number of specific controls. Usually, such selection is made on the assumption that control frameworks are rigid and inflexible. Instead, the strategist should take a different approach: select a control framework based on industry alignment and then institute a risk management process for developing additional controls based on the results of risk assessments. This is precisely the approach described in ISO/IEC 27701, as well as in the NIST Privacy Framework. Start with a well-known control framework and then create additional controls, if needed, to address risks specific to the organization. When assessing the use of a specific framework, there may be occasions where a specific control area may not be applicable. In those cases, do not just ignore the section. Document both the business and technical reasons why the organization chose not to use the control area. This will assist when a question is raised at a point in the future as to why the decision was made not to implement the control area. The date and those involved in the decision should also be documented.

Several standard privacy and security frameworks are discussed in the remainder of this section:

- ISO/IEC 27701
- HIPAA
- NIST Privacy Framework
- NIST Cybersecurity Framework (CSF)
- ISO/IEC 27001
- NIST SP 800-53
- NIST SP 800-122
- Center for Internet Security Critical Security Controls (CIS CSC)
- Payment Card Industry Data Security Standard (PCI DSS)

 EXAM TIP CDPSE candidates are not expected to memorize the contents of any privacy or security control frameworks for the exam, but you should be generally aware of them and their purposes.

ISO/IEC 27701

ISO/IEC 27701:2019, *Security techniques – Extension to ISO/IEC 27001 and ISO/IEC 27002 for privacy information management – Requirements and guidelines* is an international standard that directs the formation and management of a Privacy Information Management System (PIMS), including the controls and processes to ensure privacy by design and proper ongoing monitoring and management of personal information. The first version of this standard was published in August 2019.

ISO/IEC 27701 follows a similar structure to ISO/IEC 27001 and is divided into three main sections: requirements, guidance, and controls.

Requirements　The requirements section describes required activities included in effective PIMS. The structure of the requirements section uses the same seven sections as ISO/IEC 27001.

Guidance　The guidance section provides direction on how privacy programs can utilize ISO/IEC 27002 within the PIMS and provides specific considerations for controllers and processors. The guidance section is broken out into three groups: PIMS-specific guidance related to ISO/IEC 27002, additional ISO/IEC 27002 guidance for personally identifiable information (PII) controllers, and additional ISO/IEC 27002 guidance for processors.

PIMS-specific guidance related to ISO/IEC 27002 expands on the 14 control categories to enable an organization to evaluate the control objectives and controls in the context of risks to information security as well as risks to privacy.

The sections on additional ISO/IEC 27002 guidance for PII controllers and additional ISO/IEC 27002 guidance for processors create specific guidance on privacy management for controllers and processors.

Controls The controls section contains a baseline set of controls for controllers and processors that can be used in context with ISO/IEC 27001.

The controls for controllers and processors in ISO/IEC 27701 are described in these four categories:

- Conditions for collection and processing
- Obligations to PII principals
- Privacy by design and privacy by default
- PII sharing, transfer, and disclosure

NOTE ISO/IEC 27701 is available from https://www.iso.org/standard/71670 .html.

HIPAA

The Health Insurance Portability and Accountability Act established requirements for the protection of ePHI. These requirements apply to virtually every corporate or government entity (known as a *covered entity*) that stores or processes ePHI. HIPAA requirements fall into three main categories:

- Administrative safeguards
- Physical safeguards
- Technical safeguards

Several controls reside within each of these three categories. Each control is labeled as Required or Addressable. Required controls must be implemented by every covered entity. Addressable controls are considered optional in each covered entity, meaning the organization does not have to implement an Addressable control if it does not apply or if there is negligible risk if the control is not implemented.

NOTE HIPAA is available from https://www.gpo.gov/fdsys/pkg/CRPT-104hrpt736/pdf/CRPT-104hrpt736.pdf.

NIST Privacy Framework

The NIST Privacy Framework is a guide for organizations that need to protect and properly handle personal information. The Privacy Framework is deliberately organized similarly to the NIST CSF to facilitate the parallel use of both tools. The framework consists of three parts:

- **Core**   A set of privacy protection activities that facilitate the communication of protection activities. There are five core activities: Identify, Govern, Control, Communicate, and Protect.

- **Profile**   An organization's current set of activities used to protect personal information. You can think of the profile as a baseline that can be referenced at a future date to gauge progress, as well as a foundation for defining the desired future state of a privacy program.

- **Implementation Tiers**   These are maturity levels, from least to most mature: Partial, Risk Informed, Repeatable, and Adaptive.

Similar to the extension of ISO/IEC 27001 with ISO/IEC 27701, the integration with the NIST CSF is highlighted in the Privacy Framework Core with a key that identifies whether control objectives are identical to the CSF, or if they align with the CSF but the descriptions have been adapted for privacy programs. This approach reinforces the idea that an effective privacy program requires the integration with information security and risk management programs.

Organizations seeking to adopt the framework will find a wealth of information, including crosswalks (mappings to other standards), profiles, guidance, and tools to build and improve their privacy practices.

NOTE   All of this information is available at https://www.nist.gov/privacy-framework.

NIST Cybersecurity Framework

The NIST CSF is a risk-based life-cycle methodology for assessing risk, enacting controls, and measuring control effectiveness, unlike ISO/IEC 27001. The components of the NIST CSF are as follows:

- **Framework Core**   This set of functions—Identify, Protect, Detect, Respond, Recover—makes up the life cycle of high-level functions in an information security program. The Framework Core includes a complete set of controls (known as *references*) within the four activities.

- **Framework Implementation Tiers**   These are maturity levels, from least mature to most mature: Partial, Risk Informed, Repeatable, Adaptive.

- **Framework Profile**   This is an alignment of elements of the Framework Core (the functions, categories, subcategories, and references) with an organization's business requirements, risk tolerance, and available resources.

Organizations implementing the NIST CSF would first perform an assessment by measuring its maturity (Implementation Tiers) for each activity in the Framework Core. Next, the organization would determine the desired levels of maturity for each activity in the Framework Core. The differences found would be gaps that would need to be filled through several means, which could include the following:

- Hiring additional resources
- Training resources
- Adding or changing business processes or procedures
- Changing system or device configuration
- Acquiring new systems or devices

NOTE The NIST CSF is available from https://www.nist.gov/cyberframework.

ISO/IEC 27001

ISO/IEC 27001, *Information technology – Security techniques – Information security management systems – Requirements*, is an international standard for information security and risk management. This standard contains a requirements section that outlines a properly functioning information security management system (ISMS) and a comprehensive control framework.

ISO/IEC 27001 is divided into two sections: requirements and controls. The requirements section describes required activities found in effective ISMSs. The controls section contains a baseline set of controls that serve as a starting point for the organization. The standard is updated periodically; the latest version was released in 2015 and is known as ISO/IEC 27001:2015. The requirements in ISO/IEC 27001 are described in seven sections: Context of the organization, Leadership, Planning, Support, Operation, Performance evaluation, and Improvement.

While ISO/IEC 27001 is a highly respected control framework, its adoption has been modest, partly because a single copy of the standard costs more than US$100 (unlike NIST standards, which are free of charge); this makes it unlikely that students or professionals will pay this much for a standard to learn more about it. Despite this, ISO/IEC 27001 is growing in popularity in organizations throughout the world.

NOTE ISO/IEC 27001 is available from https://www.iso.org/standard/54534 .html (registration and payment required).

NIST SP 800-53 and NIST SP 800-53A

NIST SP 800-53, *Security and Privacy Controls for Federal Information Systems and Organizations*, is one of the most well-known and adopted security control frameworks. NIST SP 800-53 is required for all US government information systems, as well as all information systems in private industry that store or process information on behalf of the US government.

Even though the NIST 800-53 control framework is required for US federal information systems, many organizations that are not required to employ the framework have utilized it, primarily because it is a high-quality control framework with in-depth implementation guidance and because it is available without cost.

NIST SP 800-53A, *Assessing Security and Privacy Controls in Federal Information Systems and Organizations: Building Effective Assessment Plans*, is the companion standard to NIST SP800-53 that defines techniques for auditing or assessing each control in NIST SP 800-53.

 **NOTE** NIST SP 800-53 and NIST SP 800-53A are available from http://csrc.nist .gov/publications/PubsSPs.html.

NIST SP 800-122

NIST SP 800-122, *Guide to Protecting the Confidentiality of Personally Identifiable Information (PII)*, contains directives for protecting personal information. While the guidelines are required of US government agencies, many other organizations employ them, as they are considered good practices.

 **NOTE** NIST SP 800-122 is available from http://csrc.nist.gov/publications/ PubsSPs.html.

Center for Internet Security Critical Security Controls

The CSC framework from CIS, or CIS CSC, is a control framework that traces its lineage to the SANS Institute. The framework is still commonly referred to as the "SANS Top 20" or "SANS 20 Critical Security Controls."

 **NOTE** CIS CSC controls available from https://www.cisecurity.org/critical-controls/ (registration required).

PCI DSS

PCI DSS is a global control framework specifically for the protection of credit card numbers and related information when stored, processed, and transmitted on an organization's networks. The PCI DSS was developed by the PCI Standards Council, a consortium of the world's dominant credit card brands, namely, Visa, MasterCard, American Express, Discover, and JCB.

PCI DSS is mandatory for all organizations that store, process, or transmit credit card data. Organizations with larger volumes of card data are required to undergo annual onsite audits. Many organizations use the controls and the principles in PCI DSS to protect other types of financial and personal data such as account numbers, Social Security numbers, and dates of birth.

NOTE PCI DSS is available from https://www.pcisecuritystandards.org (registration and license agreement required).

Risk Objectives

A vital part of strategy development is the determination of desired risk levels. One of the inputs to strategy development is the understanding of the current level of risk, and the desired future state may also have a level of risk associated with it.

It is quite difficult to quantify risk, even for the most mature organizations. Getting risk to a reasonable "high-medium-low" is simpler, though less straightforward, and difficult to do consistently across an organization. In specific instances, the costs of individual controls can be known, and the costs of theoretical losses can be estimated, but doing this across an entire risk-control framework is tedious and yet uncertain, because the probabilities of occurrence for threat events amount to little more than guesswork.

NOTE A key part of a security strategy may well be the reduction of risk (it could also be cost reduction or compliance improvement). When this is the case, the strategist will need to employ a method for determining before-and-after risk levels that are reasonable and credible. For the sake of consistency, a better approach would be the use of a methodology— however specific or general—that fits with other strategies and discussions involving risk.

Strategy Resources

A strategy describes the process by which goals and objectives are to be met. Before an organization can develop a privacy and security strategy, it must first understand what privacy and security measures are currently in place. Existing resources paint a picture of an organization's current capabilities, including behaviors, skills, practices, and posture. The gap between the current state and future state can then be filled via technologies, skills, policies, or practices.

Two types of inputs must be considered: those that will influence the development of strategic objectives and those that define the current state of privacy and security programs and their protective controls. The following inputs must be considered before objectives are developed:

- Risk assessments
- Threat assessments

When suitable risk and threat assessments have been completed, a privacy or security strategist can then develop strategic objectives; if objectives have already been created, the strategist can determine whether these strategic objectives will satisfactorily address risks and threats identified in those assessments.

Privacy and security strategists can examine several other inputs to help them understand the workings of the current privacy and security program. Many of these activities are more security-centric than privacy-centric because a successful privacy program requires an effective security program as a foundation. These activities including the following:

- **Program charter**   The organization may have a privacy program charter that defines strategy, roles and responsibilities, objectives, and other matters.

- **Risk assessments**   A risk assessment can reveal privacy and security risks present in the organization. This helps the strategist understand threat scenarios and their estimated impacts and frequency of occurrence. Risk assessment results provide the strategist with valuable information about the types of resources required to bring risks up to acceptable levels. This is vital for developing and validating strategic objectives.

- **Threat assessments**   A threat assessment offers information about the types of threats most likely to have an impact on the organization. It provides an additional perspective on risk, because the assessment focuses on external threats and threat scenarios, regardless of the presence or effectiveness of preventive or detective controls.

NOTE   A threat assessment is an essential element of strategy development. Without a threat assessment, there is a possibility that strategic objectives may fail to address important threats. This would result in a privacy or security strategy that would not adequately protect the organization.

- **Vulnerability assessments**   A vulnerability assessment helps the strategist better understand the current privacy and security postures of the organization's processes and infrastructure. The vulnerability assessment may target personnel, business processes, network devices, appliances, operating systems, subsystems such as web servers and database management systems, and applications—or any suitable combination thereof.

- **Maturity assessments**   A maturity assessment provides valuable information about the maturity of business processes so that the strategist will better understand whether processes are orderly, organized, consistent, measured, examined, and periodically improved.

- **Audits**   Internal and external audits can tell the strategist quite a bit about the state of the organization's privacy and security programs. A careful examination of audit findings can potentially provide significant details on regulatory compliance, control effectiveness, vulnerabilities, disaster preparedness, or other aspects of the program—depending on the objectives of those audits.

NOTE   The topic of audits is discussed in considerable detail in the book *CISA Certified Information Systems Auditor All-In-One Study Guide.*

- **Policies**   An organization's privacy and security policies, as well as its practices with regard to these policies, may say a great deal about its desired current state. Privacy and security policies can be thought of as an organization's internal laws and regulations with regard to the protection and proper use of personal information and other assets. Examining current privacy and security policies can reveal a lot about what behaviors are required in the organization. Assessments, discussed earlier in this list, helps a strategist understand the organization's compliance with its policies.

NOTE   Many organizations align their privacy and security policies with the privacy and security control frameworks they have adopted.

- **Standards**   Privacy and security standards describe, in detail, the methods, techniques, technologies, specifications, brands, and configurations to be used throughout the organization. As with privacy and security policies, it is important that privacy and security managers understand the breadth of coverage, strictness, compliance, and last review and update of the organization's standards. These all indicate the extent to which an organization's privacy and security standards are used—if at all.

- **Guidelines**   The very presence of current and actionable guidelines may signal a higher than average maturity level. Most organizations don't get any further than creating policies and standards, so the presence of proper guidelines means that the organization may have (or had, in the past) sufficient resources or prioritization to make documenting guidance on policies important enough to do. According to their very nature, guidelines are typically written for personnel who need assistance on compliance with policies and standards.

- **Processes and procedures**   An organization's processes and procedures may speak volumes about its level of discipline, consistency, risk tolerance, and the maturity of not only its privacy and security programs but also of IT and the business in general. Like other types of documents discussed in this section, the relevance, accuracy, and thoroughness of process and procedure documents are indicators of maturity and commitment to robust privacy and security programs. Strategists need to confirm whether processes and procedures are actually followed, or if they are merely written artifacts.

- **Architecture**   An organization's documentation of systems, networks, data flows, and other aspects of its environment gives privacy and security strategists a lot of useful information about how the organization has implemented its information systems and the business processes it supports. Documentation in the form of architecture diagrams is as important as written policies, standards, guidelines, and other artifacts. The strategist needs to determine whether the organization's architecture supports the organization's goals, objectives, and operations.

- **Controls**   The strategist should look for artifacts and interview personnel to determine whether specific controls are in place. The presence of documentation alone may not indicate whether controls are being utilized or whether documentation is just more shelfware. Interviewing personnel and observing controls in action is a better way to determine whether controls are in use. Internal and external audits also help the strategist understand the controls' effectiveness. A strategist will also need to understand whether the controls in place are part of a control framework such as ISO/IEC 27701, ISO/IEC 27001, NIST 800-53, CIS CSC, GLBA, HIPAA, or PCI.

- **Skills and knowledge**   An inventory of skills gives the strategist an idea of what staff members are able to accomplish. Understanding skills at all levels helps the strategist understand the types of work that the current staff is able to perform, where minor skills gaps exist, and where the strategist may recommend additional staff through hiring, contracting, or professional services. A key consideration to keep in mind is the potential for a major shift in practices and technologies. A good example is if the organization has been "playing it loose" with personal information and has not yet adopted data governance and data management practices that are required in modern privacy programs. If the staff lacks knowledge about these practices, the organization will struggle to put them into place to comply with applicable privacy regulations.

- **Metrics**   Properly established metrics will serve as a guide for the long-term effectiveness of privacy and security controls and processes. Evaluating metrics helps the strategist understand what works well and where improvement opportunities reside. The strategist can then design end states with more certainty and confidence than if metrics didn't exist.

- **Assets**   The strategist needs to determine whether the organization has sufficient formal asset management practices and records to keep track of its hardware (including virtual machines and other virtual assets), software, and data. Asset management is a key activity for both privacy and security programs—a quote often used among information security professionals is, "You cannot protect what you cannot find."

- **Risk ledger** The presence of a risk ledger can give the strategist a great deal of insight into risk management and risk analysis activities in the organization. Depending on the detail available in the risk ledger, a strategist may be able to discern the scope, frequency, quality, and maturity of risk assessments; the presence of a risk management and risk treatment process; and whether there are records of incidents.

- **Risk treatment decision records** When available, risk treatment records reveal what issues warranted attention, discussion, and decisions. Coupled with the risk ledger, this information can provide a record of issues tackled by the organization's risk management process.

- **Insurance** The privacy or security strategist may want to know whether the organization has cybersecurity insurance or any general insurance policy that covers some types of cyber events and incidents. As important as having cyber insurance is, equally important is the reason the organization purchased it, such as compliance requirements, customer requirements, prior incidents, or a risk treatment decision. It is vitally important to understand the terms of any cyber-insurance policy. While the amounts of benefits are important, the most important aspects of a cyber-insurance policy are its terms, conditions, and exclusions.

- **Data management practices** The strategist needs to understand whether the organization implements formal data management practices, including but not limited to a data classification policy, an internal privacy policy, and whether any tooling exists (such as data loss prevention [DLP] in its many forms) to provide visibility and control over the movement and use of personal information and other sensitive data.

- **Critical data** Privacy and security strategists need to understand the nature and use of an organization's critical data. But first it's important to understand the term *critical*. There are at least three common uses of the term when associated with data: operational criticality, highly sensitive (including personal information), and market criticality (including intellectual property and other competitive data).

- **Business impact analysis** A business impact analysis (BIA) identifies an organization's business processes, the interdependencies between processes, the resources required for process operation, and the impact on the organization if any business process is incapacitated for a time for any reason. It is also useful for privacy and security professionals aside from business continuity purposes, because it gives the security strategist a better idea of which business processes and systems warrant the greatest protection.

NOTE The presence of a recent BIA provides a strong indication of the organization's maturity through its intention to protect its most critical processes from disaster scenarios. Correspondingly, the absence of a BIA suggests that the organization does not consider business continuity and disaster recovery (BCDR) as having strategic importance.

- **Privacy and security incident logs**   Privacy and security incident logs provide the strategist with a history of privacy and security incidents that have occurred in the organization. Depending on the information captured in the incident logs, the strategist may be able to discern the maturity of the organization's privacy and security programs, especially its incident response program. The lack of incident logs is a good indicator of the lack of an incident response process.

- **Outsourced services**   The degree to which a particular organization has outsourced its business applications to the cloud is not the concerning matter. Instead, what's important is the amount of due care exercised in the process of outsourcing, namely whether a formal third-party risk management (TPRM) program is in place.

- **Culture**   The culture of an organization can tell the strategist a lot about the state of privacy and security. Many people mistakenly believe that privacy and information security are all about the technology. While technology is part of privacy and security, people are the most important aspect of a privacy and security program. No amount of technology can adequately compensate for an incorrect attitude and understanding about protecting an organization's information assets or for mishandling of personal information. People are absolutely key.

CAUTION   When considering an organization's culture, the strategist needs to rely more upon the organization's actual operations, as opposed to its statements of culture and values. The actual organizational culture may not align with the organization's claims.

- **Maturity**   The characteristics of privacy and security management programs discussed in this list all contribute to the overall maturity of the organization's program. By itself, the maturity level of the program doesn't tell the strategist anything about the program's details. The strategist's observations of the overall program will provide a visceral feeling for its overall maturity.

- **Risk appetite**   Undocumented in most organizations, risk appetite can be discerned through the record of risk treatment decisions and observation of an organization's executive culture. Even then, however, the attitude and culture of risk appetite may differ from an organization's actual practices.

Privacy Program Strategy Development

After the strategist has performed risk and threat assessments and carefully reviewed the state of privacy and security programs through the examination of artifacts, the strategist can develop strategic objectives. Generally speaking, strategic objectives will fall into one or more of these categories:

- Improvements in data management processes
- Improvements in protective controls
- Improvements in incident visibility
- Improvements in incident response
- Reductions in risk, including compliance risk
- Reductions in cost
- Increased resiliency of key business systems

These categories all contribute to strategic improvements in an organization's privacy and security programs. Depending on the current and desired future state of privacy and security, objectives may represent large projects or groups of projects implemented over several years to develop broad new capabilities, or they may be smaller projects focused on improving existing capabilities.

Here are some examples of broad, sweeping objectives for developing new privacy and security capabilities:

- Define and implement a data loss prevention (DLP) system to provide visibility and control over the movement of personal information.
- Define and implement a SIEM system to provide visibility into privacy, security, and operational events.
- Define and implement a privacy incident response program.
- Define and implement a security awareness learning program.

Here are examples of objectives for improving existing capabilities:

- Integrate vulnerability management and GRC systems.
- Link privacy awareness and access management programs so that staff members must successfully complete privacy awareness training to retain access to systems containing personal information.

Once one or more objectives have been identified, the strategist will undertake several activities that are required to meet the objectives. These activities are explained in the remainder of this section.

 **NOTE** The strategist must consider many inputs before developing objectives and strategies to achieve them. These inputs serve a critical purpose: to help the strategist understand the organization's current state. The journey to developing and achieving a strategy is not possible without understand the journey's starting point. These are discussed in the previous section, "Strategy Resources."

Gap Analysis

In developing a privacy and security strategy and objectives, privacy and security professionals may often spend too much time focusing on the end goals and not enough time on the current state of the organization's privacy and security program. Without having sufficient knowledge of the current state, however, the strategist will find that accomplishing objectives will be more difficult, and achieving success will be less certain.

A gap analysis helps the strategist understand missing capabilities and augment existing capabilities to achieve the desired end state. When performing a gap analysis, the strategist examines the present condition of processes, technologies, and people.

A gap analysis focuses on several aspects of a privacy or security program, including one or more of the items found in the earlier section, "Strategy Resources."

When examining all of this and other information about an organization's privacy and security programs, the strategist should bring the appropriate measure of skepticism. There is much to know about what information is found, but the absence of information may speak volumes as well. Here are some considerations:

- **Absence of evidence is not evidence of absence** This time-honored adage applies to artifacts in any program. For instance, a sparse or nonexistent incident log may be an indication of several things: the organization may not have the required visibility to know when an incident has taken place, the organization's staff may not be trained in the recognition of incidents, or the organization may be watching only for "black swan" events and may be missing routine incidents.

- **Freshness, usefulness, and window dressing** When it comes to policy, process, and procedure documentation, it is important to find out whether documents are created for appearances only (in which case they may be well-kept secrets except by their owners) or whether they are widely known and utilized. A look at these documents' revision histories tells part of the story, while interviewing the right personnel completes the picture by revealing how well the documents' existence is made known and whether they are really used.

- **Scope, turf, and politics** In larger organizations, privacy and security managers need to understand current and historical practices with regard to roles and responsibilities for privacy, security, and related activities. For example, records for a global security program may instead reflect only what is occurring in the Americas, even though there may not be anything found in writing to the contrary.

- **Reading between the lines**   Depending upon the organization's culture and the ethics of current or prior privacy and security personnel, records may not accurately reflect goings-on in the program. In other words, there may be overemphasis, underemphasis, distortions, or simply "look-the-other-way" situations that may result in records being incomplete.

- **Off the books**   For various reasons, certain activities and proceedings in a privacy or security program may not be documented. For example, certain incidents may conveniently *not* be present in the incident log—otherwise, external auditors might catch the scent and go on a foxhunt, causing all manner of unpleasantries.

- **Regulatory requirements**   When examining each aspect of a privacy or security program, the program's manager needs to ask one important question: Is that activity included because it is required by regulations (with hell and fury from regulators if absent) or because the organization is managing risk and attempting to reduce the probability and/or impact of potential threats?

A common approach to determining the future state in a gap analysis is to determine the current maturity of a process or technology and compare that to the desired maturity level. Continue reading in the next section for a discussion on maturity levels.

Strengths, Weaknesses, Opportunities, and Threats Analysis

Strengths, weaknesses, opportunities, and threats (SWOT) analysis is a tool used in support of strategy planning. SWOT involves introspective analysis, where the strategist asks questions about the four components of the object of study:

- **Strengths**   What characteristics of the business give it an advantage over others?

- **Weaknesses**   What characteristics of the business put it at a disadvantage?

- **Opportunities**   What elements in the environment could the business use to its advantage?

- **Threats**   What elements in the environment threaten to harm the business?

SWOT analysis involves the use of a matrix of the four elements, shown in Figure 1-2.

Capability Maturity Models

The Software Engineering Institute (SEI) at Carnegie Mellon University accomplished a great deal with its development of the Capability Maturity Model Integration for Development (CMMi-DEV). Maturity models in other technology disciplines have also been developed, such as the Systems Security Engineering Capability Maturity Model (SSE-CMM) developed by the International Systems Security Engineering Association (ISSEA).

The CMMi-DEV uses five levels of maturity to describe the formality of a process:

- **Level 1: Initial**   This represents a process that is ad hoc, inconsistent, unmeasured, and unrepeatable.

- **Level 2: Repeatable**   This represents a process that is performed consistently and with the same outcome. It may or may not be well-documented.

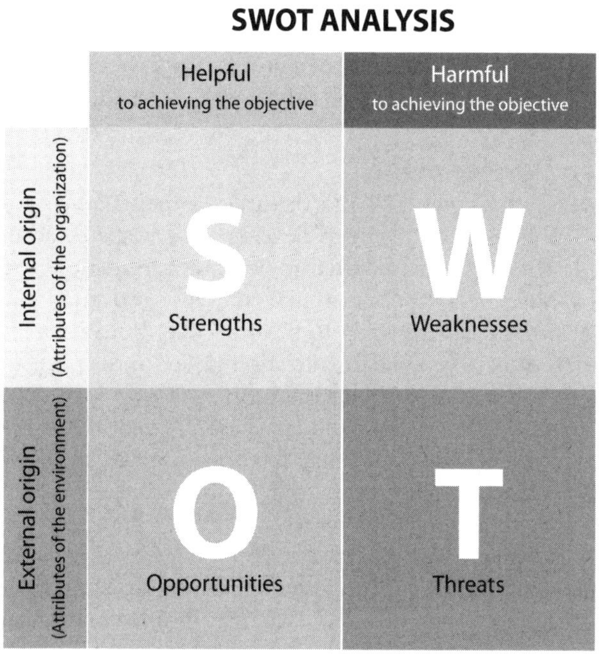

Figure 1-2
A SWOT matrix
with its four
components
(Courtesy of
Xhienne)

SWOT ANALYSIS

	Helpful to achieving the objective	Harmful to achieving the objective
Internal origin (Attributes of the organization)	**S** Strengths	**W** Weaknesses
External origin (Attributes of the environment)	**O** Opportunities	**T** Threats

- **Level 3: Defined**   This represents a process that is well-defined and documented.
- **Level 4: Managed**   This represents a quantitatively measured process with one or more metrics.
- **Level 5: Optimizing**   This represents a measured process that is under continuous improvement.

Not all strategists are familiar with maturity models. Strategists unaccustomed to capability maturity models need to understand two important characteristics of the models and how they are used:

- *Level 5 is not the ultimate objective.* Most organizations' average maturity level targets range from 2.5 to 3.5. There are few organizations whose mission justifies level 5 maturity. The cost of developing a level 5 process or control is often prohibitive and out of alignment with risks.

- *Each control or process will have its own maturity level.* It is neither common nor prudent to assign a single maturity level target for all controls and processes. Instead, organizations with skilled strategists can determine the appropriate level of maturity for each control and process. They need not all be the same. Instead, it is more appropriate to use a threat-based or risk-based model to determine an appropriate level of maturity for each control and process. Some will be 2, some will be 3, some will be 4, and a few may even be 5.

TIP The common use of capability maturity models is the determination of the current maturity of a process, together with analysis, to determine the desired maturity level process-by-process and technology-by-technology.

Road Map Development

Once strategic objectives, risk and threat assessments, and gap analyses have been completed, the strategist can begin to develop road maps to accomplish each objective. A road map is a list of steps required to achieve a strategic objective. The term *road map* is an appropriate metaphor because it represents a journey that, in the details, may not always appear to be contributing to the objective. But in a well-designed road map, each task and each project gets the organization closer to the objective.

A road map is just a plan, but the term *road map* is often used to describe the steps required by an organization to undertake and accomplish a long-term, complex, and strategic objective. Often a road map be thought of as a series of projects—some running sequentially, others concurrently—that an organization uses to transform its processes and technology to achieve the objective.

Figure 1-3 depicts a road map for an 18-month identity and access management project.

A roadmap should be a top-down endeavor, following the usual hierarchy of control of an organization's operations. The roadmap may contain one or more of the following elements:

- **Policy development** Sweeping changes in organization practices around data protection and data management will probably require policy changes to codify expected behavior and system characteristics. While not generally required in most industries, structuring the organization's security policy with one or more relevant standards or frameworks is nonetheless a common practice. Common standards and frameworks used as a structure for security policy include NIST CSF, NIST SP 800-53, ISO/IEC 27001, HIPAA/HITECH, PCI DSS, and CIS CSC. Privacy controls may be adopted from ISO/IEC 27701 or the NIST Privacy Framework.

- **Controls development** The strategist may need to enact one or more controls in specific business processes to ensure desired outcomes related to data management and data protection. Generally, controls are developed (and retired) as a result of a risk assessment, and this may be the case when developing a privacy program strategy.

EXAM TIP CDPSE candidates are not expected to memorize the contents of control frameworks for the exam but are expected to understand their purpose and use.

- **Standards development** Changes in policies, controls, or underlying technologies may necessitate the development or update of one or more standards. While standards are often developed with regard to topics such as passwords and encryption, privacy-related standards can be developed on topics such as aggregation and de-identification.

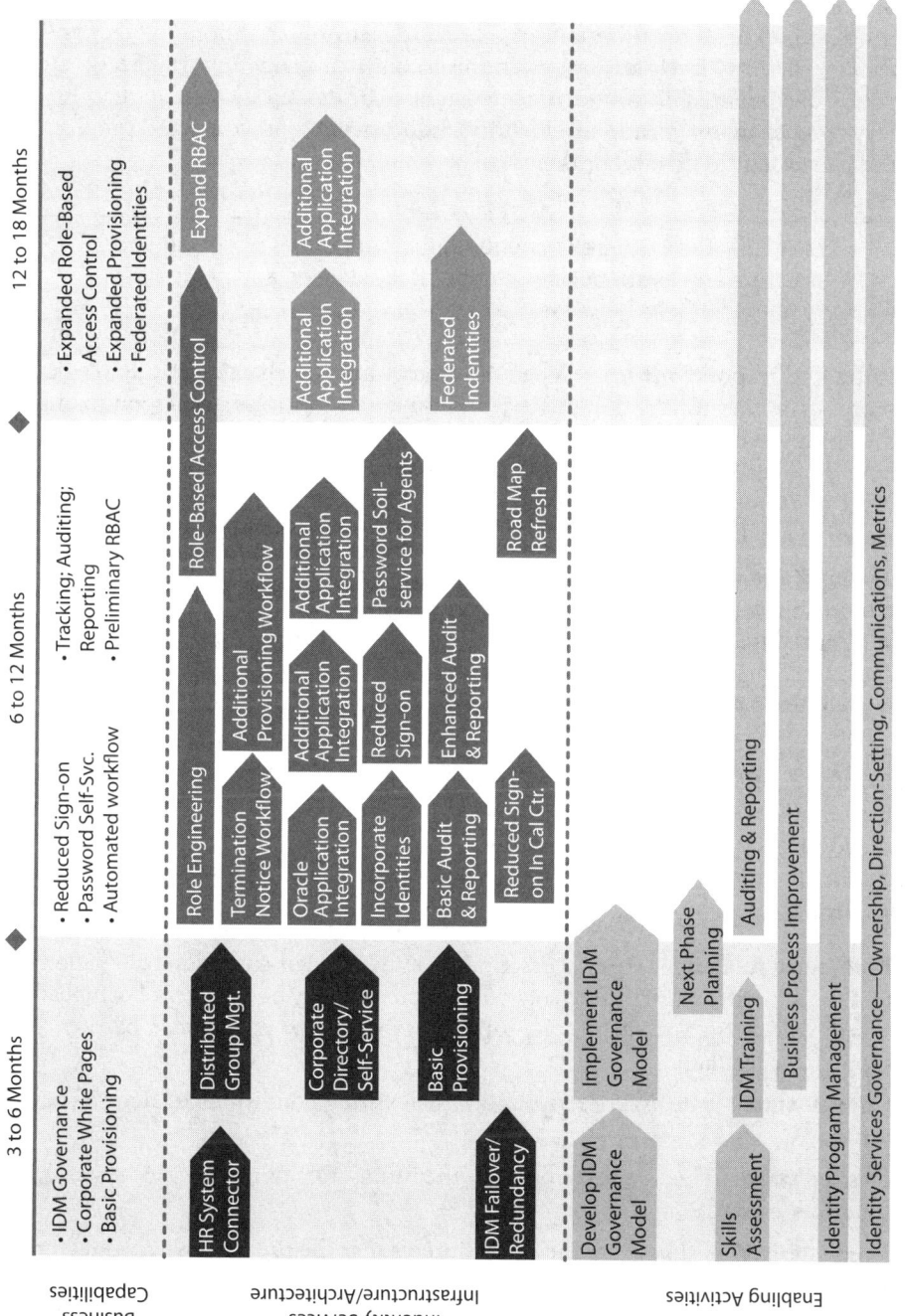

Figure 1-3 Sample road map for identity and access management initiative (Courtesy *Hi-Tech Security Solutions* magazine)

- **Processes and procedures**   Often, the purpose of a new privacy or security strategy is to increase in maturity of privacy- or security-related technologies and activities in an organization. And because many organizations' privacy and security maturity levels are low, often this means that many important tasks are poorly documented or not documented at all. The desired increase in maturity may compel an organization to identify undocumented processes and procedures and assign staff to document them.

 **EXAM TIP**   The CDPSE exam requires that candidates understand the structure and use of policies, standards, guidelines, and procedures.

- **Roles and responsibilities**   When the strategy involves changes in technologies or processes (as they usually do), this may in turn impact the roles and responsibilities for privacy and security personnel, IT workers, and perhaps other staff. Where business processes are added or changed, this often means that changes need to be made to the roles and responsibilities of personnel. There may also be new positions, requiring the development of charter documents and job descriptions.

- **Training and awareness**   Execution of a new privacy or security strategy often has broad reach, impacting technology as well as policies, standards, processes, and procedures. This results in new information, in many forms and for several audiences, including general privacy and security awareness, updated policies and procedures, and new information systems.

Developing a Business Case

Many organizations require the development of a business case prior to approving significant expenditures on privacy or security initiatives. A *business case* is a written statement that describes the initiative and describes its business benefits. Following are the typical elements included in a business case:

- **Problem statement**   This is a description of the business condition or situation that the initiative is designed to solve. The condition may be a matter of compliance, a finding in a risk assessment, or a capability required by a customer, partner, supplier, or regulator.

- **Current state**   This is a description of the existing conditions related to the initiative.

- **Desired state**   This is a description of the future state of the relevant systems, processes, or staff.

- **Success criteria**   These are the defined items that the program will be measured against.

- **Requirements**   This is a list of required characteristics and components of the solution that will remedy the current state and bring about the desired future state.

- **Approach**   This is a description of the proposed steps that will result in the desired future state. This section may include alternative approaches that were considered, with reasons why they were not selected. If the initiative requires the purchase of products or professional services, business cases may include proposals from vendors. Alternatively, the business case may include a request for proposal (RFP) or request for information (RFI) that will be sent to selected vendors for additional information.

- **Plan**   This will include costs, timelines, milestones, vendors, and staff associated with the initiative.

Mature organizations utilize an executive steering committee that evaluates business cases for proposed initiatives and makes go/no-go decisions for initiatives. Business cases are often presented to a steering committee in the form of an interactive discussion, providing business leaders with the opportunity to ask questions and propose alternative approaches.

Characteristics of business cases should include the following:

- **Alignment with organization**   The business case should align with the organization's goals and objectives, risk appetite, and culture.

- **Alignment with regulations**   A business case should cite and align with applicable privacy and data protection regulations.

- **Statements in business terms**   Problem statements, current state, and future state descriptions should all be expressed in business terms.

Establishing Communications and Reporting

Effective communications and reporting are critical elements of successful privacy and security programs. Because success depends mainly on people, in the absence of effective communications, they won't have the required information to make good privacy- and security-related decisions. Without regard for privacy or information security, the results of decisions may include the emergence of unacceptable risks and even harmful incidents.

These are common forms of communications and reporting that are related to privacy and information security:

- **Board of directors meetings**   Discussions of strategies, objectives, risks, incidents, and industry developments keep board members informed about privacy and security in the organization and elsewhere.

- **Governance and steering committee meetings**   Discussions of privacy and security strategies, objectives, assessments, risks, incidents, and developments guide decision-makers as they discuss strategies, objectives, projects, and operations.

- **Privacy and security awareness**   Periodic communications to all personnel help keep them informed on changes in privacy and security policies and standards, good privacy and security practices, and risks they may encounter such as phishing and social-engineering attacks.

- **Privacy and security advisories**   Communications on potential threats help keep affected personnel aware of developments that may require them to take steps to protect the organization from harm.

- **Privacy and security incidents**   Communications internally as well as with external parties during an incident keep incident responders and other parties informed. Organizations typically develop privacy and security incident plans and playbooks in advance, which include business rules on internal communications as well as with outside parties including customers, regulators, and law enforcement.

- **Metrics**   Key metrics are reported upward in an organization, keeping management, executives, and board members informed as to the effectiveness and progress in the organization's privacy and security programs.

When building or expanding a privacy or security program, it's best to utilize existing communications channels and add relevant privacy and security content to those channels, as opposed to building new, parallel channels. Effective privacy and security programs make the best use of existing processes, channels, and methods in an organization.

Obtaining Management Commitment

The execution of a privacy or security strategy requires management commitment. Without that commitment, the strategist will be unable to obtain funding and other resources to implement the strategy.

Getting management commitment is not always straightforward. Often, executives and board members are unaware of their fiduciary responsibilities as well as the potency of modern threats. Many organizations mistakenly believe they are an unlikely target of hackers and cyber-criminal organizations because they are small or uninteresting. Further, the common perception of executives and senior managers is that privacy and security tactical problems are solved with "firewalls and antivirus software" and that privacy and information security are in no way related to business issues and business strategy.

A privacy or security strategist in a situation where top management lacks a strategic understanding about privacy and security will need to embark on efforts to inform top management on one or more aspects of modern privacy or information security management. When success is elusive, it may be necessary to bring in outside experts to convince executives that their privacy or security manager is not attempting to build a kingdom, but instead is just trying to build a basic program to keep the organization out of trouble. As part of developing an effective communication approach, the strategist should not use fear, uncertainty, or doubt in an attempt to move the leadership team toward adopting the strategy. The better approach, as noted in this section, is to relate it to the leadership team in business terms and opportunities to improve business functions.

Strategy Constraints

While the development of a new strategy may bring hope and optimism to the privacy or security team, there is no guarantee that changes in an organization can be implemented without friction and even opposition. Instead, the privacy and security manager should anticipate and be prepared to maneuver around, over, or through many constraints and obstacles.

No privacy or security manager plans to fail. However, the failure to anticipate obstacles and constraints may result in the failure to execute even the best strategy. The presence of an excellent strategy, even with executive support, does not mean that obstacles and constraints will simply step out of the way. Instead, obstacles and constraints represent the realities of human behavior, as well as structural and operational realities that may present challenges to the privacy and security manager and the organization as a whole. There is apt meaning to the phrase "the devil's in the details."

Typical constraints, obstacles, and other issues include the following:

- **Basic resistance to change** It is basic human nature to be suspicious of change, particularly when we as individuals have no control over it and have no say about it. Change is bad, or so we tend to think. "We've always done it this way" is a common refrain. Strategists need to consider methods of involving management and staff members in anticipated changes, such as town hall meetings, surveys, and cross-functional committees.

- **Culture** Organizational culture can be thought of as the collective consciousness of all workers, regardless of rank. Privacy and security strategists should not expect to change the culture significantly but instead should work with the culture when developing and executing the privacy or security strategy.

- **Organizational structure** The strategist must understand the organization's command-and-control structure, which is often reflected by the organizational chart. However, there may be an undocumented aspect to the org chart, which is actually more important: this indicates who is responsible for what activities, functions, and assets.

- **Staff capabilities** A strategy cannot be expected to succeed if the new or changed capabilities do not align with what staff members are able to do. A gap analysis to understand the present state of the organization's privacy or security program (discussed earlier in this chapter) needs to consider staff knowledge, skills, and capabilities. Where gaps are found, the strategy needs to include training or other activities to impart the necessary skills and language to staff.

NOTE When an organization lacks staff with specific knowledge about privacy or security techniques or tools, it may look to external resources to augment internal staff. The strategist needs to consider the costs and availability of with these resources. Consultants and contracts in many skill areas are difficult to find; even larger firms may have backlogs of several months as a result.

- **Budget and cost** The strategist must determine, with a high degree of precision, all of the hard and soft costs associated with each element of a strategy. Often, executive management will want to see alternative approaches; for example, if additional labor is required, the strategist may decide to determine the costs of hiring additional personnel versus the cost of retaining consultants or contractors.

- **Time**   Realistic project planning is needed so that everyone will know when project and strategy milestones will be completed. Project and strategy timelines must take into account all business circumstances, including peak period and holiday production freezes (where IT systems are maintained in a more stable state) and external events such as regulatory deadlines, audits, and other significant events that may impact schedules.

- **Legal and regulatory obligations**   An organization may include items in its strategy that represent business capabilities that are required to exist for legal or regulatory reasons. The enactment of new privacy laws are specifically the subject of this book and represent considerable changes in practices in many organizations. The extraterritorial nature of some new privacy laws complicates this further.

- **Acceptable risk**   Initiatives in the privacy or security strategy need to align with executive management's risk appetite. However, increased pressure from privacy regulations may also be impacting risk appetite and forcing organizations to build more structures and defenses than they would otherwise choose to do.

The Obstacle of Organizational Inertia

Every organization has a finite capacity to undergo change. This is a fact that is often overlooked by overly ambitious strategists who want to accomplish a great deal in too short a time. I have coined the term *organizational inertia* to represent an analogy to Newton's laws of motion: an object either remains at rest or continues to move at a constant velocity, unless acted upon by a force. In an organization, this means that things will be done in the same way until some force requires the organization to change what is done or how things are done. The greater amount of change that is needed, the greater the outside force is required to implement the change.

The nature of organizational inertia, or its resistance to change, is threefold:

- Operational people changing their processes and procedures
- Learning curve
- Human resistance to change

Chapter Review

Privacy and security governance involves the top-down management and control of privacy, security, and risk management in an organization. Governance is usually undertaken through a steering committee that consists of executives from throughout the organization. The steering committee is responsible for setting overall strategic direction and policy, ensuring that security and security strategies are in alignment with one another and also with the organization's IT and business strategy and objectives. The wishes of the steering committee are carried out through projects and tasks that steer the privacy and security organizations toward strategic objectives.

For privacy and information security programs to be successful, they must align with the business and its overall mission, goals, objectives, and strategy. Privacy and security programs must take into account the organization's notion of asset value, culture, risk tolerance/appetite, regulations, legal obligations, and market conditions. Successful and aligned privacy and security programs do not lead the organization, but enable and support it as it carries out its mission and pursues its goals.

Risk appetite refers to the level of risk that an organization is willing to accept while in the pursuit of its mission, strategy, and objectives. Risk treatment and risk acceptance decisions should be assigned to and made by associated business owners and executives who are accountable for those decisions. The chief privacy officer and the chief information security officer facilitate and communicate the information and only in specific instances will own a risk item.

Privacy and security governance is accomplished using the same means as IT governance: it begins with board-level involvement that sets the tone for risk appetite and is carried out through the CPO and CISO, who develop privacy and security policies, as well as strategic privacy and security programs, including data management, software assurance, change management, vendor management, configuration management, incident management, vulnerability management, privacy and security awareness training, and identity and access management.

Privacy and security governance is used to establish roles and responsibilities for privacy- and security-related activities throughout all layers of the organization, from the board of directors to individual staff.

The board of directors is responsible for overseeing all activities in an organization. Boards select and manage a chief executive officer who is responsible for developing a governance function to manage assets, budgets, personnel, processes, and risk.

Privacy and security steering committees are responsible for privacy and security strategic planning. The privacy and security steering committees will develop and approve privacy and security policies and appoint managers to develop and maintain processes, procedures, and standards, all of which should align with each other and with the organization's overall mission, strategy, goals, and objectives.

The CPO is responsible for the protection and proper use of sensitive personal information (often referred to as personally identifiable information, or PII). The CPO's information protection responsibilities are sometimes shared with the CISO, who has overall information protection responsibilities.

The CISO develops business-aligned security strategies that support the organization's overall mission and goals and is responsible for the organization's overall security program, including policy development, risk management, and perhaps some operational activities such as vulnerability management, incident management, access management, and security awareness training. In some organizations, the topmost security executive has the title of chief security officer or chief information risk officer.

Virtually all other roles in IT have privacy and/or security responsibilities, including software development and integration, data management, network management, systems management, operations, service desk, internal audit, and all staff.

A formal metrics program provides both qualitative and quantitative data on the effectiveness of many elements of an organization's privacy and security programs and operations. Metrics can be developed via the SMART method: specific, measurable, attainable, relevant, and timely. Metrics must align with the organization's mission, strategy, and objectives. Some metrics can be used to report on results in the recent past, but some metrics should serve as leading indicators or drive calls to action by the leadership team.

A common shortcoming of a metrics program is its failure to provide relevant metrics for various audiences. For instance, reporting the number of packets dropped by a firewall or the number of viruses detected by antivirus to the board of directors provides no value for that audience. As an organization develops its metrics program, it must take care to develop metrics that matter for each audience.

A strategy is a plan to achieve a defined set of objectives to enable the vision of the organization to be successfully achieved. Objectives are the desired future states in an organization and, in the context of privacy, in the organization's privacy program. A strategy should be business-aligned, should deliver value, should optimize resources, and should be measurable.

The development of an overall privacy strategy includes the adoption of or alignment with a control framework. Most organizations select an industry-standard control framework such as COBIT, NIST 800-53, ISO/IEC 27001, CIS CSC, HIPAA, or PCI DSS. Then, as part of the organization's risk management processes, additional controls are added as identified risks warrant. Evolving privacy frameworks such as the NIST Privacy Framework or ISO/IEC 27701 can be used to identify privacy controls that should be added to an organization's security control tool bag.

Strategy development may include establishing desired risk levels. This may be expressed in qualitative or quantitative terms, depending upon an organization's maturity. Many resources are needed for the development of a strategy. These resources include several types of information that reveal the current state of the organization, including risk assessments, vulnerability assessments, business impact analyses, metrics, risk ledgers, and incident logs. Several other inputs are required that define the structure of a privacy or security program, including policies, standards, guidelines, processes and procedures, insurance, and outsourced services.

To develop a strategy, the strategist must first understand the organization's present state and then define one or more desired future states. A gap analysis helps the strategist understand missing capabilities. The development of a road map defines the steps to develop missing capabilities and augment existing capabilities so that the strategy will be realized.

The strategist may choose to use the strengths, weaknesses, opportunities, and threats (SWOT) analysis model in support of strategic planning. The strategist may also employ capability maturity models to help determine appropriate future states of key security processes.

Often it is necessary to build a business case so that executive management will agree to support and fund a strategy. A business case typically includes a problem statement, followed by a description of the current state, the desired future state, requirements, an approach, and a plan to achieve the strategy. Often a business case is reviewed by a business or IT steering committee consisting of business stakeholders.

A strategist must be aware of potential obstacles to achieving strategic objectives, including organizational culture, organizational structure, existing staff capabilities, budgets, time, and legal and regulatory obligations. A business-aligned strategy should take these obstacles into account and minimize them if the strategy is to be approved and achieved.

Quick Review

- The addition of privacy and information security as part of fiduciary duty by board members and executives is an important and growing trend in business today.

- Privacy and security executives and the board of directors are responsible for implementing a privacy and security governance model encompassing privacy strategy, security strategy, and other mandates. As a result, the industry is seeing a shift from a passive to a more active board when it comes to privacy and cybersecurity issues.

- Privacy and security programs should be in alignment with the organization's overall mission, goals, and objectives. This means that the chief privacy officer, chief information security officer, and others should be aware of, and involved in, strategic initiatives and the execution of the organization's strategic goals.

- Risk appetite is generally expressed in qualitative terms such as "very low tolerance for risk" and "no tolerance for risk." Different activities in an organization will have different risk appetites.

- An organization's definitions of roles and responsibilities may or may not be in sync with its culture of accountability. For instance, an organization may have clear definitions of responsibilities documented in policy and process documents and yet may rarely hold individuals accountable when preventable security events occur.

- Ideally, an organization's board of directors will be aware of privacy and security risks and may direct that the organization enact safeguards to protect the organization. However, in many organizations, the board of directors is still uninvolved in privacy and security matters; in these cases, it is still possible to have successful risk-based privacy and security programs, provided they are supported by senior executives.

- Privacy and information security is the responsibility of every person in an organization; however, the means for assigning and monitoring privacy and security responsibilities to individuals and groups vary widely.

- The NIST Cybersecurity Framework and Privacy Framework are emerging as foundations for the development of high-level metrics. Still, it is up to every privacy and security leader to develop metrics that are meaningful and applicable to various audiences with an interest in receiving them.

- The methodology for calculating return on security investment is widely discussed but not widely practiced, mainly because it is difficult to calculate the benefit of security controls designed to detect or prevent incidents that occur infrequently.

- Privacy and security managers developing strategies in an organization without a program will need to rely on their past experiences, anecdotal accounts of practices, and policies in the organization.

- Privacy and security strategists should be mindful of each organization's tolerance for change within a given period of time. While much progress may be warranted, the amount of change that can be reasonably implemented within a short amount of time is limited.

- Many organizations ruminate over the selection of a control framework. Instead, the organization should select a framework and then make adjustments to its controls to suit the business. A control framework should generally be considered a starting point, not a rigid and unchanging list of controls—except in cases where regulations stipulate that controls may not be changed.

- While it is important for privacy and security strategists to understand the present state of the organization when developing a strategic road map, the strategist must proceed knowing that there can never be a sufficient level of understanding. Besides, if even the most thorough snapshot has been taken, the organization is slowly (or perhaps quickly) changing anyway. Execution of a strategic plan is intended to accelerate changes in certain aspects of an organization that is slowly changing anyway.

- Capability maturity models are useful tools for understanding the maturity level of a process and developing desired future states. The maturity of processes in the organization will vary, as it is appropriate for some processes to have high maturity while it is acceptable for others to have lower maturity. The right question to ask about each separate process is "What is the appropriate level of maturity for this process?"

- Each organization has its own practice for the development of business cases for the presentation, discussion, and approval for strategic initiatives.

- Privacy and security strategists must anticipate obstacles and constraints affecting the achievement of strategic objectives and consider refining those objectives so that they can be realized.

Questions

1. Privacy governance is most concerned with:

    A. Privacy policy

    B. Security policy

    C. Privacy strategy

    D. Security executive compensation

2. A gaming software startup company does not employ penetration testing of its software. This is an example of:

    A. High tolerance of risk

    B. Noncompliance

    C. Irresponsibility

    D. Outsourcing

3. An organization's board of directors wants to see quarterly metrics on risk reduction. What would be the best metric for this purpose?

    A. Number of data subject requests received

    B. Viruses blocked by antivirus programs

    C. Packets dropped by the firewall

    D. Time to patch vulnerabilities on critical servers

4. Which of the following metrics is the best example of a leading indicator?

    A. Average time to mitigate security incidents

    B. Increase in the number of attacks blocked by the intrusion prevention system (IPS)

    C. Increase in the number of attacks blocked by the firewall

    D. Percentage of critical servers being patched within service level agreements (SLAs)

5. The term *legitimate interest* refers to what privacy activity?

    A. The basis for a user access request

    B. Whether data collection is allowed by law

    C. The legal basis for processing personal information

    D. An alternative to lawful processing of personal information

6. The best definition of a strategy is:

    A. The objective to achieve a plan

    B. The plan to achieve an objective

    C. The plan to achieve business alignment

    D. The plan to reduce risk

7. The primary factor related to the selection of a control framework is:

   A. Industry vertical

   B. Current process maturity level

   C. Size of the organization

   D. Compliance level

8. As part of understanding the organization's current state, a privacy strategist is examining the organization's privacy policy. What does the policy tell the strategist?

   A. The level of management commitment to privacy

   B. The compliance level of the organization

   C. The maturity level of the organization

   D. None of these

9. While gathering and examining various privacy-related business records, the privacy officer has determined that the organization has no privacy or security incident log. What conclusion can the privacy officer make from this?

   A. The organization does not have privacy or security incident detection capabilities.

   B. The organization has not yet experienced a privacy or security incident.

   C. The organization is recording privacy or security incidents in its risk register.

   D. The organization has effective privacy policies.

10. One primary difference between GDPR and CCPA is:

    A. GDPR requires an opt out while CCPA requires an opt in.

    B. Only GDPR asserts extraterritorial jurisdiction.

    C. Only CCPA asserts extraterritorial jurisdiction.

    D. GDPR requires an opt in while CCPA requires an opt out.

11. A privacy strategist has examined a business process and has determined that personnel who perform the process do so consistently, but there is no written process document. The maturity level of this process is:

    A. Initial

    B. Repeatable

    C. Defined

    D. Managed

12. A privacy strategist has examined several business processes and has found that their individual maturity levels range from Repeatable to Optimizing. What is the best future state for these business processes?

    A. All processes should be changed to Repeatable.

    B. All processes should be changed to Optimizing.

C. There is insufficient information to determine the desired end states of these processes.

D. Processes that are Repeatable should be changed to Defined.

13. In an organization using HIPAA as its control framework, the conclusion of a recent risk assessment stipulates that additional controls not present in HIPAA but present in ISO/IEC 27001 should be enacted. What is the best course of action in this situation?

A. Adopt ISO/IEC 27001 as the new control framework.

B. Retain HIPAA as the control framework and update process documentation.

C. Add the required controls to the existing control framework.

D. Adopt NIST SP 800-53 as the new control framework.

14. A privacy strategist is seeking to improve the privacy program in an organization with a strong but casual culture. What is the best approach here?

A. Conduct focus groups to discuss possible avenues of approach.

B. Enact new detective controls to identify personnel who are violating policy.

C. Implement security awareness training that emphasizes new required behavior.

D. Lock users out of their accounts until they agree to be compliant.

15. A privacy strategist recently joined a retail organization that operates with slim profit margins and has discovered that the organization lacks several important privacy capabilities. What is the best strategy here?

A. Insist that management support an aggressive program quickly to improve the program.

B. Develop a risk ledger that highlights all identified risks.

C. Recommend that the biggest risks be avoided.

D. Develop a risk-based strategy that implements changes slowly over an extended period of time.

Answers

1. **C.** Privacy governance is the mechanism through which a privacy strategy is established, controlled, and monitored. Long-term and other strategic decisions are made in the context of privacy governance.

2. **A.** A software startup in an industry like gaming is going to be highly tolerant of risk: time to market and signing up new customers will be its primary objectives. As the organization achieves viability, other priorities such as security will be introduced.

3. **D.** The metric on time to patch critical servers will be the most meaningful metric for the board of directors. The other metrics, while potentially interesting at the operational level, do not convey business meaning to board members.

4. **D.** The metric of percentage of critical servers being patched within SLAs is the best leading indicator because it is a rough predictor of the probability of a future security incident. The other metrics are trailing indicators because they report on past incidents.

5. **C.** *Legitimate interest* refers to a legal basis for the collection and processing of personal information where the interests of the organization collecting and processing personal information is balanced with the data subject's interests.

6. **B.** A strategy is the plan to achieve an objective. An objective is the "what" that an organization wants to achieve, and a strategy is the "how" the objective will be achieved.

7. **A.** The most important factor influencing a decision of selecting a control framework is the industry vertical. For example, a healthcare organization would likely select HIPAA as its primary control framework, whereas a retail organization might select PCI DSS.

8. **D.** By itself, privacy policy tells someone little about an organization's privacy practices. An organization's policy is only a collection of statements; without examining business processes, business records, and interviewing personnel, a privacy professional cannot develop any conclusions about an organization's privacy practices.

9. **A.** An organization that does not have a privacy or security incident log probably lacks the capability to detect and respond to an incident. It is not reasonable to assume that the organization has had no incidents, since minor incidents occur with regularity. Claiming that the organization has effective controls is unreasonable, as it is understood that incidents occur even when effective controls are in place (because not all types of incidents can reasonably be prevented).

10. **D.** GDPR requires that organizations provide data subjects an opportunity to opt in to be included in the collection and use of personal information. CCPA requires organizations to provide an opportunity to opt out for data subjects who no longer want their personal data used by organizations.

11. **B.** A process that is performed consistently but is undocumented is generally considered to be Repeatable.

12. **C.** There are no rules that specify that the maturity levels of different processes need to be the same or at different values relative to one another. In this example, each process may already be at an appropriate level based on risk appetite, risk levels, and other considerations.

13. **C.** An organization that needs to implement new controls should do so within its existing control framework. It is not necessary to adopt an entirely new control framework when a few controls need to be added.

14. **A.** Organizational culture is powerful, as it reflects how people think and work. In this example, there is no mention that the strong culture is bad, only that it is casual. Punishing people for their behavior may cause resentment, a revolt, or the loss of good employees who decide to leave the organization. The best approach here is to try to understand the culture and work with people in the organization to figure out how a culture of privacy and security can be introduced successfully.

15. **D.** A privacy strategist needs to understand an organization's capacity to spend its way to lower risk. It is unlikely that an organization with low profit margins is going to agree to an aggressive improvement plan. Developing a risk ledger that depicts these risks may be a helpful tool for communicating risk, but by itself, it involves is no action to change anything. Similarly, recommending risk avoidance may mean discontinuing the very operations that bring in revenue.

Management

In this chapter, you will learn about
- Privacy-related roles and responsibilities
- Privacy training and awareness
- Third-party risk management
- Auditing privacy processes and systems
- Privacy and security incident response

This chapter covers Certified Data Privacy Solutions Engineer (CDPSE) job practice 1, "Privacy Governance," part B, "Management." The entire Privacy Governance domain represents 34 percent of the CDPSE) examination.

Privacy management typically focuses on business operations concerned with the collection, use, and disposal of personal information. The scope of privacy management includes business processes and the information systems supporting them.

The processes discussed in this chapter include IT and security roles and responsibilities and their support of privacy, privacy policy, data monitoring, working with data subjects and authorities, privacy training, third-party risk management, auditing privacy operations, privacy incident management, and continuous improvement of a privacy program.

Privacy Roles and Responsibilities

Privacy and information security governance are most effective when every person in the organization knows what is expected of them. More mature organizations develop formal roles and responsibilities that establish clear expectations for personnel with regard to their part in all matters related to the protection and proper use of systems and personal information.

In the context of organizational structure and behavior, a *role* is a description of normal activities that employees are obliged to perform as part of their employment. Roles are typically associated with a *job title* or *position title*, a label assigned to each person that designates his or her place in the organization. Organizations strive to adhere to more or less standard position titles so that other people in the organization, upon knowing someone's position title, will have at least a general idea of a person's role in the organization.

Typical roles include the following:

- IT auditor
- Systems engineer
- Privacy analyst
- Accounts receivable manager
- Individual contributor

Often a position title also consists of a person's *rank,* which denotes a person's seniority, placement within a command-and-control hierarchy, span of control, or any combination of these. Typical ranks include the following, in order of increasing seniority:

- Supervisor
- Manager
- Senior manager
- Director
- Senior director
- Executive director
- Vice president
- Senior vice president
- Executive vice president
- President
- Chief executive officer
- Member, board of directors
- Chairman, board of directors

This should not be considered a complete listing of ranks. Larger organizations also include the modifiers *assistant* (as in assistant director), *general* (general manager), *associate* (a junior position), and *first* (first vice president).

A *responsibility* is a statement of outcomes that a person is expected to support. Like roles, responsibilities are typically documented in position descriptions and job descriptions. Typical responsibilities include the following:

- Perform monthly corporate expense reconciliation
- Troubleshoot network faults and develop solutions
- Audit internal privacy controls and prepare exception reports

In addition to specific responsibilities associated with individual position titles, organizations typically also include general responsibilities in all position titles. Examples include the following:

- Understand and conform to information security policy, data protection policy, harassment policy, and other policies
- Understand and conform to a code of ethics and behavior

In the context of privacy and information security, an organization assigns roles and responsibilities to individuals and groups to meet the organization's privacy and security strategies and objectives.

RACI Charts

Many organizations utilize Responsible-Accountable-Consulted-Informed (RACI) charts to denote key responsibilities in business processes, projects, tasks, and other activities. A RACI chart assigns levels of responsibility to individuals and groups. The development of a RACI chart helps personnel determine roles for various business activities. A typical RACI chart is shown in the following table:

Activity	Responsible	Accountable	Consulted	Informed
Request user account	End user	End user manager	IT service desk, end user manager	Asset owner, security team
Approve user account	Asset owner	Chief operating officer	End user manager, security team	End user, internal audit, IT service desk
Provision user account	IT service desk	IT service manager	Asset owner	End user, end user manager, security team, privacy team
Audit user account	Internal auditor	Internal audit manager	Asset owner, privacy manager	IT service desk, IT service manager, end user manager, privacy team

The same RACI chart can also be depicted as a second example in the next table. This RACI chart specifies the roles carried out by several parties in the user account access request process:

Activity	End User	Manager	IT Service Desk	IT Service Manager	Asset Owner	COO	Internal Audit	Audit Manager	Security Team	Privacy Team
Request user account	R	A	I		I			I		
Approve user account	I	C	I	I	R	A	I		C	
Provision user account	I	I	R	A	C				I	I
Audit user account		I	I	I	C		R	A	I	IC

The meanings of the four roles in a RACI chart are as follows:

- **Responsible (R)** The person or group that performs the actual work or task.
- **Accountable (A)** The person who is ultimately answerable for complete, accurate, and timely execution of the work. This person often manages those in the Responsible role.

(continued)

- **Consulted (C)** One or more people or groups who are consulted for their opinions, experience, or insight. People in the Consulted role may be subject-matter experts for the work or task, or they may be owners, stewards, or custodians of an asset associated with the work or task. Communication with the Consulted role is two-way.
- **Informed (I)** One or more people or groups who are informed by those in other roles. Depending on the process or task, the Informed role may be told of an activity before, during, or after completion. Communication with Informed is one-way.

Several considerations must be taken into account when assigning roles to individuals and groups in a RACI chart, including the following:

- **Skills** Some or all individuals in a team assignment, as well as specifically named individuals, need to have the skills, training, and competence to carry out tasks as required.
- **Segregation of duties** Critical tasks, such as the user account provisioning the RACI chart depicted earlier, must be free of segregation-of-duties conflicts. This means that two or more individuals or groups are required to carry out a critical task. In this example, the requestor, approver, and provisioner cannot be the same person or group.
- **Conflict of interest** Critical tasks must not be assigned to individuals or groups when such assignments will create conflicts of interest. For example, a user who is an approver cannot approve a request for his or her own access. In this case, a different person must approve the request—while also avoiding a segregation-of-duties conflict.

There are some variations of the RACI model, including PARIS (Participant, Accountable, Review Required, Input Required, Sign-off Required) and PACSI (Perform, Accountable, Control, Suggest, Informed).

Board of Directors

The board of directors in an organization is a body that oversees activities in an organization. Depending on the type of organization, board members may be elected by shareholders or constituents, or they may be appointed. This role can be either paid or voluntary in nature.

Activities performed by the board of directors, as well as directors' authority, are usually defined by a constitution, bylaws, or external regulation. The board of directors is typically accountable to the owners of the organization or, in the case of a government body, to the electorate.

In many cases, board members have a *fiduciary duty*. This means they are accountable to shareholders or constituents to act in the best interests of the organization with no appearance of impropriety, conflict of interest, or ill-gotten profit.

In nongovernment organizations, the board of directors is responsible for appointing a chief executive officer (CEO) and possibly other executives. The CEO, then, is accountable to the board of directors and carries out the board's directives. Board members may also be selected for any of the following reasons:

- **Investor representation**   One or more board members may be appointed by significant investors to give them control over the organization's strategy and direction.

- **Business experience**   Board members bring outside business management experience, which helps them develop successful business strategies for the organization.

- **Access to resources**   Board members bring business connections, including additional investors, business partners, suppliers, or customers.

Often, one or more board members will have business finance experience to bring financial management oversight to the organization. In the case of US public companies, the Sarbanes-Oxley Act requires board members to form an audit committee; one or more audit committee members are required to have financial management experience. External financial audits and internal audit activities are often accountable directly to the audit committee to perform direct oversight of the organization's financial management activities. As the issues of privacy and information security become more prevalent in discussions at the executive level, some organizations have added a board member who is technically savvy or have formed an additional committee often referred to as the technology risk committee.

The board of directors is generally expected to require that the CEO and other executives implement a corporate *governance* function to ensure that executive management has an appropriate level of visibility and control over the operations of the organization. Executives are accountable to the board of directors to demonstrate that they are effectively carrying out the board's strategies.

Many, if not most, organizations are highly dependent upon information technology for their daily operations. Many also process personal information for their workforce and often for their customers or constituents. As a result, privacy and information security are important topics to boards of directors. Today's standard of due care for corporate boards requires that they include privacy and information security considerations in the strategies they develop and the oversight they exert on their organizations. In its publication *Cyber-Risk Oversight*, the National Association of Corporate Directors (NACD) has developed five principles about the importance of information security:

- Principle 1: Directors need to understand and approach cybersecurity as an enterprise-wide risk management issue, not just an IT issue.

- Principle 2: Directors should understand the legal implications of cyber risks as they relate to their specific circumstances.

- Principle 3: Boards should have adequate access to cybersecurity expertise, and discussions about cyber-risk management should be given regular and adequate time on board meeting agendas.

- Principle 4: Boards should set the expectation that management will establish an enterprise-wide cyber-risk management framework with adequate staffing and budget.

- Principle 5: Board management discussions about cyber risk should include identifying which risks to avoid, which to accept, and which to mitigate or transfer through insurance, as well as specific plans associated with each approach.

The wording of these information security principles makes them entirely relevant to the mission of protecting personal information and to its proper usage.

Executive Management

Executive management is responsible for carrying out directives issued by the board of directors. In the context of privacy and information security management, this includes ensuring that the organization has sufficient resources available to implement privacy and security programs and to develop and maintain controls to protect critical assets and personal information.

Executive management must ensure that priorities are balanced. In the case of IT, privacy, and security, these functions are usually tightly coupled, but are sometimes in conflict. IT's primary mission is the development and operation of business-enabling capabilities through the use of information systems. In contrast, the missions of privacy and information security include protection, compliance, and proper usage. Executive management must ensure that these sometimes-conflicting missions successfully coexist.

Typical IT, privacy, and security-related executive position titles include the following:

- Chief information officer (CIO)
- Chief technology officer (CTO)
- Chief privacy officer (CPO) or data protection officer (DPO)
- Chief information security officer (CISO)

To ensure the success of the organization's privacy and information security programs, executive management should be involved in three key areas.

- **Ratification and enforcement of corporate privacy and security policies** This may take different forms, such as formal minuted ratification in a governance meeting, a statement for the need for compliance along with a signature within the body of the privacy or security policy document, a separate memorandum to all personnel, or other visible communication to the organization's rank and file that stresses the importance of and need for compliance to the organization's privacy and information security policies.

- **Leadership by example** Executive management should lead by example and not exhibit behavior suggesting they are "above" policy. Executives should not have the appearance of enjoying special privileges of a nature that suggests that one or more policies do not apply to them. Instead, their behavior should visibly support privacy and security policies that all personnel are expected to comply with.

- **Ultimate responsibility**   Executives are ultimately responsible for all actions carried out by the personnel who report to them. Executives are also ultimately responsible for all outcomes related to organizations to which operations have been outsourced.

Privacy and Security Steering Committees

Many organizations form a security and privacy steering committee—separate or combined—consisting of stakeholders from many (if not all) of the organization's business units, departments, functions, and key locations. Some organizations will separate privacy and security into separate committees, especially if there are differences in membership or focus.

A privacy or security steering committee may have a variety of responsibilities, including the following:

- **Risk treatment deliberation and recommendation**   The steering committee may discuss relevant risks and potential avenues of risk treatment, and develop recommendations for said risk treatment for ratification by executive management.

- **Prioritization, discussion, and coordination of IT, privacy, and security projects**   The steering committee members may discuss various IT, privacy, and security projects to resolve any resource or scheduling conflicts. They may also address potential conflicts between multiple projects and initiatives and work out solutions.

- **Review of recent risk assessments**   The steering committee may discuss recent risk assessments to develop a shared understanding of their results, as well as discuss remediation of findings.

- **Discussion of new laws, regulations, and requirements**   The committee may discuss new laws, regulations, and requirements that may impose changes in the organization's operations. Committee members can develop high-level strategies that their respective business units or departments can further build out.

- **Review of recent privacy and security incidents**   Steering committee members can discuss recent privacy and security incidents and their root causes. Often this can result in changes in processes, procedures, or technologies to reduce the risk and impact of future incidents.

Reading between the lines, the primary mission of a steering committee is to identify and resolve conflicts and to maximize the effectiveness of privacy and security programs, as balanced among other business initiatives and priorities.

Business Process and Business System Owners

Business process and system owners are typically nontechnical personnel in management positions in an organization. While they may not be technology or compliance experts, in many organizations, their business processes are enhanced by IT in business applications and other capabilities. In the context of information privacy, the term "business system" includes databases containing personal information.

Remembering that IT, privacy, and information security functions serve the organization and not the other way around, business process and business system owners are accountable for making business decisions that sometimes impact the use of IT, the use of personal information, the organization's security posture, or any combination of these. A simple example is a decision on whether an individual employee should have access to specific personal information. While IT or security may have direct control over which personnel have access to what information, the best decision is a policy-backed business decision by the manager responsible for the information.

The responsibilities of business process and business system owners include the following:

- **Access grants**   Process owners decide whether individuals or groups should be given access to the system, as well as the level and type of access.
- **Access revocation**   Process owners should decide when individuals or groups no longer require access to a system, signaling the need to revoke that access.
- **Access reviews**   Process owners should periodically review access lists to see whether each person and group should continue to have their access.
- **Subject inquiries and requests**   Process owners receive privacy-related inquiries from data subjects in the form of queries about personal data usage, corrections to personal data, opt-in and opt-out requests, requests to be removed, and complaints.
- **Configuration**   Process owners determine the configuration needed for systems and applications, ensuring their proper function and support of applications and business processes.
- **Function definition**   In the case of business applications and services, process owners determine which functions will be available, how they will work, and how they will support business processes. Typically, this definition is constrained by functional limitations within an application, service, or product.
- **Process definition**   Process owners determine the sequence, steps, roles, and actions carried out in their business processes.
- **Physical location**   Process owners determine the physical location of their systems. Factors influencing location choices include physical security, proximity to other systems, proximity to relevant personnel, and data protection and privacy laws.

Often, business and system owners are nontechnical personnel, so it may be necessary to translate business needs and applicable laws and regulations into technical specifications.

 **EXAM TIP**   For the exam, do not confuse the terms *business owner* and *system owner* with persons who possess a majority of shares of the organization. Instead, these terms connote responsibility for business operations.

Custodial Responsibilities

In many organizations, system owners are not involved in the day-to-day activities related to the management of their systems, especially when those systems are applications and the data used by them. Instead, somebody (or several people) in the IT organization acts as a proxy for system owners and makes access grants and other decisions on their behalf.

Although this is a common practice, it is often carried too far, resulting in the system owner being virtually unaware, uninvolved, and uninformed. Instead, system owners should be aware of, and periodically review, activities carried out by people, groups, and departments making decisions on their behalf.

The most typical arrangement is that people in IT make access decisions on behalf of system owners based on established policies and practices. Except in cases where there is a close partnership between these IT personnel and system owners, these IT personnel often do not adequately understand the business nature of systems or the implications when certain people are given access to them. Most often, far too many staff members have access to systems, usually with higher privileges than necessary.

Privacy by Design

Privacy by design involves proactively embedding privacy into the design and operation of IT systems, networked infrastructure, and business practices. The principle of privacy by design is explicitly stated in General Data Protection Regulation (GDPR) Article 25, "Data protection by design and by default." This principle should be included in every organization's privacy policy, whether they are subject to GDPR or other privacy regulations.

This is easier said for new information systems that benefit from a "clean-sheet" design. It is more difficult and costly to retrofit existing information systems developed before modern privacy laws.

 **EXAM TIP** You should remember that privacy by design and privacy by default are key tenets of the GDPR and are only implied by other privacy regulations.

Chief Privacy Officer

Some organizations, typically those that manage large amounts of personal information related to employees, customers, or constituents, will employ a chief privacy officer (CPO). Some organizations have a CPO because applicable regulations such as the Gramm-Leach-Bliley Act (GLBA) require it. Other regulations such as the Health Information Portability and Accountability Act (HIPAA), the Fair Credit Reporting Act (FCRA), and the GLBA place a slate of responsibilities upon an organization that compels them to hire an executive responsible for overseeing compliance. Others have a CPO because they store massive amounts of personal information and have chosen to appoint an executive-level individual to be responsible for managing the privacy program.

The roles of a CPO typically include safeguarding personal information and ensuring that the organization does not misuse the personal information at its disposal. Because many organizations with a CPO also have a CISO, the CPO's duties mainly involve oversight into the organization's proper handling and use of personal information.

The CPO is sometimes seen as a customer advocate, and often this is the actual role of the CPO, particularly when regulations require a privacy officer.

Another similar title with similar responsibilities includes the data protection officer or data privacy officer (DPO). While responsibilities may be similar to the CPO, it is important to highlight that DPOs are expected to operate in strictly an oversight of governance role. In some cases, a CPO may not be able to fulfill the role of a DPO, particularly in organizations where the CPO has responsibility for implementing data processing activities or systems that will enable data processing activities.

 **NOTE** Many smaller organizations appoint an existing staff member as the acting privacy officer.

Does GDPR Require a DPO?

Much discussion and debate has ensued over GDPR's requirements for organizations to hire or retain a DPO. The GDPR is somewhat vague on the matter. Section 4, Article 37 reads:

> The controller and the processor shall designate a data protection officer in any case where:
>
> (a) the processing is carried out by a public authority or body, except for courts acting in their judicial capacity;
>
> (b) the core activities of the controller or the processor consist of processing operations which, by virtue of their nature, their scope and/or their purposes, require regular and systematic monitoring of data subjects on a large scale; or
>
> (c) the core activities of the controller or the processor consist of processing on a large scale of special categories of data pursuant to Article 9 or personal data relating to criminal convictions and offences referred to in Article 10.

The key language is included in subsections (a) and (b), which have some subjectivity. When it comes to GDPR, most companies are obliged to assign a DPO.

Chief Information Security Officer

The CISO is the highest ranking information security title in an organization. A CISO will develop business-aligned security strategies that support present and future business initiatives and will be responsible for the development and operation of the organization's information risk program, the development and implementation of security policies, security incident response, and perhaps some operational security functions.

In some organizations, the CISO reports to the COO or the CEO. In other organizations, the CISO may report to the CIO, chief legal counsel, or another executive in the organization.

Other similar titles with similar responsibilities include the following:

- **Chief security officer (CSO)**   A CSO often is responsible for physical security and workplace safety in addition to cybersecurity.

- **Chief information risk officer (CIRO)**   Generally, this represents a change of approach to the CISO position, from being protection-based to being risk-based.

- **Chief risk officer (CRO)**   This position is responsible for all aspects of risk, including information risk, business risk, compliance risk, and market risk. This role is separate from IT.

Many organizations do not have a CISO but instead have a director or manager of information security who reports further down in the organization chart. There are several possible reasons for organizations not having a CISO, but generally, it can be said that the organization does not consider information security as a strategic function. This will hamper the visibility and importance of information security and often results in information security being a tactical function concerned with basic defenses such as firewalls, antivirus software, and other tools. In such situations, responsibility for strategy-level information security implicitly lies with some other executive, such as the CIO. This situation often results in the absence of a security program and the organization's general lack of awareness of relevant risks, threats, and vulnerabilities.

For small to medium-sized organizations, a full-time strategic security leader may not be cost-effective. In these situations, it is advisable to contract with a virtual CISO (vCISO) to assist with strategy and planning. The benefit of this type of approach for organizations that may not require or cannot afford a full-time person is that it enables the organization to benefit from the knowledge of a seasoned security professional to assist in managing the information security program.

Software Development

Positions in software development are involved in the design, development, and testing of software applications and often include the following:

- **Systems architect**   This position is usually responsible for the overall information systems architecture in the organization. This may or may not include overall data architecture as well as interfaces to external organizations.

- **Systems analyst**   A systems analyst is involved with the design of applications, including changes in any application's original design. This position may develop technical requirements, program design, and software test plans. If an organization licenses applications developed by other companies, the systems analyst designs interfaces to other applications.

- **Software engineer/developer**   This position develops application software. Depending upon their level of experience, people in this position may also design programs or applications. In organizations that utilize purchased application software, developers often create custom interfaces, application customizations, and custom reports.

- **Software tester**   This position tests changes in programs made by software engineers/developers.

While the trend to outsourcing applications has resulted in organizations infrequently developing their own applications from scratch, software development roles persist in organizations. Developers are needed to create customized modules within software platforms, as well as integration tools to connect applications. Still, most organizations have a smaller number of developers than they did decades ago.

 **EXAM TIP**   You should remember that the regulatory requirements of privacy *by design* usually rest with the systems architect for organizations that develop software that processes personal information.

Rank Sets Tone and Gives Power

A glance at the highest ranking privacy and information security positions in an organization reveals much about executive management's opinion of privacy and information security in larger organizations. Executive attitudes about privacy and security are reflected in the privacy and security leaders' titles, which may resemble the following:

- **Privacy manager or security manager**   Privacy and information security are tactical only and often viewed as consisting only of basic tactical controls. The privacy and security managers have no visibility into the development of business objectives. Executives consider privacy and security as unimportant and based on simple practices only.

- **Privacy director or security director**   Privacy and information security are essential, and the director has moderate decision-making capability but little influence on the business. A director in a larger organization may have little involvement in overall business strategies and little or no access to executive management or the board of directors.

- **Vice president**   Privacy and information security are strategic objectives but do not influence business strategy and objectives. The vice president will have some access to executive management and possibly the board of directors.

- **CISO/CIRO/CSO/vCISO/CPO/DPO**   Privacy and information security are strategic objectives, and business objectives are developed with full consideration for risk. The C-level security and privacy personnel have free access to executive management and the board of directors.

Data Management

Positions related to data management are responsible for developing and implementing database designs and for maintaining databases. These personnel will be carrying out some of privacy's design principles. These positions are concerned with data within applications, as well as data flows between applications:

- **Data manager**   This position is responsible for data architecture and data management in larger organizations.

- **Database architect**   This position develops logical and physical designs of data models for applications. With sufficient experience, this person may also design an organization's overall data architecture.

- **Big data architect**   This position develops data models and data analytics for large, complex data sets.

- **Database administrator (DBA)**   This position builds and maintains databases designed by the database architect and those databases that are included as part of purchased applications. The DBA monitors databases, tunes them for performance and efficiency, and troubleshoots problems.

- **Database analyst**   This position performs tasks that are junior to the database administrator, carrying out routine data maintenance and monitoring tasks.

- **Data scientist**   This position applies scientific methods, builds processes, and implements systems to extract knowledge or insights from data.

 **EXAM TIP**   CDPSE candidates need to understand that the roles of data manager, big data architect, database architect, database administrator, database analyst, and data scientist are distinct from data owners. The former are IT department roles for managing data models and data technology, whereas the latter role governs the business use of, and access to, data in information systems.

Network Management

Positions in network management are responsible for designing, building, monitoring, and maintaining voice and data communications networks, including connections to outside entities and the Internet:

- **Network architect**   This position designs data and voice networks and designs changes and upgrades to networks as needed to meet new organization objectives.

- **Network engineer**   This position implements, configures, and maintains network devices such as routers, switches, firewalls, and gateways.

- **Network administrator**   This position performs routine tasks in the network, such as making configuration changes and monitoring event logs.

- **Telecom engineer**   Positions in this role work with telecommunications technologies such as telecom services, data circuits, phone systems, and conferencing systems,.

Systems Management

Positions in systems management are responsible for architecture, design, building, and maintenance of servers and operating systems. This may include desktop operating systems as well. Personnel in these positions also design and manage virtualized environments as well as microsegmentation.

- **Systems architect**   This position is responsible for the overall architecture of systems (usually servers), in terms of both the internal architectures and the relationships between systems.
- **Systems engineer**   This position designs, builds, and maintains servers and server operating systems.
- **Storage engineer**   This position designs, builds, and maintains storage subsystems.
- **Systems administrator**   This position performs maintenance and configuration operations on systems.

Operations

In larger organizations, positions in operations are responsible for day-to-day operational tasks that may include networks, servers, databases, and applications:

- **Operations manager**   This position is responsible for overall operations that are carried out by others. Responsibilities will include establishing operations shift schedules.
- **Operations analyst**   This position may be responsible for developing operational procedures; examining the health of networks, systems, and databases; setting and monitoring the operations schedule; and maintaining operations records.
- **Controls analyst**   This position monitors batch jobs and performs data entry work and other tasks to make sure they are operating correctly.
- **Systems operator**   This position monitors systems and networks, performs backup tasks, runs batch jobs, prints reports, and performs other operational tasks.
- **Data entry**   This position is responsible for keying batches of data from hard copy or other sources.
- **Media manager**   This position maintains and tracks the use and whereabouts of backup tapes and other media.

Privacy Operations

Though few organizations have personnel in a privacy operations function, the staff in many business departments have access to personal information of the organization's workforce or to its customers or constituents. These business functions include

- Human resources
- Sales and marketing
- Customer support
- Warranty or assurance services
- Business operations

Workers in these and other business functions need to be aware of the implications of having access to personal information and the organization's privacy policy, so that their day-to-day work does not run afoul of privacy policy or applicable laws.

 **NOTE** For the most part, it's more important to know that an organization has assigned various privacy responsibilities to designated personnel. The structure of the organization (or "org chart") is less important.

Security Operations

Positions in security operations are responsible for designing, building, and monitoring security systems and security controls to ensure the confidentiality, integrity, and availability of information systems:

- **Security architect**   This position designs security controls and systems such as authentication, audit logging, intrusion detection systems (IDSs), intrusion prevention systems (IPSs), and firewalls.

- **Security engineer**   This position designs, builds, and maintains security services and systems that are designed by the security architect. Such systems include firewalls, IDSs and IPSs, WAFs, web content filters, cloud access security brokers (CASBs), and others.

- **Security analyst**   This position examines logs from firewalls, IDSs, and audit logs from systems and applications. A security analyst could also have other responsibilities, such as performing security reviews, performing risk analyses, and maintaining security-related business records. This position may also be responsible for issuing security advisories to others in IT.

- **Access administrator**   This position is responsible for accepting approved requests for user access management changes and performing the necessary changes at the network, system, database, or application level. Often, this position is carried out by personnel in network and systems management functions; in larger organizations, user account management is performed by information security or in a separate user access department.

Privacy Audit

Positions in privacy audit are responsible for examining process design and for verifying the effectiveness of privacy policies and controls:

- **Privacy audit manager**   This position is responsible for audit operations and scheduling and managing audits.
- **Privacy auditor**   This position performs internal audits of privacy controls to ensure that they are being operated properly.

The topic of auditing privacy is discussed later in this chapter.

Security Audit

Positions in security audit are responsible for examining process design and for verifying the effectiveness of security controls:

- **Security audit manager**   This position is responsible for audit operations and scheduling and managing audits.
- **Security auditor**   This position performs internal audits of IT controls to ensure that they are being operated properly.

NOTE   While a privacy and security audit may not be a formal internal audit function, those performing security audits need to be able to exercise independence from the functions they audit.

Service Desk

Positions at the service desk are responsible for providing frontline support services to IT and IT customers:

- **Service desk manager**   This position serves as a liaison between end users and the IT service desk department.
- **Service desk analyst**   This position provides frontline user support services to personnel in the organization. This is sometimes known as a help desk analyst.
- **Technical support analyst**   This position provides technical support services to other IT personnel and perhaps also to IT customers.

Quality Assurance

In larger organizations, positions in quality assurance (QA) are responsible for evaluating IT systems and processes to confirm their accuracy and effectiveness:

- **QA manager**   This position facilitates quality improvement activities throughout the IT organization.
- **QC manager**   This position tests IT systems and applications to confirm whether they are free of defects.

Other Roles

Other roles in IT organizations include the following:

- **Third-party risk management manager** This position assesses third-party service providers to ensure that their practices do not result in unacceptable risks to the protection of sensitive information, particularly personal information of customers, constituents, or employees.

- **Vendor manager** This position is responsible for maintaining business relationships with external vendors, measuring their performance, and handling business issues.

- **Program manager** This position manages teams of project managers and oversees larger and more complex projects.

- **Project manager** This position creates project plans and manages IT projects.

General Staff

The rank and file in an organization may or may not have explicit privacy or information security responsibilities. This is determined in part by executive management's understanding of the broad capabilities of information systems and the personnel who use them. It also determines executives' understanding of the human role in privacy and information security.

Typically, general staff privacy- and security-related responsibilities include the following:

- Understanding and compliance with organization privacy and security policy

- Acceptable use of organization assets, including information systems and personal information

- Proper judgment, including proper responses to people who request access to personal information or request that staff members perform specific functions (the primary impetus for this is the phenomenon of social engineering and its use as an attack vector)

- Reporting of privacy- and security-related matters and incidents to management

Organizations with a more mature privacy and security culture have standard language in job descriptions that specify general responsibilities for the protection of assets, systems, and personal information.

Building a Privacy Operation

A privacy operation consists of activities that ensure that the collection and use of personal information complies with privacy policies and applicable regulations. Privacy operations are often implemented in an oversight capacity or as an "overlay" in an organization. Often this is implemented through the formation of a privacy office, led by the organization's privacy leader (often, but not always, the CPO or DPO), and a team of one or more analysts.

The privacy office functions as a catalyst to ensure that the intake, processing, and disposal of personal information throughout the organization is done properly—according to the organization's privacy policy, which should align with applicable privacy laws as well as other contractual and legal obligations.

It is said that, similar to information security, privacy is "everyone's job." This means that the procedures and practices followed by all persons involved in the processing of personal information include steps to ensure that personal information is used only in officially sanctioned ways. The main function of the privacy office is to ensure this ongoing outcome.

An organization establishing a privacy office needs to define the scope of its responsibilities. In smaller organizations, the scope would typically include all business operations in all locations. In larger organizations, particularly those with a presence in one or more countries with strict privacy laws (such as the GDPR), the organization may appoint local privacy personnel in each local country. This can help better align local business operations with local laws and requirements.

NOTE There is no single approach to implementing privacy operations in an organization. Some will designate separate staff, while others will appoint existing staff with various privacy-related responsibilities.

Identifying Privacy Requirements

Before the privacy office can begin enforcing privacy-related activities in an organization, it must first identify and document requirements that define the specifics regarding the collection, protection, and use of personal information.

Culture and Values

At the risk of implying a "motherhood and apple pie" sentiment, it's necessary to consult with the organization's culture and stated values as a starting point. Culture and values define the personality and uniqueness of an organization; the way that the organization values its assets, including the personal information of its customers, constituents, and employees, should be reflected in its policies and requirements.

Applicable Regulations

All regulations that apply to the organization need to be identified. This includes industry-specific regulations such as GLBA and HIPAA, as well as geographically related regulations such as Personal Information Protection and Electronic Documents Act (PIPEDA), California Consumer Privacy Act (CCPA), and GDPR. These and other regulations are cited and described in Chapter 1.

Privacy regulations are developing and changing at high velocity. Organizations need to have an established system in place to keep them informed about new and changing regulations on the topics of information privacy and cybersecurity. Cybersecurity professionals and lawyers each have their industry news sources; at present, the best sources appear to be newsletters and paid subscriptions from legal sources that keep their subscribers up to date on new laws and related developments. It is recommended that organizations keep an official inventory of applicable laws and regulations related to cybersecurity and privacy; depending upon the organization's industry sector, this inventory may extend to other industry-specific topics as well.

Legal Interpretation

It's a wise practice to employ internal or outside legal counsel to provide an interpretation of privacy and cybersecurity laws. Legal counsel experienced in these fields should first provide guidance on the applicability of these laws. For those laws deemed to be applicable, legal counsel should then guide the organization on the meaning of applicable portions and how they should be implemented.

As the de facto risk officer, an organization's legal counsel is responsible for identifying legal and regulatory requirements and risks. It is in this capacity that legal counsel determines which laws are applicable and what the organization should do to comply with them. Like individuals in other professions, legal counsel will often confer with their industry peers and outside experts to get an idea of the consensus of opinion on the applicability and compliance approach to new and existing laws.

NOTE Legal interpretation of applicable regulations is a key function that organizations need to acquire with in-house or external legal counsel.

Cybersecurity Policies, Requirements, and Regulations

As is often cited in this book, it's impossible to implement privacy successfully without also implementing effective cybersecurity. For the protective aspect of information privacy, organizations also need to identify their cybersecurity practices, policies, requirements, and applicable regulations and do their best to implement them in the form of cyber risk management and cybersecurity operations. If the protective side of privacy is unable to succeed, the proper handling side of privacy will be in danger of failure.

NOTE Organizations operating in the United States and the European Union may find it easier to apply updated GDPR-compliant operations to the entire organization, versus operating one way for EU citizen data and another way for non-EU citizen data.

Developing Privacy Policies

Privacy policies are statements that describe the collection and use of personal information, as well as actions that persons can take to inquire and make requests about their personal information. If the organization collects personal information only from its internal workers, this policy may not need to be publicly available. If the organization collects information from customers and constituents who are external to the organization, its privacy policy will most likely be published externally, often on a public web site. Similarly, at organizations that collect information from customers in person, privacy notices are often visibly posted.

Organizations that collect personal information from persons outside the organization often also have internal, nonpublic privacy policies that affect their employees and other workers. Such policies, like cybersecurity policies, will describe the required characteristics and expectations of staff members as well as information systems used by the organization.

If an organization that processes personal information employs any service providers that also store, process, or have access to the personal information used by the organization, the organization will also impose requirements upon those service providers. This aspect is discussed later in this chapter in the section, "Third-Party Risk Management."

Internal Privacy Policy

Virtually all organizations possess information about their employees and other workers that is used in their role as employer. In many countries, those organizations are required to disclose to their employees that the organization has a policy of confidentiality and that its workers' personal information is used only in its role as employer.

Organizations lacking a formal privacy program often stop there and make no further attempts to define the meaning or enforcement of "for sanctioned business purposes only." This can, however, lead to improper use of personal information through tactical decisions made by nearly any staff members who may have some or no knowledge about applicable laws on this topic.

Organizations with more mature security and privacy programs will have detailed privacy policies that define expected behaviors of its workers and the required characteristics of their information systems. In many cases, their privacy policies will exist as a part of their information security policies for the protection of said information. On the minimum side, security or privacy policies will include general statements about sensitive or personal information that shall be used "for sanctioned business purposes only" without further detail.

Whether a part of an organization's information security policy or separate, an internal privacy policy should include content on the following topics:

- *Roles and responsibilities for the organization's privacy program.* This should include
  - Those who have data management responsibilities
  - Those who approve and review access to personal information
  - Those who review and approve of new uses for personal data
  - Those who receive and process subject data requests
  - Those who have responsibility for monitoring uses of personal data
  - Those who have responsibility for responding to incidents that represent the misuse of personal information
  - Those who review privacy business process
  - Those who audit privacy business processes
- *Business processes governing the use of personal information.* This should include periodic reviews of business processes to ensure that they remain compliant with applicable laws and regulations.

- *Protection of personal information.* Generally, this will fall back to the organization's information security policy. However, this language can be included in its privacy policy (organizations are cautioned against duplicating this language, since it will require care to keep both policies in sync).

- *Consequences for violations of privacy policy.* Like an information security policy and other policies, this will describe the range of possible outcomes when persons are found to violate privacy policy. Typically, this ranges from verbal warnings to written warnings and even termination of employment.

- *Review and audit of privacy business processes.* This will describe reviews that take place within privacy business processes, as well as audits of privacy business processes. Reviews will typically be performed by process owners to confirm that all necessary actions take place as required. Audits of privacy business processes will be performed by persons outside of the organization's privacy office.

- *Measurements of privacy business processes.* Any statistics, metrics, key performance indicators (KPIs), and key risk indicators (KRIs) required in the privacy program will be described here.

- *Citations of applicable regulations and other obligations.* Privacy policy may be a place where applicable laws, regulations, and other obligations (such as terms and conditions in contracts with other organizations) may be cited in an organization's privacy policy.

 **NOTE** Some organizations will develop a privacy program charter. This high-level document describes the privacy program's goals, roles and responsibilities, authorities, key business processes, and other program-related matters. Other organizations will include these items within the internal privacy policy itself.

External Privacy Policy

Organizations that collect personal information about persons outside the organization develop privacy policies accessible by those persons. In the case of organizations that sell products or services to the general public, the privacy policy will often be publicly available on a public web site. It may also be posted on the premises where these products or services are sold.

Privacy laws enacted in many regions of the world have brought about the near universality of public-facing privacy policies posted on web sites or in the form of visible notices at business locations. Driven by these privacy laws, public-facing privacy policies generally include the following:

- Descriptions of the methods used to collect personal information
- Descriptions of the methods used to protect personal information
- Descriptions of primary and secondary uses of personal information
- Descriptions of any international transfers of personal information

- Descriptions of any third parties that may store or process personal information on behalf of the organization

- Descriptions of any tracking or logging of activities performed by the organization, such as the use of web browser cookies

- Any procedures that may be available for persons to understand how their personal information has been used or is being used

- Statements describing the legal rights of data subjects concerning the organization's use of their personal data, often including several subsections for specific regions or localities

- Contact information or contact procedures for persons who want to initiate subject data requests, inquiries, or lodge complaints regarding the organization's use of their personal information

- Contact information or contact procedures for regulators or other authorities to whom persons can make inquires or lodge complaints regarding the organization's collection or use of their personal information

- The date that the privacy policy was last updated

Some privacy laws require that an organization inform employees, customers, or constituents if the organization changes its privacy policy. For regulations that require organizations to collect an acknowledgment stating that data subjects have received privacy policy, those organizations will be required to collect a new acknowledgment for the updated privacy policy. In business applications, this often comes in the form of a brief statement that reads, "Continued use of this application constitutes consent to our updated privacy policy," with a link to the privacy policy content. Alternatively, organizations may send e-mail notices to users of a system regarding the updated privacy policy; often, the policy itself is not included in the message, but means are provided to access the policy.

 **NOTE** Like other aspects of service and service levels, organization privacy practices may soon become competitive differentiators.

Developing and Running Data Protection Operations

The protection of personal data is one of the primary responsibilities of an organization's information security function. In most organizations, this function is separate from the privacy office or privacy operations.

- Data protection operations are generally built upon a framework of security controls. Security control frameworks are discussed in Chapters 4, 5, and 6.

- Data protection technologies are typically operated by the IT department, although in some organizations, a separate security operations function will manage these. The technology of data protection is discussed more fully in Chapter 4.

- Decisions regarding the ongoing development of security privacy and controls are a part of the larger risk management life cycle, described in Chapter 1.

 **NOTE** The management of information security is discussed in great detail in the book *CISM Certified Information Security Manager All-In-One Exam Guide* (McGraw Hill, 2018).

Developing and Running Data Monitoring Operations

Monitoring the use of personal data is at the core of many organizations' privacy programs. Since privacy is concerned with the protection and use of personal information, some privacy operations are uniquely different from operational security processes.

Data Discovery Scanning

To determine whether personnel are complying with privacy and data classification policies, organizations will conduct data discovery scans of its data storage. Typically performed by automated DLP (data loss prevention) scanning tools, these scans generally target file shares and other structured and unstructured repositories and employ rules to identify specific types of information.

Examples of scan targets include account numbers, credit card numbers, and government-issued identification numbers, in an attempt to discover whether files containing personal information have been stored on file shares in violation of policy. Scans can also include nonpersonal information such as source code, financial information, and intellectual property.

Upon receiving the results of discovery scans, security or privacy analysts will investigate the presence of these files and attempt to determine why those files are there, who put them there, and why. Where such files are used as a part of sanctioned business processes, security or privacy analysts will confirm that access rights comply with access policies, including least privilege and need-to-know principles. Where such files are not a part of legitimate business processes, corrective action is taken to prevent such security or privacy issues from recurring.

Data discovery often identifies undocumented procedures as well as improper behaviors. Over time, corrective actions will gradually lift the maturity of related business processes and inform staff of proper data handling policies and procedures.

 **CAUTION** Privacy regulations are not always explicitly clear on which data fields are considered personal information. Legal counsel may be needed to clarify this, so that privacy operations can be sure to monitor effectively.

Data Movement Monitoring

Information systems can be supplemented with DLP tooling that will monitor the movement of sensitive and personal information in real time. Monitoring agents placed in key information systems can detect the creation, movement, and deletion of specific information and generate alerts that are sent to security or privacy personnel for investigation and follow-up.

Following are examples of data movement monitoring:

- **E-mail**   Agents on endpoints can detect sensitive information in the contents of incoming or outgoing e-mail.

- **Endpoint storage**   Agents on endpoints can detect the local storage of information.

- **File servers**   Agents on file servers and other storage systems can detect the creation and movement of information.

- **USB storage**   Agents on endpoints can detect the movement of information to and from external USB storage devices.

- **Internet ingress/egress**   Agents on endpoints and network ingress/egress points (including Internet connections) can monitor data movement.

In all of these cases (and more that are not mentioned here), alerts can be sent to security or privacy analysts who would investigate these events to determine whether the data movement is legitimate (in which case, alerts can be adjusted to reduce the number of "false positives") or whether corrective action is warranted.

In addition to monitoring the movement of sensitive information, DLP monitoring agents can be configured to intervene and prevent the data movement that is attempted. When implementing these DLP systems, organizations often configure them initially to operate in passive monitoring mode to help them understand and distinguish legitimate business processes from activities that violate policy. Then, carefully, organizations can configure DLP agents to intervene in circumstances where data movement is a policy violation.

On end-user systems, in many cases, DLP agents can display a window to the end user that asks for confirmation of the intended data movement. While the agents do not overtly block such data movement, asking for confirmation can remind users that some data movement may violate policy. Still, users can be empowered in some circumstances to confirm that the intended data movement is legitimate. This action will still produce an event or an alert that can be investigated to determine whether the user is abusing policy. If the movement is legitimate, privacy and security analysts can determine whether they want users to continue to confirm such data movement or whether the movement can be permitted without intervention.

DLP systems, whether used to perform discovery scans or monitor data movement, require a good deal of "tuning" to ensure that they do not interfere with legitimate business processes but properly alert personnel when potential violations of data classification or privacy policies occur. Because business processes typically change slowly over time, the task of tuning DLP is never finished and is an ongoing activity.

Working with Data Subjects

Modern data privacy laws require transparency not only with regard to the collection and use of personal data, but also to provide one or more means for data subjects to make inquiries and requests regarding the use of their personal information. The procedures for making such subject data requests are typically spelled out in an organization's privacy policy (laws such as GDPR require privacy policy to describe this). Occasionally, such procedures may be located elsewhere, such as in a user guide or in system documentation.

Inquiries for Data Usage

Data subjects may send an inquiry regarding the presence and usage of their personal information. Their request may be general or quite specific. For instance, a data subject may ask whether any of his personal information is present in the organization's systems. Or a data subject may ask about specific personal information, such as a home address.

Smaller organizations may provide only an inquiry form, an e-mail address, a telephone number, or a surface mail address where such inquiries may be sent. These organizations must train personnel to manage these inquiries properly and respond to data subjects within specific timeframes (which are sometimes spelled out in regulations). Personnel who handle these requests will need to have access to systems and applications containing personal information so that they may respond to these requests accurately.

Larger organizations automate inquiries in some cases. For instance, a data subject with an existing account on an organization's systems can log in and click a link to learn how and where personal information is used. Often, such tools provide the means for data subjects to make changes to some of their information. This is discussed more fully in the next section.

Organizations typically maintain a log of inquiries, including the subject's name (or other identifying information), so that management can better understand the frequency of requests and the workload incurred. Privacy personnel will recognize that these logs themselves may also contain protected personal information.

Requests for Corrections

In some circumstances, a data subject may ask for changes in their personal information used by an organization. For instance, a data subject may change residences and need to update a mailing or shipping address. Or they may make changes in a payment method, family status, or service provider such as insurance. Finally, sometimes personal information is mistyped, and spelling and other corrections are needed.

Organizations are required to provide one or more means through which data subjects can request these corrections. Data subjects often can make these changes through self-service programs, but sometimes they must request that personnel in the organization make the changes on their behalf.

Privacy policies often provide one or more methods that can be used by data subjects to make these requests. Whether the means are automated or manual, organizations typically log these events as a part of routine systems and activity measurements. Like other mature business processes, this logging will sometimes compel management to make changes or improvements to systems and processes. For example, if the organization is receiving numerous requests that personnel must make manually, the organization may provide more self-service tools for data subjects to make some changes themselves.

PART I

Requests for Removal

GDPR made famous the notion of "the right to be forgotten," meaning the outright removal of a data subject from an organization's records. Not a new concept, data subjects often want to opt out of an activity a particular organization may be conducting that involves a data subject. As with other subject data requests, privacy policy will provide specific means for such requests to be made.

Organizations accepting opt-out or data removal requests must understand the nature of such data and other laws requiring the retention of records. For example, a former employee may request that her employment records be removed; however, employment law may require that employment records be retained for many years after the end of the employment. Similarly, a similar request made to a bank or credit union may conflict with laws requiring the retention of banking transaction records. However, marketing organizations that facilitate mail or telephone marketing campaigns may have few or no retention requirements and are compelled to remove a subject's data on request. The same can be said of social networking organizations that have few statutory requirements for retaining subject data. Finally, privacy laws cite specific exclusions to data removal requests: GDPR, for instance, does not require courts or prison systems to expunge a person's criminal history. Nice try, though!

Complaints

To improve customer service, organizations may include a means for permitting data subjects to lodge complaints regarding the organization's use of their personal information. A data subject may be venting in the complaint, or the complaint may be an implicit request for a change in the person's relationship with the organization.

Personnel in the organization will need to consider complaints carefully, including whether a complaint describes an activity that could violate the organization's privacy policy. For this reason alone, organizations should pay close attention to data subject complaints, as they may be the only way organizations can become aware of privacy or security incidents.

Working with Authorities

Many privacy laws provide for the creation of government authorities that act in a supervisory capacity as a part of the enforcement of these laws. For instance, Articles 51 through 54 of the GDPR define supervisory authorities and their responsibilities. (Although this book does focus on organizations that will, from time to time, work with supervisory authorities, details on work performed by supervisory authorities are beyond its scope.)

In my experience as a privacy and security professional for more than 20 years, I have learned that the most important ingredient to successful relationships and encounters with external parties, including auditors, regulators, and supervisory authorities, is this: completeness of business records. This includes the following:

- Up-to-date process information
- Data flow diagrams (or detailed descriptions of data flows)

- Effective processes
- Complete business records

Nothing frustrates these external parties more than an organization that is disorganized and out of control. When such organizations do produce information, it will be regarded with skepticism, as external parties will wonder if the data was conjured up at the last minute or "cooked" (altered in an attempt to avoid accountability). This could even be regarded as a lack of cooperation with a supervisory authority. (GDPR Article 31 reads, "The controller and the processor and, where applicable, their representatives, shall cooperate, on request, with the supervisory authority in the performance of its tasks.")

Privacy and cybersecurity laws often require organizations that store or process personal information to have privacy and security breach procedures. Further, organizations should identify, in advance of any incident or breach, all applicable laws, regulations, and other obligations, and all instances where notification to regulators, supervisory authorities, and affected parties are required should a breach occur. Then, at the onset of an incident or breach, the organization simply carries out its procedures, which are known and practiced in advance.

Privacy Training and Awareness

Personnel are the primary weak point in an organization's privacy and cybersecurity status. Such breaches are mainly caused by lapses in judgment, inattentiveness, fatigue, work pressures, a shortage of skills, and a lack of awareness of corporate policies. Personnel are generally considered the largest and most vulnerable portion of an organization's attack surface, and for good reason: most breaches start with a social engineering attack, often via e-mail.

Many organizations conduct security awareness training so that personnel are aware of these common attacks, as well as several other topics that mainly fall into the category known as *Internet hygiene*, which is the safe use of computers and mobile devices while accessing the Internet. Fewer organizations provide privacy awareness training that enables personnel to be aware of the expectations regarding the proper use of customer, constituent, and employee personal information.

Training Objectives

The primary objective of privacy and security awareness programs is the keen awareness, on the part of all personnel, of the proper handling of personal information, the reality of the different types of attacks that they may be subject to, and what they are expected to do in various situations. Further, personnel must understand and comply with an organization's acceptable use policy, privacy policy, security policy, code of ethics or code of conduct, and other applicable policies.

Better privacy and security awareness training programs include opportunities to practice skills, with testing at the end of training sessions. In computer-based training, users should be required to pass the test successfully with a minimum score—70 percent is a typical minimum score to complete the course.

The best privacy and security awareness training courses, whether in-person or online, are engaging and relevant. While some organizations conduct privacy and security awareness training for compliance purposes, many organizations have a genuine interest in their personnel getting the most value out of the training. The point of privacy and security awareness training is, after all, the reduction of risk.

Business records should be created, recording when each employee receives training. Many organizations are subject to privacy and security regulations that require personnel to complete awareness training; business records provide ample evidence of users' completion of their training.

Creating or Selecting Content

Privacy and security managers need to develop or acquire awareness training content for personnel in the organization. The content that is selected or developed should be

- **Understandable**   The content should make sense to all personnel. A common mistake that security and privacy managers make is to create or select content that is overly technical and difficult for many nontechnical personnel to understand.

- **Relevant**   The content should be applicable to the organization and its users. For example, training on the topic of cryptography would be irrelevant to the vast majority of personnel in most organizations. Irrelevant content can cause personnel to disengage from further training.

- **Actionable**   The content should ensure that personnel know what to do (and not do) in common scenarios.

- **Memorable**   The best content will give personnel opportunities to practice their skills at some of the basic tasks important to privacy and security, including selecting and using passwords, reading and responding to e-mail, making good decisions about the use of personal information, and interacting with persons inside and outside the organization.

Audiences

When planning an awareness training program, privacy and security managers need to understand the entire worker population and their various roles in the organization. This helps managers understand what training subject matter is relevant to which groups of workers. Managers need to ensure that all workers get all the training they need and not overburden workers with training that is not relevant to their jobs.

For example, workers in a large retail organization fall into four categories:

- **Corporate workers**   These persons all use computers, and most of them use mobile devices. Most have access to sensitive information, including personal information about customers and/or employees.

- **Retail floor managers**   These persons work in retail store locations and use computers daily in their jobs.

- **Retail floor cashiers**   These persons work in retail store locations. They do not use computers, but they do collect payments by cash, check, and credit card.
- **Retail floor workers**   These persons work in retail store and warehouse locations and do not use computers.

Privacy and security managers should package awareness training so that each audience receives relevant training. In this example, retail floor workers probably need little Internet or computer-related security awareness training, but instead would receive training on physical security and workplace safety topics. Cashiers need training on fraud techniques (counterfeit currency, currency counting fraud, and matters related to credit card payments such as skimming). Corporate workers and retail floor managers should probably receive full-spectrum privacy and security training since they all use computers, and many have access to personal information and sensitive information. Retail floor managers should also receive all of the training delivered to retail floor workers and cashiers because they also work at retail locations and supervise these personnel.

 **NOTE**   Privacy training ensures that all personnel understand the organization's privacy policies, practices, and expectations.

Information Workers

Workers in an organization who have contact with the personal information of employees, customers, or constituents should receive privacy awareness training. Information workers need to be aware of the organization's policies on the protection and proper use of personal information, so that the organization is less likely to suffer a privacy breach caused by poor judgment.

Technical Workers

Technical workers in an organization, typically IT personnel, should be trained in security techniques relevant to their positions. Technical workers are responsible for system architecture, system and network design, implementation, and administration. Without security training, these workers may unknowingly have lapses in judgment that could result in significant vulnerabilities that could lead to compromises.

Technical workers also need privacy awareness training to be aware of the proper handling of personal information. This is especially important for the organization, so that information systems will be designed and configured to bring about the greatest possible protection and sound handling of personal information.

Software Developers

Software developers typically receive little or no education on privacy by design or secure software development in colleges, universities, and tech schools. The art and science of privacy by design and of secure coding, then, is new to many software developers.

Training for software developers helps them to be more aware of the common mistakes made by software developers, including these:

- Vulnerabilities that permit injection attacks
- Broken authentication and session management that can lead to attackers accessing other user sessions
- Cross-site scripting
- Broken access control
- Security misconfiguration
- Sensitive data exposure
- Insufficient attack protection
- Cross-site request forgery
- Using components with known vulnerabilities
- Underprotected APIs

This list, which changes from time to time, is published by the Open Web Application Security Project (OWASP; at www.owasp.org), an organization that helps software developers better understand the techniques needed for secure application development and deployment.

Privacy and security training for software developers should also include protection of the software development process itself. Topics in secure software development generally include the following:

- Protection of source code
- Source code reviews
- Care when using open source code
- Testing of source code for vulnerabilities and defects
- Archival of changes to source code
- Protection of systems used to store source code, edit and test source code, build applications, test applications, and deploy applications

Note that some of these aspects are related to the architecture of development and test environments and may not be needed for all software developers.

Third Parties

Privacy and security awareness training needs to be administered to all personnel who have access to personal information through any means. Because this may include personnel who are employees of other organizations, those workers need to participate in the organization's privacy and security awareness training. In larger organizations, the curriculum for third-party personnel may be altered somewhat, since there may be portions of the privacy and security awareness training content that do not apply to outsiders.

New Hires

New employees, as well as consultants and contractors, should be required to attend privacy and security training as soon as possible. There is a risk that new employees could make mistakes early in their employment and prior to their training, since they would not be familiar with all of the practices in the organization.

Better organizations link access control with privacy and security training: new employees are not given access to systems until after they have completed their privacy and security training. This gives new workers added incentive to complete their training quickly, since they want to be able to get access to corporate applications and get to work.

Annual Training

Most privacy and security awareness programs include annual refresher training for all workers. Required by some regulations, annual training is highly recommended, as this helps keep privacy, security, and Internet safety a part of every worker's day-to-day thinking process and helps them to avoid common mistakes. Further, because handling procedures, protective techniques, and attack techniques change quickly, annual refresher training helps workers be aware of these developments.

Training takes time, and people tend to put it off for as long as possible. Workers can be offered incentives to complete their training through various types of rewards. For example, all workers who complete their training in the first week can win gift cards or other prizes.

Organizations generally choose one of two options for annual training:

- Train the entire organization all at once.
- Train one-twelfth of the organization on workers' hire month anniversaries

Communication Techniques

Privacy and security awareness training programs often utilize a variety of means for imparting information-handling procedures, Internet hygiene, and safe computing information to its workers. Communication techniques often include

- **E-mail**   Privacy and security managers may occasionally send out advisories to affected personnel to inform them of recent developments, such as a new phishing attack. Occasionally, a senior executive will send a message to all personnel to emphasize that security is every worker's job and that security is to be taken seriously.

- **Internal web site**   Organizations with internal web sites or web portals may, from time to time, include privacy and security content.

- **Video monitors, posters, and bulletins**   Sometimes a privacy or security message can be delivered on monitors, posters, or bulletins on various topics keeps people thinking about privacy and security.

Third-Party Risk Management

Third-party risk management (TPRM) refers to activities used to discover and manage risks associated with external organizations performing operational functions for an organization. Many organizations outsource some of their information processing to third-party organizations, often in the form of cloud-based software as a service (SaaS) and platform as a service (PaaS), and often for economic reasons: it is less expensive to pay for software in a leasing arrangement than developing, implementing, integrating, and maintaining software internally. Similarly, many organizations prefer to lease server operating systems versus purchasing their own hardware.

TPRM involves the extension of techniques used to identify and treat privacy and security risk within the organization. The same risks present in third parties' services are present within an organization's processing environment. The discipline of third-party risk exists because of the complexities associated with identifying risks in third-party organizations, as well as risks inherent in doing business with specific third parties. At its core, third-party risk is like other risk management; the difference lies in acquiring relevant information to identify risks outside of the organization's direct control.

 **NOTE** Organizations lacking a mature TPRM program should implement a process that incorporates both security and privacy requirements to relevant service providers.

Cloud Service Providers

Organizations moving to cloud-based environments often assume that those cloud service providers have taken care of many or all information security functions, when often this is not the case at all. This often results in security and privacy breaches, often a result of each party believing that the other was performing key data protection tasks. Many organizations are unfamiliar with the shared responsibility model that delineates which party is responsible for which operations and security functions. Tables 2-1 and 2-2 depict shared responsibility models in terms of operations and security, respectively.

 **NOTE** The specific responsibilities for operations and security between an organization and any specific service provider may vary somewhat from these tables. It is vital that an organization clearly understand its precise responsibilities for each third-party relationship, so that no responsibilities that may introduce risks to the organization are overlooked or neglected.

The privacy officer should recognize that third-party service providers generally play little or no role in the data-handling aspect of privacy. These functions are fully the responsibility of the organization using third-party services, not the third party itself.

TPRM has been the subject of many standards and regulations that compel organizations to be proactive in discovering security risks present in their critical third-party providers. Historically, many organizations were not voluntarily assessing these

Operation	On-premises	IaaS	PaaS	SaaS
Applications	Org	Org	Org	Provider
Data	Org	Org	Org	Provider
Runtime	Org	Org	Provider	Provider
Middleware	Org	Org	Provider	Provider
Operating system	Org	Org	Provider	Provider
Virtualization	Org	Provider	Provider	Provider
Servers	Org	Provider	Provider	Provider
Storage	Org	Provider	Provider	Provider
Networking	Org	Provider	Provider	Provider
Data center	Org	Provider	Provider	Provider

Table 2-1   Operational Shared Responsibility Model

third parties. Statistical data about breaches over several years has revealed that more than half of all breaches have a nexus in third parties. This statistic has illuminated the magnitude of the third-party risk problem and has resulted in the enactment of laws and regulations in many industries that now require organizations to build and operate effective third-party risk programs in their organizations.

Activity	On-premises	IaaS	PaaS	SaaS
Human Resources	Org	Shared	Shared	Provider
Privacy	Org	Org	Org	Shared
Application Security	Org	Org	Shared	Provider
Identity and access management	Org	Org	Shared	Provider
Log management	Org	Org	Shared	Provider
System monitoring	Org	Org	Shared	Provider
Incident response	Org	Org	Shared	Shared
Data Governance	Org	Org	Shared	Shared
Data Encryption	Org	Org	Shared	Provider
Host intrusion detection	Org	Org	Shared	Provider
Host hardening	Org	Org	Shared	Provider
Asset management	Org	Org	Shared	Provider
Network intrusion detection	Org	Org	Provider	Provider
Network security	Org	Org	Provider	Provider
Security policy	Org	Shared	Shared	Provider
Physical security	Org	Provider	Provider	Provider

Table 2-2   Security and Privacy Shared Responsibility Model

Privacy Regulation Requirements

In GDPR parlance, organizations that use third-party service providers are often, but not always, considered *data controllers*, which are entities that determine the purposes and means of the processing of personal data, which can include directing third parties to process personal data on their behalf. The third parties that process data for data controllers are known as *data processors*.

HIPAA requires that covered entities (organizations subject to HIPAA regulations) establish business associate agreements (BAAs) with every service provider that has access to the covered entity's information or information systems. Sarbanes-Oxley requires that organizations perform up-front and periodic due diligence on financially relevant service providers.

TPRM Life Cycle

The management of business relationships with third parties is a life-cycle process. The life cycle begins when an organization contemplates the use of a third party to augment or support the organization's operations. The life cycle continues during the ongoing relationship with the third party and concludes when the organization no longer uses the third party's services: all connections are severed, and all data stored at the third party is removed or destroyed.

Initial Assessment

Before establishing a business relationship with a third party, an organization will assess and evaluate the third party for suitability. Often this evaluation is competitive, where two or more third parties are vying for the formal relationship. During the evaluation, the organization will require that each third party provide information describing its services, generally in a structured manner through a request for information (RFI) or a request for proposal (RFP) process.

In their RFIs and RFPs, organizations often include sections on privacy and security, so that the organizations can better understand how each third party protects the organization's information. This, together with information about the services themselves, pricing, and other information, reveals details that the organization uses to select the third party that will provide services.

Legal Agreement

Before services can begin, the organization and the third party will negotiate a legal agreement that describes the services provided, along with service levels, quality, pricing, and other terms included in typical legal agreements. Based on the details discovered in the assessment phase, the organization can develop a section in the legal agreement that addresses privacy and security. This part of the legal agreement will typically cover these subjects:

- **Privacy and/or security program** Require the third party to have a formal privacy and/or security program, including but not limited to governance, policy, risk management, annual risk assessment, internal audit, vulnerability management, incident management, secure development, privacy and security awareness training, data protection, and third-party risk.

- **Security and/or privacy controls**   Require the third party to have a controls framework, including linkages to risk management and internal audit.

- **Vulnerability assessments**   Require the third party to undergo penetration tests or vulnerability assessments of its service infrastructure and applications, performed by a competent security professional services firm, with reports made available to the organization upon request.

- **External audits and certifications**   Require the third party to undergo annual SOC 1 and/or SOC 2 Type 2 audits (SOC stands for System and Organization Controls), TrustArc audits, ISO/IEC 27001 certifications, HITRUST certifications, Payment Card Industry Reports on Compliance (PCI ROCs), or other industry-recognized and applicable external audits, with reports made available to the organization upon request.

- **Privacy and security incident response**   Require the third party to have a formal privacy and security incident capability that includes testing and training.

- **Privacy and security incident notification**   Require the third party to notify the organization in the event of a suspected and confirmed breach, within a specific timeframe, typically 24–48 hours. The language around "suspected" and "confirmed" needs to be developed very carefully so that the third party cannot sidestep this responsibility.

- **Right to audit**   Require the third party to permit the organization to conduct an audit of the third-party organization without cause. If the third party does not want to permit this, one fallback position is to insist on the right to audit in the event of a suspected or confirmed breach or other circumstances. Further, include the right to have a competent security professional services firm perform an audit of the third party privacy and security environment on behalf of the organization (useful for several reasons, including geographic location; the external audit firm will be more objective).

- **Periodic review**   Require the third party to permit an annual review of its operations, privacy, and security. This can improve confidence in the third party's privacy and security.

- **Third-party disclosures**   Require the third party to list any contracted parties it uses to perform services, and a suitable third-party due diligence process.

- **Annual due diligence**   Require the third party to respond to annual questionnaires and evidence requests as a part of the organization's third-party risk program.

- **Cyber insurance**   Require the third party to carry a cyber-insurance policy with minimum coverage levels. Require the third party to comply with all requirements in the policy so that the policy will pay out in the event of a privacy or security event. A great option is to have the organization be a named beneficiary on the policy, in the event of a widespread breach that could result in a large payout to many customers being diluted.

Organizations with many third parties may consider developing standard privacy and security clauses that include all of these provisions. Then, when a new third-party service is being considered, the organization's privacy and security teams can perform their upfront examination of the third party's privacy and security environment, and then make adjustments to the privacy and security clauses as needed.

During the vetting process, organizations will often find one or more shortcomings in the third party's privacy or security program that the third party is unwilling or unable to remediate right away. There are still options, however: The organization can compel the third party to enact improvements in a reasonable period after starting the business relationship. For example, a third-party service provider may not have an external audit such as a SOC 1 or SOC 2 audit, but it may agree to undergo such an audit one year later. Or a third-party service provider that has never had external penetration testing could be compelled to begin testing at regular intervals. Alternatively, the third party could be required to undergo a penetration test and be required to remediate all critical- and high-level issues before the organization will begin using the third party's services.

Classifying Third Parties

Organizations utilizing third parties often discover a wide range of risks: Some third parties may have access to large volumes of operationally critical or personal information, while others may have access to small volumes of personal information, and still others do not access data associated with critical operations at all. Because of this wide span of risk levels, many organizations choose to develop a scheme consisting of levels of risk, based on criteria important to the organization. Typically, this risk scheme will have two to four risk levels, with each third party assigned to a risk level.

Organizations need to assess their third parties periodically to ensure that they remain at the right level of classification. Third parties that provide a variety of services may initially be classified as low risk, but in the future, if the third party is retained to provide additional services, this could result in reclassification at a higher level of risk.

The purpose of this classification is explained in the following sections on questionnaires and assessing third parties.

Questionnaires and Artifacts

Organizations that utilize third parties need to assess those third parties periodically. Generally, this consists of the creation of a privacy and/or security questionnaire that is sent to the third party, with the request to answer all of the questions and return to the organization within a reasonable amount of time. The organization may choose not to rely simply on the answers provided by the third party. In that case, the organization can also request that the third party furnish specific artifacts that serve as evidence that support the responses in the questionnaire. Here are some typical artifacts that an organization will request of its third party:

- Privacy and security policies
- Privacy and security controls
- Privacy and security awareness training records
- New-hire checklists
- Details on employee background checks (not necessarily actual records but a description of the checks performed)

- Nondisclosure and other agreements signed by employees (not necessarily signed copies, but blank copies)
- Vulnerability management process
- Secure development process
- Copy of general insurance and cyber-insurance policies
- Incident response plan and evidence of testing

An organization that uses many third parties may find that it utilizes various types of services: some store or process large volumes of personal or critical data, others are operationally critical but do not access personal information, and other categories. Often it makes sense for an organization to utilize different versions of questionnaires, one or more for each category of third party, so that the majority of questions asked of each third party are relevant. Organizations that don't do this risk having large portions of questionnaires being irrelevant, which could be frustrating to third parties that would rightfully complain of wasted time and effort.

As described earlier on the classification of third parties, organizations often use different questionnaires for different risk levels of third parties. For example, third parties in categories of high risk would be asked to complete very extensive questionnaires that include requests for many pieces of evidence. In contrast, third parties of medium risk would receive shorter questionnaires, and low-risk third parties would receive very short questionnaires. While it is courteous to send questionnaires of appropriate length to various third parties (mainly to avoid overburdening low-risk third parties with huge questionnaires), remember that this practice also increases the burden on the organization, since someone has to review the questionnaires and attached evidence. An organization with hundreds of third parties does not want to overburden itself with the task of analyzing hundreds of questionnaires, each with hundreds of questions, when most of the third parties are lower risk and warrant shorter questionnaires.

Assessing Third Parties

To discover risks, organizations need to assess their third parties, not only at the onset of the business relationship (before the legal agreement is signed, as explained earlier), but periodically thereafter. Business conditions and operations often change over time, necessitating that third parties be assessed throughout the relationship.

Organizations assessing third parties often recognize that IT, privacy, and security controls are not the only forms of risk that require examination. Instead, organizations generally will seek other forms of information about its more important third parties, including

- Financial risk
- Geopolitical risk
- Inherent risk
- Recent security breaches
- Lawsuits

These and other factors can influence the overall risk to the organization and can manifest in various ways, including degradations in overall privacy and security, failures to meet production or quality targets, and even failure of the business.

Because of the effort required to collect information on these other risk areas, organizations often rely on outside service organizations that collect information on companies and make it available on a subscription basis. Of course, these are also third-party organizations that require an appropriate measure of due diligence.

Risk Mitigation

When assessing its third parties, organizations that carefully examine the information provided by the parties often discover some unacceptable aspects. In these cases, the organization will analyze the issues and decide on a course of action.

For instance, a highly critical third party indicates that it does not perform annual privacy and security awareness training for its employees, and the organization finds this unacceptable. To remedy this, the organization needs to analyze the risk (in a manner not unlike any risk found internally) and decide on a course of action. In this example, the organization contacts the third party and attempts to compel them to institute annual privacy and security awareness training for its employees.

Sometimes, a deficiency problem in a third party is not so easily solved. For example, a third party that has been providing services for many years indicates in its annual questionnaire that it does not employ encryption of stored personal information. At the onset of the third-party business relationship, this was not a common practice, but over time it has become a common practice in the organization's industry. The service provider, when confronted with this, explains that it is not operationally feasible to implement encryption of personal information in a manner acceptable to the organization, mainly for financial reasons: because of the significant impact of cost on its operations, the third party would have to increase its prices to cover these costs. In this example, the organization and the third party would need to discover the most pragmatic course of action, so that the organization can be satisfied with the level of risk and control its own costs.

Auditing Privacy Operations

The purpose of any audit is to confirm, using objective means, the effectiveness of controls and processes. An audit may be performed by a customer, regulator, audit firm, or internal audit staff. An audit of an organization's privacy processes and underlying information system will be performed using established audit practices.

This section contains a summary of audit practices in the context of information privacy. For more detailed information about audit planning and audits, refer to *CISA Certified Information Systems Auditor All-In-One Exam Guide, Fourth Edition,* particularly Chapter 3, "The Audit Process." (Be sure to use the latest edition.)

Privacy Audit Scope

An audit is planned for good reason. Because audits are disruptive and often expensive, one or more compelling business drivers should be identified to help management determine whether an audit is needed and what the desired outcome of the audit is likely to be.

For privacy, the *scope* of a privacy audit is likely to be the controls, processes, and systems used to protect personal information, or the controls, processes, and systems used to collect, process, and use personal information. Just as privacy itself has two perspectives (the protection of personal information and the use of personal information), a privacy audit is likely to focus on one or both of these perspectives.

Privacy Audit Objectives

The term *audit objectives* refers to the specific goals for a privacy audit. Generally, the objectives of a privacy audit are to determine whether privacy controls exist and are effective in some specific aspect of business operations in an organization. Generally, a privacy audit is performed to comply with applicable privacy regulations or related legal obligations. An audit may also be performed as the result of a recent privacy incident or event.

Depending on the subject and nature of the audit, the auditor may examine privacy controls and related evidence herself, or she may instead focus on the business content that is processed by the controls. For example, if the focus of a privacy audit is an organization's subject data request process, the auditor may focus on requests in the system to see whether they comply with the organization's privacy policy or applicable regulations. Or the auditor could focus on the information systems processes that support the processing of personal information. Formal audit objectives should make such a distinction so that the auditor has a sound understanding of the objectives. Objectives tell the auditor what to examine during the audit. Of course, knowing the type of audit to be undertaken helps too; this is covered in the next section.

Types of Privacy Audits

The scope, purpose, and objectives of a privacy audit will determine the type of audit that will be performed. Auditors need to understand each type of audit, including the procedures that are used for each:

- **Operational audit** This type of audit involves an examination of privacy controls, security controls, or business controls to determine the controls' existence and effectiveness. The focus of an operational audit is usually the operation of one or more controls, and it could concentrate on the management of a business process or on the business process itself.

- **Information systems audit** This type of audit involves a detailed examination of an IT department's operations related to the storage and processing of personal information. An IS audit looks at IT governance to determine whether the IT department is aligned with overall organization goals, objectives, privacy policy, security policy, and applicable regulations. The audit may also look closely at all of the major IT processes, including service delivery, change and configuration management, security management, systems development life cycle, business relationship and supplier management, and incident and problem management. The integrity of a privacy process ultimately depends upon the integrity of the underlying IT systems and processes. This audit will determine whether each control objective and control is effective and operating properly.

- **Integrated audit** This type of audit combines an operational audit and an information systems audit to help the auditor fully understand the entire environment's integrity. Such an audit will closely examine privacy operations processes, procedures, and records, as well as the IT applications used to store and process personal information.

- **Administrative audit** This type of audit involves an examination of the operational efficiency of privacy-related business processes.

- **Compliance audit** This type of audit is performed to determine the level and degree of compliance with one or more applicable privacy regulations, other legal requirements, or internal policies and standards. If a particular privacy law requires an external audit, the compliance audit may have to be performed by approved or licensed external auditors and/or be performed using specific audit standards. If, however, the law does not explicitly require audits, the organization may still decide to perform one-time or regular audits to determine the level of compliance to the law. Internal or external auditors may perform this type of audit, and it typically is performed to give management a better understanding of the level of compliance risk.

- **Forensic audit** This type of audit is usually performed by an IS auditor or a forensic specialist in support of an anticipated or active legal proceeding. This is typically part of the investigation of a privacy breach. To withstand cross-examination and to avoid having evidence being ruled inadmissible, strict procedures must be followed in a forensic audit, including the preservation of evidence and a chain of custody of evidence.

- **Service provider audit** Because many organizations outsource parts of their operations, third-party service organizations will undergo one or more external audits to increase customer confidence in the integrity of the third-party's services. In the United States, a Statement on Standards for Attestation Engagements No. 18, *Reporting on Controls at a Service Organization* (SSAE 18) can be performed on a service provider's operations and the audit report transmitted to customers of the service provider.

Privacy Audit Planning

The auditor must obtain information about the privacy audit that will enable her to establish the audit plan. Information needed includes

- Location or locations that need to be visited
- A list of the business processes and supporting applications to be examined
- The personnel to be interviewed
- The technologies supporting each application
- Privacy policies, security policies, standards, and data flow diagrams that describe the environment and the personal data stored and processed there

This and other information will enable the auditor to determine the resources and skills required to examine and evaluate privacy-related business processes and information systems. The auditor will be able to establish an audit schedule and a good idea of the types of evidence needed. The auditor may be able to make advance requests for certain other types of evidence even before the onsite phase of the audit begins.

For an audit with a risk-based approach, the auditor has a couple of options:

- Precede the audit itself with a risk assessment to determine which privacy processes or controls warrant additional audit scrutiny.

- Gather information about the organization and historic events to discover risks that warrant additional audit scrutiny.

Audit Statement of Work

For an external audit, the auditor may need to develop a statement of work or engagement letter that describes the audit purpose, scope, duration, and costs. The auditor may require written approval from the client before audit work can officially begin.

Establish Audit Procedures

Using information obtained regarding audit objectives and scope, the auditor can develop procedures for the audit. For each privacy process, control, and objective to be tested, the auditor can specify the following:

- A list of people to interview

- Inquiries to make during each interview

- Documentation (policies, procedures, and other documents) to request during each interview

- Audit tools to use

- Sampling rates and methodologies

- How and where evidence will be archived

- How evidence will be evaluated

- How findings will be reported

Communication Plan

The auditor will develop a communication plan to keep the auditor's management, and the auditee's management, informed throughout the audit project. The communication plan may contain one or more of the following:

- A list of evidence requested, usually in the form of a PBC (provided by client) list, which is typically a worksheet that lists specific documents or records and the names of personnel who can provide them (or who provided them in a prior audit).

- Regular written status reports that include activities performed since the last status report, upcoming activities, and any significant findings that may require immediate attention.

- Regular status meetings where audit progress, issues, and other matters may be discussed in person or via conference call.
- Contact information for both auditor and auditee so that both parties can contact each other quickly if needed.

Report Preparation

The auditor needs to develop a plan that describes how the audit report will be prepared. This will include the format and the content of the report, as well as how findings will be established and documented.

The auditor will need to ensure that the audit report complies with all applicable audit standards, including applicable regulations and ISACA IS audit standards.

Wrap-up

The auditor must perform some tasks at the conclusion of the audit, including the following:

- Deliver the report to the auditee.
- Schedule a closing meeting so that the results of the audit can be discussed with the auditee and so that the auditor can collect feedback.
- For external audits, send an invoice to the auditee.
- Collect and archive all work papers. Enter their existence in a document management system so that they can be retrieved later if needed and to ensure their destruction when they have reached the end of their retention life.
- Update PBC documents if the auditor anticipates that the audit will be performed again in the future.
- Collect feedback from the auditee and convey to audit staff as needed.

Post-audit Follow-up

After a given period (which could range from days to months), the auditor should contact the auditee to determine what progress the auditee has made to remedy any audit findings. This establishes a tone of concern for the auditee organization and helps to establish a dialogue whereby the auditor can help auditee management work through any needed process or technology changes as a result of the audit.

Privacy Audit Evidence

Evidence is the information collected by the auditor during the course of the audit project. The contents and reliability of the evidence obtained are used by the auditor to reach conclusions on the effectiveness of privacy controls and control objectives. The auditor needs to understand how to evaluate various types of evidence and how (and if) it can be used to support audit findings.

The auditor will collect many kinds of evidence during an audit, including observations, written notes, correspondence, independent confirmations from other auditors, process and procedure documentation, and business records.

When the auditor examines evidence, he needs to consider several characteristics that will contribute to its weight and reliability, including the following:

- **Independence of the evidence provider** Evidence provided by the process owner may be tainted (to influence audit results); an independent evidence provider may be preferred.

- **Qualifications of the evidence provider** The evidence provider should be a person qualified to represent the process or system being audited.

- **Objectivity** The evidence should be objective. Digital evidence is more objective than the opinion of a process owner, for instance.

- **Timing** The evidence should be timely and should be appropriate to the issue at hand. Some evidence, such as system logs, may be available only for a short period.

 **NOTE** Evidence collected in a privacy audit is likely to contain personal information. The auditor and auditee will need to understand the protective measures required to protect this evidence, whether it can be anonymized, and how long it must be retained.

Gathering Evidence

The privacy auditor must understand and be familiar with the methods and techniques used to gather evidence during an audit. The methods and techniques used most often in audits include reviews of the organization chart, department and project charters, third-party contracts and service level agreements (SLAs), policies and procedures, risk register, incident log, standards, system documentation, interviews of personnel, re-performance (where auditors will confirm that the organization's processes and systems calculate results properly), and passive observation.

 **NOTE** The privacy auditor should pay attention to what department charters, policies, and procedure documents *do* say, as well as what they *don't* say. He should perform corroborative interviews to determine if these documents define the organization's behavior or if they're just window dressing. This will help the auditor understand the maturity of the organization, a valuable insight that will be helpful when writing the audit report.

Sampling

Sampling refers to a technique used when it is not feasible to test an entire *population* of privacy events or transactions. The objective of sampling is to select a portion of a population so that the characteristics observed will reflect the characteristics of the entire population. There are several methods for sampling, including the following:

- Statistical sampling
- Judgmental sampling (also known as nonstatistical sampling)
- Attribute sampling

- Variable sampling
- Stop-or-go sampling
- Discovery sampling
- Stratified sampling

 EXAM TIP CDPSE candidates are not expected to memorize specific sampling techniques or their purposes, although it is important that you understand the general concept.

Relying on the Work of Other Auditors

Audit departments and external auditors, like other IT service organizations, are challenged to find qualified audit professionals who understand all aspects of organizations' technologies in use. Increased specialization in IT is resulting in auditors who have increased technical knowledge in certain areas and fewer auditors with all of the knowledge required to perform an audit. Third-party service providers usually do not permit customers to audit them, but instead rely on external auditors to perform audits, and then make those audit reports available to the customer. These and other factors are putting increasing pressure on organizations to outsource some auditing tasks (or entire audits) to third-party organizations and to rely upon audit reports from other sources.

For example, it's unlikely that Amazon Web Services or Microsoft Azure will permit any customer to audit them. Amazon and Microsoft will instead commission a number of types of external audits, such as SOC 1, SOC 2, ISO/IEC 27001, or PCI, and make those audit reports (or summaries) available to their customers upon request. The auditors in customer organizations often have little choice but to rely upon them.

Reporting Privacy Audit Results

The work product of a privacy audit project is the *audit report*, a written report that describes the entire audit project, including audit objectives, scope, controls evaluated, opinions on the effectiveness and integrity of those controls, and recommendations for improvement.

While an auditor or audit firm will generally use a standard format for an audit report, some privacy laws and standards require that an audit report regarding those laws or standards contain specific information or be presented in a particular format. Still, there will be some variance in the structure and appearance of audit reports created by different audit organizations.

The auditor is typically asked to present findings in a closing meeting, explaining the audit and its results and being available to answer questions about the audit. The auditor may include an electronic presentation to guide discussions of the audit.

Auditing Specific Privacy Practices

An audit of an organization's privacy is likely to focus on one or more key aspects of an organization's privacy policy and operations. Several are discussed here.

Auditing Privacy Policy

An audit of an organization's privacy policy is going to focus on one or more of these:

- **Compliance with applicable privacy regulations**   Does the organization's privacy policy align with privacy regulations that the organization is required to comply with?

- **Compliance with privacy policies**   Does the organization's practices align with its internal and external privacy policies? For example, if the organization's external privacy policy claims that the organization does not sell personal information, the auditor will determine whether this is really true.

- **Alignment with security policy and practices**   Is the organization's security policy content adequate for protecting personal information? And do the organization's practices align with its policies?

Auditing Data Management

An organization's data management audit, which should include the protection and management of personal information, is likely to focus on one or more of these:

- **Data classification**   The auditor will examine the organization's data classification policy and handling procedures. Since most organizations lack automation, auditors will want to interview workers to see whether there is a pervasive awareness of the classification policy and to determine how often it is applied in practice. If automation is present in the form of DLP or other solutions, the auditor will examine those systems to understand their capabilities and how they are managed.

- **Data protection**   The auditor will examine one or more facets of information security to see how effectively the organization protects personal information. This potentially covers a large variety of topics, from system hardening to identity and access management.

- **Data flows**   The auditor will examine data flow diagrams and supporting documentation to see whether the organization truly knows where personal information flows and resides in its environment.

- **Data loss prevention**   The auditor will examine any DLP systems to see whether they effectively identify the presence and flows of personal information.

Auditing Data Collection

Auditors looking at an organization's practice of collecting personal information will examine several aspects of data collection:

- **Security**   The auditor will look for secure protocols to ensure the protection of personal information in transit and upon arrival on the organization's systems.

- **Alignment with privacy policy**   The auditor will compare data collection practices with privacy policy to see whether they align. For instance, if privacy policy states that only name, address, and phone number are collected, the auditor will compare that policy with systems that collect data to confirm whether these are the only items collected from data subjects.

- **Alignment with applicable regulations** The auditor will confirm whether data collection practices are compliant with specific privacy regulations.

- **Consent** The auditor will examine privacy policy and data collection practices to understand how the organization obtains consent from the data subject at the time of collection. Note that an absence of consent is not necessarily a violation of policy or regulations, as there are circumstances in which it is infeasible or unnecessary for an organization to obtain consent.

- **Data aggregation** The auditor will seek to understand the organization's practices for aggregating personal information collected from the data subject with data obtained from other sources.

Auditing Data Subject Requests

The auditor will examine business processes and supporting information systems to understand how the organization receives data subject requests and how the organization responds to them. Aspects of an audit will include the following:

- **Data subject authentication** The auditor will examine the procedures used by the organization to authenticate the user. The authentication process itself may include the collection of data that authenticates the user; the auditor will need to understand how that information is used, whether it is retained, and if that information is disclosed in subsequent requests.

- **Effectiveness of response** The auditor looks for the procedures undertaken by the organization to see whether they identify all areas where a subject's data resides and whether all instances are disclosed to the data subject.

- **Accuracy of response** The auditor examines procedures to determine whether the organization correctly processes the request, including changes or corrections.

- **Completeness of response** The auditor checks to see whether the organization's response to a data subject includes all instances of storage and use of personal information in its response to the data subject.

- **Timeliness of response** The auditor examines records to see how long the organization takes to respond to requests.

- **Recordkeeping** The auditor examines business records to see what information about the data subject request is retained.

- **Compliance with policy and applicable regulations** Finally, the auditor will confirm whether the organization's processing of data subject requests aligns with its privacy policy and applicable regulations.

Auditing Data Minimization

The auditor will examine data collection and aggregation practices and compare those with the organization's services to see whether it appears that the organization is collecting more items of personal information than are necessary for the organization to provide its services. This will include a comparison with language in the organization's privacy policy to see whether these, too, are in alignment.

Auditing Anonymization and Pseudonymization

The auditor will examine the organization's anonymization and pseudonymization practices to see whether these practices are effective. Here, the auditor will need to "think outside the box" to see whether there are ways in which anonymized and pseudonymized data can be reconstituted and associated with specific natural persons. Whether through ineptness, simple misconfiguration of a system, or outright deception, some organizations tend to fulfill only the *appearance* of anonymization and pseudonymization, but not the *fact* of anonymization and pseudonymization. For instance, merely removing the name of a person otherwise identified as a "46-year-old male electrical engineer with a family of four living on Main Street" is probably an insufficient practice of anonymization.

Auditing Privacy Incident Management

Auditing privacy-related incident management and investigative procedures requires attention to several key activities, including these:

- **Investigation policies and procedures**   The auditor should determine whether there are any policies or procedures regarding privacy and security investigations. This would include who is responsible for performing investigations, where information about investigations is stored, and to whom the results of investigations are reported. Where subject data is examined in an investigation, the auditor will seek to understand how that subject data is protected and used and whether this aligns with the organization's privacy policy and applicable regulations.

- **Computer crime investigations**   The auditor should determine whether there are policies, processes, procedures, and records regarding computer crime investigations. The auditor should understand how internal investigations are transitioned to supervisory authorities, regulators, or law enforcement.

- **Security incident response**   Because security incident response is relevant to privacy in instances where personal data is involved, the auditor should examine security incident response policies, procedures, and plans to determine whether they are up to date. Interviewing incident responders to gauge their familiarity with incident response procedures can indicate the effectiveness of training and tabletop exercises. The auditor should examine some of the records from actual security incidents to see whether the responses were effective and whether the organization conducted post-incident reviews to identify process improvements.

- **Privacy incident response**   For incidents that involve the misuse of personal information, the auditor will examine incident response plans and incident records to see how the organization responds to these incidents. If any personal information becomes a part of the incident response record, the auditor will see whether the organization complies with privacy policy and applicable regulations for the storage and use of those records.

Figure 2-1
Relationship
between ISACA
audit standards,
audit guidelines,
and Code of
Professional
Ethics

Audit Guidelines
(optional)

Audit Standards
(mandatory)

Code of Ethics
(mandatory)

- **Computer forensics** The auditor should determine whether there are procedures for conducting computer forensics. The auditor should also identify tools and techniques available to the organization for the acquisition and custody of forensic data. The auditor should identify whether any employees in the organization have received computer forensics training and are qualified to perform forensic investigations. Because some organizations employ outside firms for forensics assistance, the auditor should examine any contract in place to see whether this prearranged capability was properly established.

Audit Standards

ISACA has published its Information Technology Assurance Framework in the *ITAF: A Professional Practices Framework for IS Audit/Assurance* (currently in its third edition and available free of charge at www.isaca.org/ITAF). ITAF consists of the ISACA Code of Professional Ethics, IS audit and assurance standards, IS audit and assurance guidelines, and IS audit and assurance tools and techniques. The relationship between these is illustrated in Figure 2-1.

Privacy Incident Management

A privacy incident is an event in which one or more data subjects' personal information has been inappropriately used or disclosed in a manner contrary to applicable laws or regulations. A privacy incident can also be thought of as an event that represents a violation of an organization's privacy and/or security policy. For instance, if an organization's privacy policy states that it is not permitted to copy personal information to an external data storage device, then such use would be considered a privacy incident.

NOTE This section focuses primarily on privacy breaches. For a more detailed explanation of security breaches, read Chapter 5, "Information Security Incident Management," in *CISM Certified Information Security Manager All-In-One Exam Guide.*

Phases of Incident Response

An effective response to a privacy incident is organized, documented, and rehearsed. The phases of a formal incident response plan are explained in this section.

For incident response to be effective, organizations must anticipate that incidents will occur and, accordingly, develop incident response plans, test those plans, and train personnel so that incident response will be effective and timely.

Briefly, the phases of incident response, in order, are

- Planning
- Detection
- Initiation
- Analysis
- Containment
- Eradication
- Recovery
- Remediation
- Closure
- Post-incident review
- Retention of evidence

Planning

This step involves the development of written response procedures that are followed when an incident occurs. These procedures are created once the organization's practices, processes, and technologies are well understood. This helps to ensure that incident response procedures align with privacy policy, security policy, business operations, the technologies in use, and practices in place regarding its architecture, development, management, and operations.

Detection

Detection represents the time when an organization is initially aware that a privacy incident is taking place or has taken place. Because of the variety of events that characterize a privacy incident, an organization can become aware of an incident in several ways, including

- Application or network slowdown or malfunction
- Alert from the IDS, IPS, DLP system, web filter, CASB, and other detective and preventive security systems
- Alert from a security incident and event management system (SIEM)
- Alerts from media outlets and their investigators and reports
- Notification from an employee or business partner
- Anonymous tip
- Notification from a regulator

Initiation

This is the phase where a response to the incident begins. Typically, this will include a declaration of an incident, followed by notifications that are sent to response team members so that response operations may begin. Notifications are also typically sent to business executives so that they may also be informed.

 **NOTE** While each organization's privacy incident response plan will vary, an incident is typically confirmed either in the initiation or analysis phases. At that time, organizations may be required to notify regulators, supervisory authorities, and/or affected parties.

Analysis

In this phase, response team members analyze available data to understand the cause, scope, and impact of the incident. This may involve the use of forensic analysis tools to understand activities on individual systems.

Containment

Here, incident responders perform or direct actions that halt the progress or advancement of an incident. The steps required to contain an incident will vary according to the means used by the attacker.

Eradication

In this phase of incident response, responders take steps to remove the source of the incident. This could involve removing malware, blocking incoming attack messages, or removing an intruder.

Recovery

When the incident has been evaluated and eradicated, there is often a need to recover systems or components to their pre-incident state. This may include restoring data or configurations or replacing damaged or stolen equipment.

Remediation

This activity involves any necessary changes that will reduce or eliminate the possibility of a similar incident occurring in the future. This may take the form of process or technology changes.

Closure

Closure occurs when eradication, recovery, and remediation are completed. Incident response operations are officially closed.

Post-incident Review

Shortly after the incident closes, incident responders and other personnel will meet to discuss the incident: its cause, its impact, and the organization's response. The discussion will range from lessons learned to possible improvements in technologies and processes to improve defense and response.

Retention of Evidence

Incident responders and other personnel will direct the retention of evidence and other materials used or collected during the incident. This may include information that may be used in legal proceedings, including prosecution, civil lawsuits, and internal investigations. A chain of custody may be required to ensure the integrity of evidence.

Several standards are available that guide organizations toward a structured and organized incident response, including NIST SP 800-61, *Computer Security Incident Handling Guide*.

Privacy Incident Response Plan Development

Effective incident response plans take time to develop. A privacy manager who is developing an incident response plan must first thoroughly understand business processes, privacy policy, data flows, and underlying information systems, and then discover resource requirements, dependencies, and failure points. A privacy manager may first develop a high-level incident response plan, which is usually followed by the development of several incident response playbooks, which are step-by-step instructions to follow when specific incidents occur.

NOTE Because many privacy incidents are also security incidents, the development of a privacy incident response plan should be performed in close cooperation with the security manager to avoid duplication of effort and to utilize existing response plan resources and practices.

Resources

Before developing privacy incident response procedures, a privacy manager needs to identify required and available resources for incident detection and response. Perhaps the most important resource is the organization's security incident response plan. A correctly designed security incident response plan will recognize and respond to incidents, including information theft and destruction. When this is in place, two elements are needed to develop a privacy incident response plan:

- Callouts to privacy incident responders, so that they may orchestrate notifications to regulators and affected parties

- Detection and response to incidents of misuse of personal information that are not themselves security incidents

Besides these, other resources that privacy managers need to identify include

- Privacy incident response personnel, beyond those workers identified as security incident responders. Privacy incident response personnel will be responsible for examining information systems to understand the nature of a "misuse of personal information" incident.

- Forensics capabilities, including chain of custody procedures. Since a privacy incident may involve notifications to outside parties, a chain of custody will ensure that evidence retention is robust.

- Attorney–client privilege, to ensure that certain aspects of incident response are protected.

- Contact information and methods for regulators and supervisory authorities.

- Prewritten notifications to regulators, supervisory authorities, affected parties, and the public.

Incident Response Playbooks

More mature organizations have developed numerous (as many as a dozen or more) "playbooks," which are detailed procedures to be followed when specific types of security incidents take place. Typical playbook scenarios include ransomware, denial of service, lost or stolen laptop or mobile device, destructive malware, compromise of a user account, and more.

Privacy incident response plans need their playbooks as well, since many privacy incidents are not security incidents per se, but instead represent the misuse of personal information. Thus, privacy managers developing privacy incident response plans need to develop their own response playbooks so that incident responders can quickly work through investigation, containment, and recovery steps. A privacy incident is not a good time to begin to learn how a specific system works, where its logs reside (and how to read them), and how to run reports to understand the steps that resulted in the incident. Better organizations develop these playbooks in advance, so that incident responders can quickly determine what happened and why it happened.

Response Plan Tabletop Testing

When privacy incident response plans (and playbooks) have been developed, they need to be tested in one or more *tabletop* exercises. These are facilitated discussions led by an experienced incident responder, who walks personnel through a typical privacy incident scenario, step by step. At the same time, participants read their privacy incident response plans and discuss the steps they'd be taking if a real incident were taking place.

Privacy Continuous Improvement

The philosophy of continuous improvement is a mainstay of quality-oriented organizations. Rather than assume that all of an organization's processes, procedures, controls, and other operations are operating at an optimum level, a more realistic approach is the idea that there is always room for meaningful improvement.

Continuous improvement is primarily concerned with the fact of process and control improvement rather than the appearance of improvement. Still, an organization should consider all of its privacy and security programs' operations as "works in progress," meaning that management and staff recognize that their processes and controls have not achieved a level of perfection, and likely never will.

An organization can improve processes and controls in the following ways:

- **Accuracy**   Organization strives to improve its controls and processes so that fewer exceptions and errors occur.

- **Efficiency**   Organization will seek opportunities to make controls and processes more efficient, meaning that they will take less effort or require fewer resources while still maintaining quality objectives.

- **Timeliness**   Organization will seek ways to make controls and processes more responsive so that routine and nonroutine tasks take less time to complete.

- **Risk**   Organization will look for ways to reduce risks in controls and processes to ensure fewer opportunities for incidents.

Continuous improvement is so important that it is officially a requirement in ISO/IEC 27001:2013. Requirement 10.2 of the standard reads, "The organization shall continually improve the suitability, adequacy and effectiveness of the information security management system." Similarly, ISO/IEC 27701 (*Security techniques – Extension to ISO/IEC 27001 and ISO/IEC 27002 for privacy information management – Requirements and guidelines*) requirement 5.8 extends this to include the privacy information management system.

Chapter Review

The management of a privacy program in an organization begins with an understanding of all of the IT, security, and privacy-related roles and responsibilities. A role is a description of expected activities that each employee is obliged to perform as part of his or her employment. A responsibility is a statement of outcomes that a person is expected to support.

Models such as RACI charts are used to determine and illustrate levels and types of responsibilities in selected business processes. RACI charts show who performs tasks, who helps, and who is to be told of them. Alternatives to RACI include PARIS (Participant, Accountable, Review Required, Input Required, Sign-off Required) and PACSI (Perform, Accountable, Control, Suggest, Informed).

The formation of a privacy steering committee helps to ensure that an organization's privacy plan will align with the organization's core business operations while fulfilling business objectives and meeting compliance requirements.

Privacy operations include many activities already present in many organizations, including identity and access management, incident monitoring and response, third-party risk management, and others.

Through the concept of privacy by design and by default, privacy affects the processes, procedures, and activities of many (if not all) personnel in IT and IT security.

Building a privacy operation requires detailed knowledge of privacy requirements—that is, the activities and outcomes the organization is required to undertake at present and in the foreseeable future. Those requirements generally stem from applicable regulations; implicit and explicit expectations from employees, customers, and constituents; and other legal obligations.

External privacy policies are the official statements describing the organization's practices for the collection, use, and ultimate disposal of personal information. Privacy policies also describe how data subjects may make inquiries and requests regarding their personal information.

Internal privacy policies describe the roles, responsibilities, practices, standards, and expectations of all staff members.

Data management is a key activity for an organization's privacy program, as this includes the protection of personal information as well as all of its uses.

Modern privacy laws require that organizations establish means for accepting incoming inquiries and requests from internal and external data subjects. Effective procedures and controls are required to ensure accurate and timely response to these incoming communiques.

Subject requests for data removal require that an organization understand whether applicable laws require the retention of such data, and whether anonymization or pseudonymization may effectively be used to fulfill some of these requests.

Supervisory authorities are government or quasi-government agencies that investigate potential wrongdoing by organizations and their management of personal information. Organizations need to be prepared to respond to their requests and inquiries, not unlike those of subject data requests.

Like security awareness training to inform the workforce of expectations in behavior and safe Internet usage, privacy awareness training informs the workforce of expectations for the access and management of personal information.

Third-party risk management (TPRM) refers to activities used to discover and manage risks associated with external organizations performing operational functions for an organization. TPRM industry practices are well developed and can be adopted in a variety of ways by each organization.

The best practice in TPRM is the practice of establishing risk tiers of an organization's service provider population and applying varying levels of rigor to each tier. This ensures a risk-based allocation of resources in upfront and ongoing due diligence of service providers.

Like other important business processes, many organizations undergo audits of some or all aspects of their privacy and data management operations. The principles of information systems auditing are no different in privacy audits: the goal is to determine the effective design and operations of key controls that ensure the right outcomes required of a typical privacy program.

Audits of privacy programs may target data protection (information security) and the proper use of personal information. Audits will also frequently cover data subject access request operations.

A privacy incident is an event in which one or more data subjects' personal information has been inappropriately used or disclosed, contrary to applicable laws or regulations. A privacy incident can also be thought of as an event representing a violation of an organization's privacy and/or security policy.

The overall techniques used for privacy incident response are similar to those for security incident response. Incidents are detected, analyzed, contained, and eradicated. Systems are recovered, and remediation steps are carried out to prevent recurrence. Incidents are closed, and post-incident reviews are performed.

Privacy incident response plans need to be periodically tested, and all incident responders periodically trained.

It is vital that an organization's culture not be satisfied with its current state of policies, practices, and operations. Instead, a spirit of continuous improvement instills a culture of "we can and should do better."

Quick Review

- An employee's title and rank together connote the employee's role in the organization, along with his or her sphere of responsibility. Industries do not follow a strict model for ranks and titles.

- The placement and structure of an organization's staff hierarchies (the org chart) is less important than the practices carried out by the staff.

- The involvement of an organization's board of directors ranges from high involvement in the organization throughout the year, to merely attending a quarterly meeting, to listening to management presentations and asking a few questions.

- In a smaller, younger organization, board members are often the organization's senior executives.

- An organization can combine its privacy, security, and even IT steering committees if the participants are largely the same individuals.

- Through the concept of privacy by design and by default, privacy affects the processes, procedures, and activities of many (if not all) personnel in IT and IT security.

- Because of the high velocity of change of applicable privacy and data protection laws, organizations are compelled to establish reliable means for knowing about changes in these laws.

- The presence and use of unstructured data is often the greatest concern for a privacy officer. The power and flexibility of endpoint operating systems and their connectivity mean that employees can do nearly anything they want with personal information, subject to internal policies and any automatic controls constraining their behavior.

- The advent of data privacy laws has compelled organizations to "discover" the presence and usage of personal information in their information systems. Lax or nonexistent data governance in many organizations is often the root cause of data breaches.

- The greater the extent of outsourcing of IT services, the greater the need for an effective TPRM program.

- For a privacy program to be effective, close partnerships are required with IT, IT security, human resources, legal, and business units or departments. Privacy cannot work in a vacuum, because all of these other entities manage key business processes and information systems manage personal information.

- Audits of a privacy program that attempt to establish that only sanctioned uses of personal information exist is tantamount to proving the absence of unsanctioned activities—a difficult undertaking.

- With relatively minor additions, an organization's security incident response plan can be augmented to serve effectively as a privacy incident response plan as well.

- It is up to management to determine where improvements in processes, procedures, and controls should be implemented.

Questions

1. Privacy responsibilities are included in which of these IT positions?
   - **A.** Security engineer
   - **B.** Application developer
   - **C.** Database administrator
   - **D.** All of these

2. The purpose of a privacy and security steering committee includes:
   - **A.** Business alignment
   - **B.** Policy approval
   - **C.** Risk decisions
   - **D.** All of these

3. An organization has received a data subject request that asks the organization to remove all personal information on file. How should the organization respond?
   - **A.** Pseudonymize the data subject's personal information.
   - **B.** Anonymize the data subject's personal information.
   - **C.** Remove or anonymize the data subject's personal information.
   - **D.** Remove or anonymize the data subject's personal information as permitted by other applicable laws.

4. Program responsibilities over the activities of managing data subject requests lie with:
   - **A.** Customer support
   - **B.** The chief marketing officer
   - **C.** The chief information security officer
   - **D.** The chief privacy officer

5. The inclusion of privacy requirements in a new software development project is a direct offshoot of which principle?
   - **A.** GDPR Article 21
   - **B.** People, process, and technology
   - **C.** Privacy by design and by default
   - **D.** Privacy by design

6. The best first step in building privacy operations is:

    A. Perform a risk assessment.

    B. Identify requirements.

    C. Perform data discovery.

    D. Conduct a penetration test.

7. What is the primary data privacy law in Canada?

    A. PIPEDA

    B. CCPA

    C. GDPR

    D. CICEDA

8. What is the purpose of data discovery scanning?

    A. Determine the presence of personal information in structured data.

    B. Determine the presence of personal information in unstructured data.

    C. Observe the movement of personal information in internal networks.

    D. Observe the movement of personal information in external networks.

9. An organization wants to limit the use of USB external storage for the storage of personal information. What is the best first step to accomplish this?

    A. Implement software to detect uses of USB storage of personal information.

    B. Implement software to block uses of USB storage of personal information.

    C. Create a policy that defines limitations of USB storage.

    D. Disable USB ports on end-user computers.

10. An organization requests that each data subject submit an image of their driver's license as a means of authentication when submitting data subject requests. Should subsequent data subject requests cite the driver's license as collected information?

    A. Yes, because authentication data is always subject to data access requests.

    B. No, because the driver's license was collected outside of the collection period.

    C. No, because information submitted as a part of authentication is exempt.

    D. Yes, because the data subject's driver's license was collected by the organization.

11. A privacy strategist is developing a privacy awareness program. What is the best method for ensuring that employees have retained important content?

    A. Measure the time it takes for employees to complete training.

    B. Include competency quizzes at the end of training sessions.

    C. Note how quickly employees complete training after being asked.

    D. Include videos in privacy training content.

12. What is the purpose of the cloud services shared responsibility model?

   A. Defines responsibilities when assigned to a project team

   B. Defines which parties are responsible for which aspects of privacy

   C. Defines which parties are responsible for which aspects of security and privacy

   D. Defines which parties are responsible for which aspects of security

13. An auditor is developing a plan for auditing privacy controls in a retail organization. What type of evidence should the auditor collect to determine whether data subject requests are recorded properly?

   A. Interview data subjects.

   B. Interview control owners.

   C. Examine business records.

   D. Examine privacy policy.

14. What is the best method for ensuring that privacy incident responders are familiar with incident response procedures?

   A. Include incident responders in tabletop testing.

   B. Direct incident responders to develop incident response plans.

   C. Direct incident responders to respond to the next incident.

   D. Direct incident responders to review incident response plans.

15. What is the best approach for developing a privacy policy in an organization subject to multiple privacy regulations?

   A. Include requirements for the regulation with the greatest number of requirements.

   B. Include only the requirements for the most recent privacy regulation.

   C. Include only the requirements common to all applicable privacy regulations.

   D. Include requirements from all applicable privacy regulations.

Answers

   1. **D.** Privacy responsibilities flow to nearly all positions in IT and IT security. The principle of privacy by design and by default means that all related information systems and supporting processes and procedures need to align with privacy policy.

   2. **D.** A privacy and security steering committee exists to ensure that the privacy and security programs in an organization are business aligned. These security committees also often approve changes to policies and participate in (or approve) top-level decision-making.

   3. **D.** The organization may proceed with the data subject's data removal request, provided that there are no other laws requiring the retention of this information. For example, banks are typically not permitted to remove financial records for current or former customers.

4. D. The chief privacy officer has primary responsibility for the organization's receipt of, processing of, and response to data subject requests. Other departments may have operational responsibilities in the management of these requests, but the ultimate accountability lies with the CPO.

5. C. The principle of privacy by design and by default leads organizations to incorporate privacy requirements into systems and process development efforts. If privacy is not a part of the initial design of a system or process, organizations will have to alter it later, usually at greater cost than they would have incurred if the effort included privacy considerations initially.

6. B. When building a new privacy operation, it is first necessary to understand what activities and characteristics are required of the organization. This will come from the text of applicable security and privacy regulations and other legal obligations such as privacy contracts. Legal counsel should be on hand to interpret these regulations and other obligations correctly.

7. A. PIPEDA, or Personal Information Protection and Electronic Documents Act, is a Canadian data privacy law that seeks to ensure consumer data privacy in the context of e-commerce.

8. B. Data discovery involves the use of tools to determine the extent of the storage of personal information, primarily in unstructured data stores. The nature of business applications and their ability to produce reports and downloadable data extracts compel data privacy officers to know the extent to which the organization's personnel obtain and store personal information.

9. A. The best first step to limiting or blocking the storage of personal information on external USB devices is to implement a tool that provides visibility into the use of USB storage. Then organizations can implement tools that block USB storage with foreknowledge of any issues that may arise. For example, there may be legitimate procedures involving USB storage and personal information that can be altered, or exceptions can be made, so that USB blocking does not disrupt these procedures.

10. D. Personal information, including the image of a driver's license or other government-issued identification, that is collected by an organization for any reason must be disclosed to a data subject who inquires about what information an organization has collected.

11. B. Competency quizzes as a part of training are an effective means for recording knowledge retention.

12. C. The cloud services shared responsibility model illustrates which parties are responsible for which aspects of security and privacy in the technology stack. For example, in an IaaS environment, the service provider will be responsible for physical security, and the customer will be responsible for server patching and end-user access controls.

13. **C.** To determine whether data subject requests are logged properly, an auditor would have to examine business records. Examining privacy policy or interviewing process owners or data subjects would not reveal this.

14. **A.** The best method for incident responders to become familiar with incident response plans is for them to participate in tabletop exercises.

15. **D.** An organization's privacy policy must comply with all requirements for all applicable regulations. To do any less would put an organization in jeopardy of failing to comply with one or more regulations.

Risk Management

In this chapter, you will learn about

- Risk management life-cycle process
- The risk register
- Privacy threats and vulnerabilities
- Privacy impact assessments

This chapter covers Certified Data Privacy Solutions Engineer job practice 1, "Privacy Governance," part C, "Risk Management." The entire Privacy Governance domain represents 34 percent of the CDPSE examination.

Because so much of privacy management has its roots in information security (remember that privacy equals protection plus usage of personal information), privacy risk cannot be separated from information risk any more than data protection can be removed from privacy.

This chapter describes risk management through three lenses:

- **Risk management life cycle** This is the "big picture" process of risk management—the cyclical, iterative process used by any organization that stores or processes personal information in information systems. Organizations need to create and implement an overall privacy and cybersecurity risk management process that may or may not be a part of an existing enterprise risk management (ERM) process.

- **Privacy impact assessments** PIAs assist in providing the instantiation of risk assessments focused on changes in business processes or information systems. A PIA is an analysis of how personally identifiable information (PII) is collected, used, shared, and maintained and is intended to identify privacy risks associated with a project or proposed change to systems or business processes.

- **Threats, vulnerabilities, and attacks** To perform a proper PIA, privacy specialists need to understand the threats, vulnerabilities, and attacks associated with privacy risk. Knowledge of these will result in a PIA that is less likely to overlook one or more factors.

Privacy programs can succeed if they include regular risk assessments within the auspices of a formal risk management program. Indeed, risk management is a core tenet of cybersecurity programs, often referred to as *information security management systems* (ISMSs). Within a privacy program, organizations will undergo PIAs as a part of their standard business change management and IT change management processes. PIAs are not much different from cybersecurity risk assessments with regard to these change-management processes: they focus on the identification of changes in privacy risk that could occur as a part of these planned changes.

The Risk Management Life Cycle

Like other life-cycle processes, risk management is a cyclical, iterative activity that is used to acquire, analyze, and treat risks. This book focuses on privacy risk, but overall the life cycle for privacy risk is functionally similar to that for information risk or even business risk: a new risk is introduced into the process, the risk is studied, and a decision is made about its outcome.

Like other life-cycle processes, risk management is formally defined in policy and process documents that define the scope, roles and responsibilities, workflow, business rules, and business records. Several frameworks and standards from US and international sources define the full life-cycle risk process. Privacy and security managers are generally free to adopt any of these standards, use a blend of different standards, or develop a custom framework.

Both privacy and information risk management rely upon risk assessments that consider valid threats against the organization's information assets, considering any present vulnerabilities. Several standards and models for risk assessments can be used. The results of risk assessments are placed into a risk register, which is the official business record containing current and historic information risk threats.

Risk treatment is the activity in which decisions about risks are made after weighing various risk treatment options. Risk treatment decisions are typically made by a business owner associated with the affected business activity and ratified by an executive steering group.

The Risk Management Process

The risk management process consists of a set of structured activities that enable an organization to manage risks systematically. Like other business processes, risk management processes vary somewhat from one organization to the next, but generally they consist of the following activities:

- **Scope definition**   The organization defines the scope of the risk management process itself. Typically, scope definitions include geographical or business unit parameters. The scope definition is not part of the iterative portion of the risk management process, although scope may be redefined from time to time. In an organization's privacy program, the scope should include
  - Business processes related to the collection, use, and transfer (or sale) of personal information
  - Information systems that support these processes

- The work centers and processing centers supporting the information systems
- Information security in support of these systems and processes
- All of the aforementioned items that are outsourced to third parties

- **Asset identification and valuation**   The organization uses various means to discover and track its information (including personal information) and information system assets. A classification scheme may be present that identifies risk and criticality levels. Asset valuation is a key part of asset management processes, and the value of assets is appropriated for use in the risk management process.

- **Risk appetite**   Developed outside of the risk management life-cycle process, risk appetite is an expression of the level of risk that an organization is willing to accept. A risk appetite that is related to information and privacy risk is typically expressed in qualitative means.

- **Risk identification**   This is the first step in the iterative portion of the risk management process, when the organization identifies a risk that comes from one of several sources, including the following:

  - **Risk assessment**   This includes an overall risk assessment or a focused risk assessment.

  - **Privacy impact assessment (PIA)**   This is an analysis of how PII is collected, used, shared, and maintained as part of planned changes to a business process or information system to identify any changes in privacy risk.

  - **Data protection impact assessment (DPIA)**   This is an analysis of how planned changes to a business process or information system will impact an organization's ability to protect specific types of data (such as PII).

  - **Vulnerability assessment**   This may be one of several activities, including a security scan, a penetration test, or a source code scan.

  - **Threat advisory**   This is an advisory from a product vendor, threat intelligence feed, or news story.

  - **Internal audit**   A routine internal audit may reveal a weakness in a business process that warrants attention in the risk management process.

  - **Control self-assessment (CSA)**   The self-assessment of an internal control may identify a weakness that needs to be managed in the risk management process.

  - **Change in regulations**   A new privacy regulation, a change in an existing regulation, or a precedent set in the enforcement or in legal proceedings may compel organizations to see their processes in a new light. Occasionally this means that a process or control once thought to be compliant (or secure) may need to be revised.

  - **Risk analysis**   This analysis is focused on information that may uncover additional risks that require attention.

  - **Incident**   A security or privacy incident may reveal risks, whether associated with the incident or not. While this is sometimes a matter of risk identification in hindsight, such risks cannot be overlooked.

NOTE Threat events include various aspects of compliance risk, including audits or examinations (with their findings), fines, penalties, sanctions, or notifications to affected parties (employees, customers, or constituents).

- **Risk analysis**   This is the second step in a typical risk management process, including a PIA or DPIA. After the risk has been identified, it is then analyzed to determine several characteristics, including the following:

  - **Probability of event occurrence**   The risk analyst studies event scenarios and calculates the likelihood that an event associated with the risk will occur. This is typically expressed in the number of likely events per year.

  - **Impact of event occurrence**   The risk analyst studies different event scenarios and determines the impact of each. This may be expressed in quantitative terms (dollars or other currency) or qualitative terms (high–medium–low or a numeric scale of 1–5 or 1–10).

  - **Mitigation**   The risk analyst studies different available methods for mitigating the risk. Depending upon the type of risk, there are many techniques to choose from, including changing a process or procedure, training staff, changing architecture or configuration, or applying a security patch.

  - **Recommendation**   After studying a risk, the risk analyst may develop a recommended course of action to address the risk. This reflects the fact that the individual performing risk analysis is often not the risk decision-maker.

- **Risk treatment**   This is the last step in a typical risk management process. Here, the privacy steering committee (or appropriate authoritative group) makes or approves a decision about a specific risk. The basic options for risk treatment are

  - **Accept**   The organization elects to take no action related to the risk.

  - **Mitigate**   The organization chooses to mitigate the risk, which takes the form of some action that serves to reduce the probability of a risk event or reduce the impact of a risk event. The actual steps taken may include business process changes, system configuration changes, the enactment of a new control, or staff training.

  - **Transfer**   The practice of transferring risk is typically achieved through an insurance policy, although other forms are available, including contract assignment.

  - **Avoid**   The organization chooses to discontinue the activity associated with the risk. This choice is typically selected for an outdated business activity that is no longer profitable or for a business activity that was not formally approved in the first place.

- **Risk communication**   This takes many forms, including formal communications within risk management processes and procedures, as well as information communications among risk managers and decision-makers.

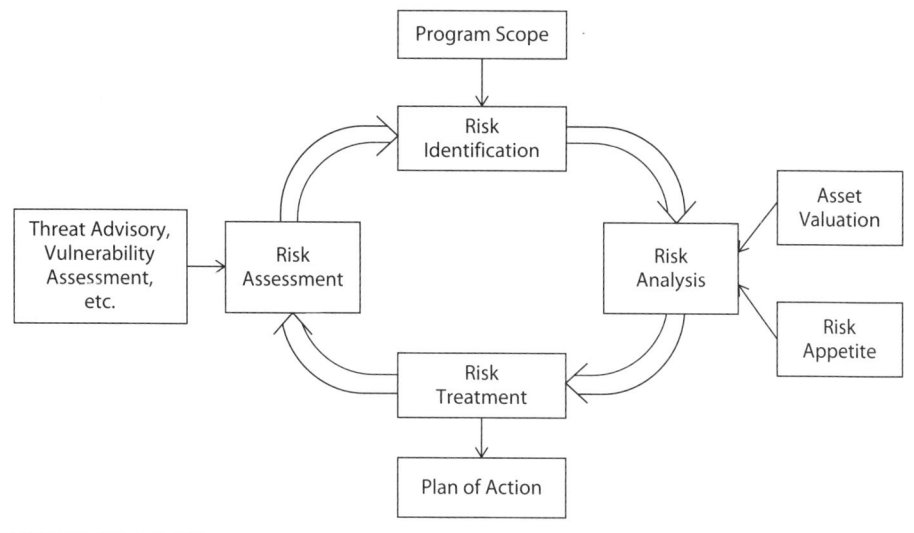

Figure 3-1   The risk management life cycle

In addition to business processes, a risk management process has business records associated with it. The *risk register*, sometimes known as a risk ledger, is the primary business record in most risk management programs. A risk register is a listing of risks that have been identified. Typically, a risk register contains many items, including a description of each risk, the level and type of each risk, and information about risk treatment decisions.

Figure 3-1 shows the elements of a typical risk management life cycle.

Risk Management Methodologies

Several established methodologies are available for organizations that want to manage risk using a formal standard. Organizations select one of these standards for a variety of reasons; they may be required to use a specific standard to address regulatory or contractual terms, they may believe that a specific standard better aligns with their overall information risk program or the business as a whole, or they may want to start with a known standard process as opposed to creating one from scratch.

NIST Standards

The National Institute for Standards and Technology (NIST) develops standards for information security and other subject matter. NIST Special Publication (SP) 800-39, *Managing Information Security Risk: Organization, Mission, and Information, System View*, describes the overall risk management process. NIST SP 800-30, *Guide for Conducting Risk Assessments*, is a detailed, high-quality standard that describes the steps for conducting risk assessments.

NIST SP 800-39   The methodology described in NIST SP 800-39 consists of multilevel risk management, at the information systems level, at the mission/business process level, and at the overall organization level. Communications up and down these levels ensure that

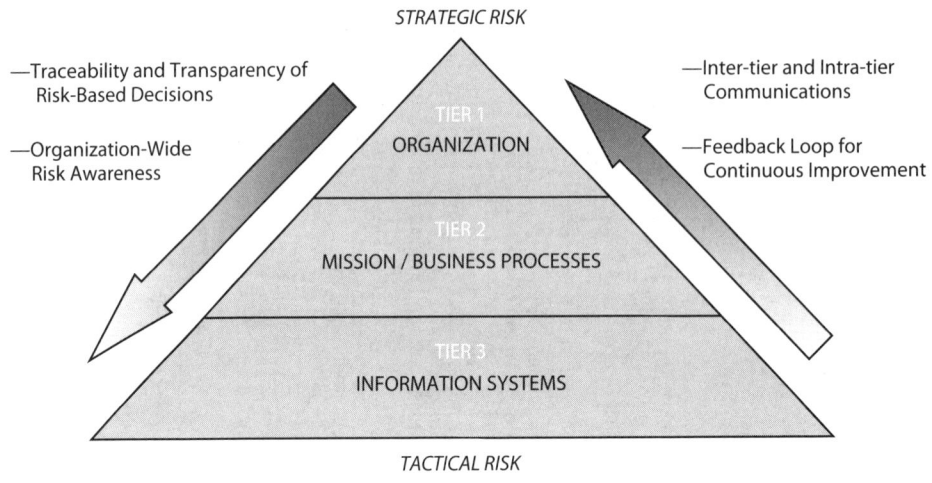

Figure 3-2 Multi-tier risk management in NIST SP 800-39 (Source: National Institute for Standards and Technology)

risks are communicated upward for overall awareness, while risk awareness and risk decisions are communicated downward for overall awareness. Figure 3-2 depicts this approach.

NOTE At first glance, this may appear to be a discussion of cybersecurity risk, but you should know that risk management and risk assessment methodology has wide application. Organizations practicing risk management can do so in the context of operationally critical information systems, manufacturing assembly lines, biomedical laboratory operations, or the protection and use of personal information. For all these, and more, the methodologies of identifying, analyzing, and treating risk are the same.

The tiers of risk management are described in NIST SP 800-39 in this way:

- **Tier 1: Organization view**   This level focuses on the role of governance, the activities performed by the risk executive, and the development of risk management and investment strategies.
- **Tier 2: Mission/business process view**   This level is all about enterprise architecture and enterprise security architecture, and ensuring that business processes are risk aware.
- **Tier 3: Information systems view**   This level concentrates on more tactical things such as system configuration and hardening specifications, vulnerability management, and the detailed steps in the systems development life cycle.

Other concepts discussed in NIST SP 800-39 include trust, the trustworthiness of systems, and organizational culture.

The overall risk management process defined by NIST SP 800-39 consists of several steps:

- **Step 1: Risk framing**   This consists of the assumptions, scope, tolerances, constraints, and priorities—in other words, the business context that is considered prior to later steps taking place.

- **Step 2: Risk assessment**   This is the actual risk assessment, where threats and vulnerabilities are identified and assessed to determine levels and types of risk.

- **Step 3: Risk response**   This is the process of analyzing each risk and developing strategies for reducing it through appropriate risk treatment for each identified risk. Risk treatment options are *accept, mitigate, avoid,* and *transfer.* This step is defined in more detail in NIST SP 800-30, described next.

- **Step 4: Risk monitoring**   This is the process of performing periodic and ongoing evaluation of identified risks to determine whether conditions and risks are changing.

NIST SP 800-30   NIST SP 800-30 describes in greater detail a standard methodology for conducting a risk assessment. The techniques in this document are quite structured and essentially involve setting up a number of worksheets where threats and vulnerabilities are recorded, along with the probability of occurrence and impact if they occur.

In this standard, these are the steps for conducting a risk assessment:

- **Step 1: Prepare for assessment**   The organization determines the purpose of the risk assessment. Primarily, it is important to know the purpose of the results of the risk assessment and the decisions that will be made as a result of the risk assessment. Next, the scope of the assessment must be determined and known. This may take many forms, including geographic and business unit boundaries or specific business processes, as well as the range of threat scenarios that are to be included. Also, any assumptions and constraints pertaining to the assessment should be identified. Further, the sources of threat, vulnerability, and impact information must be identified. (NIST SP 800-30 includes exemplary lists of threats, vulnerabilities, and impact in its appendixes.)

- **Step 2: Conduct assessment**   The organization performs the actual risk assessment. This consists of several tasks.

  a) *Identify threat sources and events.* The organization identifies a list of threat sources and events that will be considered in the assessment. The following sources of threat information are included in the standard and can be used. Organizations are advised to supplement these sources with other information as needed.

     - Table D-1: Threat source inputs
     - Table D-2: Threat sources
     - Table D-3: Adversary capabilities
     - Table D-4: Adversary intent

- Table D-5: Adversary targeting
- Table D-6: Nonadversarial threat effects
- Table E-1: Threat events
- Table E-2: Adversarial threat events
- Table E-3: Nonadversarial threat events
- Table E-4: Relevance of threat events

b) *Identify vulnerabilities and predisposing conditions.* The organization examines its environment (people, processes, and technology) to determine what vulnerabilities exist that could result in a greater likelihood that threat events may occur. The following sources of vulnerability and predisposing condition information are included in the standard and can be used in a risk assessment. Like the catalog of threats, organizations are advised to supplement these lists with additional vulnerabilities as needed.

- Table F-1: Input—vulnerability and predisposing conditions
- Table F-2: Vulnerability severity assessment scale
- Table F-4: Predisposing conditions
- Table F-5: Pervasiveness of predisposing conditions

c) *Determine the likelihood of occurrence.* The organization determines the probability that each threat scenario identified will occur. The following tables guide the risk manager in scoring each threat:

- Table G-1: Inputs—determination of likelihood
- Table G-2: Assessment scale—likelihood of threat event initiation
- Table G-3: Assessment scale—likelihood of threat event occurrence
- Table G-4: Assessment scale—likelihood of threat event resulting in adverse impact
- Table G-5: Assessment scale—overall likelihood

d) *Determine the magnitude of impact.* In this phase, the risk manager determines the impact of each type of threat event on the organization. These tables guide the risk manager in this effort:

- Table H-1: Input—determination of impact
- Table H-2: Examples of adverse impacts
- Table H-3: Assessment scale—impact of threat events
- Table H-4: Identification of adverse impacts

e) *Determine the risk level.* The organization determines the level of risk for each threat event. These tables aid the risk manager in this effort:

- Table I-1: Inputs—risk
- Table I-2: Assessment scale—level of risk (combination of likelihood and impact)

- Table I-3: Assessment scale—level of risk
- Table I-4: Column descriptions for adversarial risk table
- Table I-5: Template for adversarial risk table to be completed by risk manager
- Table I-6: Column descriptions for nonadversarial risk table
- Table I-7: Template for nonadversarial risk table to be completed by risk manager

- **Step 3: Communicate results**   When the risk assessment has been completed, the results are then communicated to decision-makers and stakeholders in the organization. The purpose of communicating risk assessment results is to ensure that the organization's decision-makers make decisions that include considerations for known risks. Risk assessment results can be communicated in several ways, including the following:
    - Publishing to a central location
    - Briefings
    - Distributing via e-mail
    - Distributing hard copies

- **Step 4: Maintain assessment**   After a risk assessment has been completed, the organization will maintain the assessment by monitoring risk factors identified in the risk assessment. This enables the organization to maintain a view of relevant risks that incorporates changes in the business environment since the risk assessment was completed. NIST SP 800-137, *Information Security Continuous Monitoring (ISCM) for Federal Information Systems and Organizations*, provides guidance on the ongoing monitoring of information systems, operations, and risks.

   **NOTE**   NIST SP 800-30 is available at https://csrc.nist.gov/publications/sp.

ISO/IEC 27005

ISO/IEC 27005, *Information technology – Security techniques – Information security risk management*, is an international standard that defines a structured approach to risk assessments and risk management. The methodology outlined in this standard is summarized here:

Step 1: Establish Context   Before a risk assessment can be performed, a number of parameters need to be established, including the following:

- **Scope of the risk assessment**   This includes which portions of an organization are to be included, based on business unit, service, line, geography, organization structure, or other means.

- **Purpose of the risk assessment**   Reasons include legal or due diligence or support of an ISMS, business continuity plan, vulnerability management plan, or incident response plan.

- **Risk evaluation criteria**   Determine the means through which risks will be examined and scored.

- **Impact criteria**   Determine how the impact of identified risks will be described and scored.

- **Risk acceptance criteria**   Specify the method that the organization will use to determine risk acceptance.

- **Logistical plan**   This includes which personnel will perform the risk assessment, which personnel in the organization need to provide information such as control evidence, and what supporting facilities are required, such as office space.

Step 2: Risk Assessment   The risk assessment is performed with the following tasks:

- **Asset identification**   Risk analysts identify assets, along with their value and criticality.

- **Threat identification**   Risk analysts identify relevant and credible threats that have the potential to harm assets, along with their likelihood of occurrence. There are many types of threats, both naturally occurring and human-caused, and accidental or deliberate. Note that some threats may affect more than one asset. ISO/IEC 27005 contains a list of threat types, as does NIST SP 800-30 (in Table D-2), described earlier.

 **NOTE**   A risk analyst should consider additional threats specific to the organization that are not contained in ISO/IEC 27005 or NIST SP 800-30.

- **Control identification**   Risk analysts identify existing and planned controls. Those controls that already exist should be examined to see whether they are effective. The criteria for examining a control includes whether it adequately reduces the likelihood or impact of a threat event. The results of this examination will conclude whether the control is effective, ineffective, or unnecessary. Finally, when identifying threats, the risk analyst may determine that a new control is warranted.

- **Vulnerability identification**   Vulnerabilities that can be exploited by threat events that cause harm to an asset are identified. Remember that a vulnerability does not cause harm, but its presence may permit a threat event to harm an asset. ISO/IEC 27005 contains a list of vulnerabilities. Note that a risk analyst may need to identify additional vulnerabilities.

- **Consequences identification**   The risk analyst will identify consequences that would occur for each identified threat against each asset. Consequences may include the loss of confidentiality, integrity, or availability of any asset, as well as a loss of human safety. Depending on the nature of the asset, consequences may take many forms, including service interruption or degradation, reduction in service quality, loss of business, reputation damage, or monetary penalties, including fines. Note that consequences may be a primary result or a secondary result of the realization of a specific threat. For example, the theft of sensitive financial information may have little or no operational impact in the short term, but legal proceedings over the long term could result in financial penalties, unexpected costs, and loss of business.

Step 3: Risk Evaluation   Levels of risk are determined according to the risk evaluation and risk acceptance criteria established in step 1. The output of risk evaluation is a list of risks, with their associated threats, vulnerabilities, and consequences.

Step 4: Risk Treatment   Decision-makers in the organization will select one of four risk treatment options for each risk identified in step 3:

- **Risk reduction (aka risk mitigation)**   The organization alters something in information technology (such as security configuration, application source code, or data), business processes and procedures, or personnel (such as training).

  In many cases, an organization will choose to update an existing control or enact a new control so that the risk reduction may be more effectively monitored over time. The cost of updating or creating a control—as well as the impact on ongoing operational costs of the control—will need to be weighed alongside the value of the asset being protected, as well as the consequences associated with the risk being treated. A risk manager remembers that a control can reduce many risks, and potentially for several assets, so the risk manager will need to consider the benefit of risk reduction in more complex terms. Chapter 6 covers a comprehensive discussion of the types of controls.

- **Risk retention (aka risk acceptance)**   The organization chooses to accept the risk and decides not to change anything.

- **Risk avoidance**   The organization decides to discontinue the activity associated with the risk. For example, an organization assesses the risks related to the acceptance of credit card data for payments and decides to change the system so that credit card data is sent directly to a payment processor so that the organization will no longer be accepting credit card data.

- **Risk transfer**   The organization transfers risk to another party. The common forms of risk transfer are insurance and outsourcing security monitoring to a third party. When an organization transfers risk to another party, there will usually be residual risk that is more difficult to treat. For example, while an organization may have had reduced costs from a breach because of cyber insurance, the organization may still suffer reputational damage in the form of reduced goodwill.

Decision-makers weigh the costs and benefits associated with each of these four options and decide the best course of action for the organization. These four risk treatment options are not mutually exclusive; sometimes, a combination of risk treatment options may be the best choice for a specific situation. For instance, a business application was found to accept weak passwords; the chosen risk treatment was a combination of security awareness training (mitigation) and acceptance (the organization elected not to modify the application as this would have been too expensive).

Further, some treatments can address more than one risk. For example, security awareness training may reduce several risks associated with end-user computing and behavior.

Often, after risk treatment, some risk—known as *residual risk*—remains. When analyzing residual risk, the organization may elect to undergo additional risk treatment to reduce the risk further, or it may accept the residual risk as is. Note that residual risk cannot be reduced to zero—there will always be some level of risk.

Because some forms of risk treatment (mainly, risk reduction and risk transfer) may require an extended period of time to be completed, risk managers usually track ongoing risk treatment activities to completion.

Step 5: Risk Communication   All parties involved in information risk—the chief information security officer (CISO) or another top-ranking information security official, risk managers, business decision-makers, and other stakeholders—need channels of communication throughout the entire risk management and risk treatment life cycle. Examples of risk communication include the following:

- Announcements and discussions of upcoming risk assessments
- Collection of risk information during risk assessments (and at other times)
- Proceedings and results from completed risk assessments
- Discussions of risk tolerance
- Proceedings from risk treatment discussions and risk treatment decisions and plans
- Educational information about security and risk
- Updates on the organization's mission and strategic objectives
- Communication about security incidents to affected parties and stakeholders

Step 6: Risk Monitoring and Review   Organizations are not static, and neither is risk. The value of assets, impacts, threats, and vulnerabilities, and the likelihood of risk occurrence should be periodically monitored and reviewed so that the organization's view of risk continues to be relevant and accurate. Monitoring should include

- Discovery of new, changed, and retired assets
- Change in business processes and practices
- Changes in technology architecture
- New threats that have not been assessed

PART I

- New vulnerabilities that were previously unknown
- Changes in threat event probability and consequences
- Security incidents that may alter the organization's understanding of threats, vulnerabilities, and risks
- Changes in market and other business conditions
- Changes in applicable laws and regulations

NOTE ISO/IEC 27005 is available at https://www.iso.org/home.html.

Factor Analysis of Information Risk

Factor Analysis of Information Risk (FAIR) is an analysis method that helps a risk manager understand the factors that contribute to risk, the probability of threat occurrence, and an estimation of potential losses. In the FAIR methodology, there are six types of loss:

- **Productivity**   Loss of productivity caused by the incident
- **Response**   Cost expended in incident response
- **Replacement**   Expense required to rebuild or replace an asset
- **Fines and judgments**   All forms of legal costs resulting from the incident
- **Competitive advantage**   Loss of business to other organizations
- **Reputation**   Loss of goodwill and future business

FAIR also focuses on the concept of asset value and liability. For example, a customer list is an asset, because the organization can reach its customers to solicit new business; however, the customer list is also a liability, because of the impact on the organization if the customer list is obtained by an unauthorized person.

EXAM TIP   CDPSE candidates are not expected to memorize risk assessment methodologies, but you should be familiar with overall risk assessment concepts.

FAIR guides a risk manager through an analysis of threat agents and the different ways in which a threat agent acts upon an asset:

- **Access**   Threat agent reads data without authorization
- **Misuse**   Threat agent uses an asset differently from its intended usage
- **Disclose**   Threat agent shares data with other unauthorized parties
- **Modify**   Threat agent modifies asset
- **Deny use**   Threat agent prevents legitimate subjects from accessing assets

FAIR is considered to be complementary to risk management methodologies such as NIST SP 800-30 and ISO/IEC 27005.

 **NOTE** You can obtain information about FAIR at https://www.fairinstitute.org.

Asset Identification

After a risk assessment's scope has been determined, the initial step in a risk assessment is the identification of assets and a determination of each asset's value. In a typical information risk assessment, assets will consist of various types of information (including intellectual property, internal operations, and personal information), the information systems that support and protect those information assets, and the business processes that are supported by these systems.

Hardware Assets

Hardware assets may include server and network hardware, user workstations, office equipment such as printers and scanners, and Wi-Fi access points. Depending on the scope of the risk assessment, assets in storage and replacement components may also be included.

Because hardware assets are installed, moved, and eventually retired, it is important to verify the information in the asset inventory periodically by physically verifying the existence of the physical assets. Depending upon the value and sensitivity of systems and data, this inventory "true-up" may be performed as often as monthly or as seldom as annually. Discrepancies in actual inventory must be investigated to verify that assets have not been stolen or moved without authorization.

Subsystem and Software Assets

Software applications such as software development tools, drawing tools, security scanning tools, and subsystems such as application servers and database management systems are all considered assets. Like physical assets, these often have tangible value and should be periodically inventoried.

Cloud-Based Information Assets

One significant challenge related to information assets lies in the nature of cloud services and how they work. An organization may have a significant portion of its information assets stored by other organizations in their cloud-based services. Unless an organization has exceedingly good business records, some of these assets will be overlooked. The main reason for this is because of the ways in which cloud services work; it's easy to sign up for a zero-cost or low-cost service and immediately begin uploading business information to the service. Unless the organization has advanced tools such as a cloud access security broker (CASB), it will be next to impossible for an organization to know all of the cloud-based services that are used.

Virtual Assets

Virtualization technology, which enables an organization to employ multiple, separate operating systems to run on one server, is a popular practice for organizations, whether on their own hardware servers located in their own data centers or in hosting facilities. Organizations employing infrastructure as a service (IaaS) are also employing virtualization technology.

 **NOTE** While IaaS and virtualization make it far easier to create and manage server assets, maintaining an accurate inventory of virtual server assets is even more challenging than it is for physical assets.

Information Assets

Information assets are less tangible than hardware assets, as they are not easily observed. Information assets take many forms:

- **Personal information** Most organizations store information about people, whether they are employees, customers, constituents, beneficiaries, or citizens. This data may include sensitive information such as contact information and personal details, transactions, order histories, and other items.

 **CAUTION** Organizations need to be mindful of the regulatory definitions of personal information within the context of applicable privacy laws and regulations.

- **Intellectual property** This type of information can take the form of trade secrets, source code, product designs, policies and standards, and marketing collateral.
- **Business operations** This generally includes merger and acquisition information and other types of business processes and records not mentioned earlier.
- **Virtual machines** Most organizations are moving their business applications to the cloud, thereby eliminating the need to purchase hardware. Organizations that use IaaS have virtual operating systems, which are another form of information. Even though these operating systems are not purchased but instead are rented, there is nonetheless an asset perspective: they take time to build and configure and therefore have a replacement cost. The value of assets is discussed more fully later in this section.

Asset Classification

Asset classification is an activity whereby an organization assigns an asset to a category that represents usage or risk. The purpose of asset classification is to determine, for each asset, its level of criticality to the organization. In an organization with a formal privacy program, asset classification will include one or more classifications for assets related to personal information.

Criticality can be related to information sensitivity. For instance, a database of customer personal information that includes contact and payment information would be considered highly sensitive and, in the event of compromise, could result in significant impact to present and future business operations.

Criticality can also be related to operational dependency. For example, a database of virtual server images may be considered highly critical. If an organization's server images were to be compromised or lost, this could adversely affect the organization's ability to continue its information processing operations.

These and other measures of criticality form the basis for information protection, system redundancy and resilience, business continuity planning, and access management. Scarce resources in the form of information protection and resilience need to be allocated to the assets that require it the most. It doesn't usually make sense to protect all assets to the same degree—the more valuable, sensitive, and critical assets should be protected more securely than those that are less valuable and critical.

NOTE The best approach to asset classification in most organizations is first to identify and classify *information* assets and then follow this with *system* classification. It should be noted that one area that is often overlooked or not addressed to a satisfactory level is dealing with unstructured data and data that resides outside of the organization's approved systems.

Data Classification

Data classification is a process whereby different sets and collections of data in an organization are analyzed for various types of sensitivity, criticality, integrity, and value. There are different ways to understand these characteristics. These are some examples:

- **Personal information**   This type of information is most commonly associated with natural persons. Examples include personal contact information, employment records, medical records, and personal financial data, including credit card and bank account numbers.

- **Sensitive information**   Information other than personal information can also be considered sensitive, including intellectual property, nonpublished financial records, merger and acquisition information, and strategic plans.

- **Operational criticality**   In this category, information must be available at all times, or perhaps the information is related to some factors of business resilience. Examples of information in this category include virtual server images, incident response procedures, and business continuity procedures. Corruption or loss of this type of information may have a significant impact on ongoing business operations.

- **Accuracy or integrity**   Information in this category is required to be highly accurate. If altered, the organization could suffer significant financial or reputational harm. Types of information include exchange rate tables, product or service inventory data, machine calibration data, and price lists. Corruption or loss of this type of information impacts business operations by causing incomplete or erroneous transactions.

- **Monetary value** This information may be more easily monetized by intruders who steal it. Types of information include credit card numbers, bank account numbers, gift certificates or cards, and discount or promotion codes. Loss of this type of information may cause direct financial losses.

Most organizations have information that falls into all of these categories, with degrees of importance within them. While this may result in a complex matrix of information types and degrees of importance or value, the most successful organizations will be those that build a fairly simple data classification scheme. For instance, an organization may develop four levels of information classification, such as Public, Confidential, Regulated, and Secret. Data classification is discussed in depth in Chapter 7.

Asset Valuation

A key part of a risk assessment is the identification of the value of an asset. In the absence of an asset's value, it is more difficult to calculate risks associated with an asset, even when qualitative risk valuation is employed. Without a known valuation, the impact of loss is more difficult to determine.

Qualitative Asset Valuation

Because many risk assessments are qualitative in nature (as will be revealed later in this chapter), establishing asset valuation in qualitative terms is common. Instead of assigning a dollar (or other currency) value figure to an asset, the value of an asset can be assigned to a low–medium–high scale or to a numeric scale such as 1–5 or 1–10.

The objective of qualitative asset valuation is to establish which assets have more value than others. Qualitative valuation gives an organization the ability to determine which assets have greater value and which have less value. This can be highly useful in an organization with a lot of assets, because it can provide a view of its high-value assets without the "noise" of comingled lower valued assets. In a privacy program, this can help identify assets associated with the processing of personal information.

Quantitative Asset Valuation

Many organizations opt to surpass qualitative asset valuation and assign a dollar (or other currency) valuation to assets. This is common in larger or more mature organizations that want to understand the actual costs that may be associated with loss events.

In a typical quantitative valuation of an asset, its value may be one of the following:

- **Replacement cost** If the asset is a hardware asset, its valuation may be determined by the cost of purchasing (and deploying) a replacement. If the asset is a database, its cost may be determined by the operational costs required to restore it from backup or the costs to recover it from its source, such as a service provider.

- **Book value** This represents the value of an asset in the organization's financial system, typically the purchase price less depreciation.

- **Net present value (NPV)** If the asset directly or indirectly generates revenue, this valuation method may be used.

- **Redeployment cost** If the asset is a virtual machine, its valuation may be determined by the cost of setting it up again. This is typically a soft cost if it is set up by internal staff, but it could be a hard cost if another company is hired to redeploy it.

- **Creation or reacquisition cost** If the asset is a database, its cost may be determined by the cost of re-creating it. If the asset is intellectual property such as software source code, its valuation may be determined according to the effort required for developers to re-create it.

- **Consequential financial cost** If the asset is a database containing personal information, its valuation may be measured in the form of financial costs that result from its theft or compromise. While the cost of recovering that database may be relatively low, the consequences of its compromise could cost hundreds of dollars per record. This is a typical cost when measuring the full impact of a privacy breach.

Risk managers must carefully determine the appropriate method for setting the value of each asset. While some instances will be fairly straightforward, others will not. In many cases, an individual asset will have more than a single valuation category. For example, a credit card database may primarily be valued on its consequential cost (because of the potential fines plus remediation costs associated with consumers who may have been harmed) and also redeployment costs, although in this case, this may be a small fraction of the total valuation.

Risk managers should document their rationale and method of valuation, particularly for sensitive and personal information assets whose valuations could vary widely depending on the method used. Better yet, larger and more mature organizations will have guidelines that specify methods and formulas to be used in information asset valuation.

 **EXAM TIP** CDPSE candidates are not expected to quantify the value of information assets, but you should be familiar with various concepts and approaches to valuation.

Threat Identification

The identification of threats is a key step in a risk assessment. A *threat* is defined as an event that, if realized, would bring harm to an asset and, thus, to the organization.

In the privacy and cybersecurity industries, the key terms involved with risk assessments are often misunderstood and misused. These terms are distinguished from one another in this way: A *threat* is an actual action that would cause harm, not the person or group (generically called an *actor* or *threat actor*) associated with it. A threat is also *not* a weakness that may permit a threat to occur; this is known as a *vulnerability*.

Threats are typically classified as external or internal, as intentional or unintentional, and as man-made or natural. The origin of many threats is outside the control of the organization but not necessarily outside of its awareness. A good privacy or security manager can develop a list of privacy- and security-related threats that are likely (more or less) to occur to any given asset.

When performing a risk assessment, the risk manager needs to develop a complete list of threats for use in the risk assessment. Because it's not always possible for a risk manager to memorize all possible threats, the security manager may turn to one or more well-known sources of threats, including the following:

- ISO/IEC 27005's Appendix C, "Examples of Typical Threats"
- NIST SP 800-30's Appendix E, "Threat Events"

Upon capturing threat events from one or both of these sources, the risk manager may well identify a few additional threats not found in these lists. These additional threats may be specific to the organization's location, business model, or other factors. A risk manager will typically remove a few of the threats from the list that do not apply to the organization. For instance, an organization located far inland is not going to be directly affected by tsunamis or hurricanes, so this threat source can be eliminated.

Internal Threats

Internal threats originate within the organization and are most often associated with employees. Quite possibly, internal employees may be the intentional actors behind these threats. This is generally known as *insider threat.*

Privacy managers need to understand the nature of internal threats and the interaction between personnel and information systems. A wide range of events can take place that constitutes threats, including the following:

- Well-meaning personnel making errors in judgment
- Well-meaning personnel making errors in haste
- Well-meaning personnel making errors because of insufficient knowledge or training
- Well-meaning personnel being tricked into doing something harmful
- Disgruntled personnel being purposefully negligent or reckless
- Disgruntled personnel deliberately bringing harm to an asset
- A trusted individual in a trusted third-party organization doing any of these

After understanding all the ways that something can go wrong, privacy and security managers may sometimes wonder that things can ever proceed as planned!

A privacy manager must understand this important concept: While employees are at the top of a short list of potential threat actors, employees are the same people who need to be given broad access to sensitive data in order for each of them to do his or her job and for the organization to function. Although there have been marginal improvements in technologies such as data loss prevention (DLP), employers must trust employees by giving them access to large sets of the organization's information, with the hope that the employees will not accidentally or deliberately abuse those privileges with potential to cause the organization great harm.

Here are some examples of employees gone rogue:

- A disgruntled internal auditor discloses salary and other personal information relating to 100,000 staff members at a large supermarket chain in an attempt to frame a colleague.

- A consulting firm for a large insurer finds that one of its consultants, who was discovered to be involved in identity theft, has e-mailed a file with more than 18,000 Medicare member details to his personal e-mail account.

- An engineer at a cloud services company breaks into and exposes millions of customer records at a large bank.

- A systems administrator at an intelligence agency acquires and leaks thousands of classified documents to the media.

A significant factor in employees gone rogue is access control policy and access management practices, which result in individual employees having access to more information than is prudent. That said, increasing the granularity of access controls is known to be time-consuming and costly, and it increases the friction of doing business; few organizations tolerate this despite identified risks.

The following list includes internal and external man-made threats that may be included in an organization's risk assessment:

Internal and External Man-Made Threats

Leak data via e-mail

Leak data via upload to unauthorized system

Leak data via external USB storage device or medium

Leak information face-to-face to unauthorized person

Perform programming error

Misconfigure system or device

Shut down application, system, or device

Perpetrate error created by any internal staff

Respond to phishing attack

Respond to social engineering attack

Share login credentials with another person

Install or run unauthorized software program

Copy sensitive data to unauthorized device or system

Destroy or remove sensitive or critical information

Retrieve discarded, recycled, or shredded information

Conduct security scan

Conduct denial-of-service attack

Internal and External Man-Made Threats

Conduct physical attack on systems or facilities

Conduct credential-guessing attack

Eavesdrop on sensitive communication

Impersonate another individual

Obtain sensitive information through illicit means

Cause data integrity loss through any action

Intercept network traffic

Obtain sensitive information through programmatic data leakage

Perform reconnaissance as part of an attack campaign

Attack via social engineering

Failure or anomaly in power

Failure in communications

Failure in heating, venting, or air conditioning

Degradation of electronic media

Damage via fire

Damage via smoke

Damage via fire retardant

Flood from water main break or drainage failure

Damage via vandalism

Damage from demonstrations/protests/picketing

Attack by terrorist

Damage via electromagnetic pulse

Damage via explosion

Damage via bombing

It may be useful to build a short list of threat actors (the people or groups that would initiate a threat event), but remember that these are not the threats themselves. However, building such a list may help the security manager identify additional threat events that may not be on the list.

The following list shows internal and external natural threats:

Internal and External Natural Threats

Forest fire or range fire

Smoke damage from a forest fire or range fire

River flood

Landslide

Internal and External Natural Threats

Avalanche

Tornado

Hurricane

Windstorm

Hailstorm

Earthquake

Tsunami

Lightning

Epidemic

Explosion of naturally occurring substances

Solar storm

External Threats

External threats originate outside of the organization. Like internal threats, they can include both deliberate and accidental actions and can be man-made or associated with naturally occurring events.

The security manager performing a risk assessment needs to understand the full range of threat actors, along with motivations. This is particularly important for organizations where specific types of threat actors or motivations are more common. For example, certain industries such as aerospace and weapons manufacturers attract industrial espionage and intelligence agencies, and certain industries attract hacktivists.

The following lists show external threat actors and the motivations behind them:

External Threat Actors

Former employees

Current and former consultants

Current and former contractors

Competitors

Hacktivists

Personnel in current and former third-party service organizations, vendors, and suppliers

Government intelligence agencies (foreign and domestic)

Criminal organizations (including individuals)

Terrorist groups (including individuals)

Activist groups (including individuals)

Armed forces (including individuals)

Threat Actor Motivations

Competitive advantage

Economic espionage

Monetary gain

Political gain

Intelligence

Revenge

Ego

Curiosity

Unintentional errors

In a risk assessment, it is essential that the assessor identify all threats that have a reasonable likelihood of occurrence. Threats that are unlikely because of geographical and other conditions are usually excluded. For example, hurricanes can be excluded for locations far from oceans, and earthquakes and volcanos can be excluded from locations where these are not known to occur. Threats such as falling meteorites and space debris are rarely included in risk assessments because of the minute chance of occurrence.

 **EXAM TIP** CDPSE candidates are not expected to memorize lists of threats, but you should be familiar with their concepts.

Advanced Persistent Threats

An *advanced persistent threat* (APT) is a particular type of threat actor, so named in the early 2000s to describe a new kind of adversary that worked slowly but effectively to compromise a target organization. Whether perpetrated by an individual or a cybercrime organization, an APT involves techniques that indicate resourcefulness, patience, and resolve. Rather than employing a "hit-and-run" or "smash-and-grab" operation, an APT actor will patiently perform reconnaissance on a target and use tools to infiltrate the target and build a long-term presence there.

APT is defined by NIST SP 800-39 as follows:

"An APT is an adversary that possesses sophisticated levels of expertise and significant resources that allow it to create opportunities to achieve its objectives using multiple attack vectors (e.g., cyber, physical, and deception). These objectives typically include establishing and extending footholds within the IT infrastructure of the targeted organizations for purposes of exfiltrating information, undermining or impeding critical aspects of a mission, program, or organization; or positioning itself to carry out these objectives in the future. The advanced persistent threat: (i) pursues its objectives repeatedly over an extended period of time; (ii) adapts to defenders' efforts to resist it; (iii) is determined to maintain the level of interaction needed to execute its objectives."

Prior to APTs, threat actors were unsophisticated and conducted operations that ran for short periods of time—a few days at most. But as more organizations put more valuable information assets online, threat actors became craftier and more resourceful; they resorted to longer term campaigns to study a target for long periods of time before attacking it. Once an attack began, it would carry on for months or longer. APTs would compromise multiple systems inside the target organization and use a variety of stealthy techniques to establish and maintain a presence using as many compromised targets as possible. Once an APT was discovered (if it is *ever* discovered), the security manager would clean up the compromised target, often not knowing that the APT had compromised many other targets, with not all of them using the same technique.

This cat-and-mouse game could continue for months or even years, with the adversary continuing to compromise targets and study the organization's systems—all the while searching for specific targets—while the security manager and others would continually chase the adversary around like the carnival game of "whack a mole."

 **NOTE** The term *advanced persistent threat* is not used as often nowadays, although its definition is largely unchanged. APTs were discussed more often when their techniques were new. But today, there are a multitude of cybercriminal organizations, along with hundreds if not thousands of talented, individual threat actors, whose techniques resemble the APTs of a dozen years ago. Today, APTs are not novel, but routine.

Emerging Threats

The theater in which cyberwarfare takes place today is constantly changing and evolving. Several forces (see Table 3-1) are at work and continually "push the envelope" in the areas of attack techniques as well as defense techniques.

The subject of emerging threats should be seen as the continuing phenomenon of new techniques, rather than as a fixed set of techniques. Often, the latest techniques are difficult to detect because they fall outside the span of attack techniques that one expects to observe from time to time. Emerging threats represent the cutting edge of

Phenomenon	Response
Emerging technologies, including bring-your-own-device (BYOD), cloud computing, virtualization, and Internet of Things (IoT)	New targets of opportunity, many of which are poorly guarded when first implemented
End user behavior analytics (EUBA) that detect anomalous end user behavior	Slower exfiltration of sensitive data in order to stay "under the radar"
Improved technologies (faster processing time)	More rapid compromise of cryptosystems
Improved technologies (faster network speeds)	More rapid exfiltration of larger data sets; easier transport of rainbow tables used to crack hash tables
Improved antimalware controls	Attack innovation—techniques evaded antimalware controls

Table 3-1 The Cascade of Emerging Threats

attack techniques that are difficult to detect and/or remediate when they are discovered. But these threats will eventually become routine, and even newer threat techniques will emerge. Privacy and security managers need to understand that, even as defensive technologies improve to help prevent and/or detect attacks of increasing sophistication, attack techniques will continuously improve in their ability to evade detection by even the most sophisticated defense techniques.

Vulnerability Identification

The identification of vulnerabilities is an essential part of any risk assessment. A *vulnerability* is any weakness in a process or system that permits an attack to compromise a target process or system successfully. In the privacy and security industries, the key terms involved with risk assessments are often misunderstood and misused. These terms are distinguished from one another in this way: A vulnerability is the weaknesses in a system that could permit an attack to occur. A vulnerability is not the attack vector or technique—this is known as a *threat*.

Vulnerabilities usually take one of these forms:

- **Configuration fault**   A system, program, or component with configuration settings has one or more settings set incorrectly, which could provide an attacker with additional opportunities to compromise a system. For example, the authentication settings on a system may permit an attacker to employ a brute-force password-guessing attack that will not be blunted by target user accounts being automatically locked out.

- **Design fault**   The relationship between components of a system may be arranged in a way that makes it easier for an attacker to compromise a target system. For instance, an organization may have placed a database server in its DMZ network instead of in its internal network, making it easier for an attacker to identity and attack.

- **Business process weakness**   A business process related to the processing of personal information may fail to prevent or detect unwanted activities in certain circumstances. For instance, an end user who extracts a large volume of personal information from a customer application and saves it directly to a personal cloud drive may bypass controls that would detect or prevent this if the file were saved locally.

- **Known unpatched weakness**   A system may have one or more vulnerabilities for which security patches are available but not yet installed. For example, a secure communications protocol may have a flaw in the way that an encrypted session is established, which could permit an attacker to take over an established communications session. A security patch may be available for the flaw, but until the security patch is installed, the flaw exists and may be exploited by anyone who understands the vulnerability and has techniques to exploit it. Sometimes, known weaknesses are made public through a disclosure by the system's manufacturer or a responsible third party. Although a patch may not yet be offered, other avenues may be available to mitigate the vulnerability, such as a configuration change in the target system.

- **Undisclosed unpatched weakness** A system may have vulnerabilities that are known only to the system's manufacturer and that are not publicized. Until an organization using one of these systems learns of the vulnerability via a security bulletin or a news article, the organization can do little to defend itself, short of employing essential security techniques such as system hardening, network hardening, and secure coding.

- **Undiscovered weakness** Security managers have long accepted the fact that all kinds of information systems have security vulnerabilities that are yet to be discovered, disclosed, and mitigated. New techniques for attacking systems are constantly being developed, and some of these techniques can exploit weaknesses no one knew to look for. As newly discovered techniques involve examining active memory for snippets of sensitive information, system and tool designers continue to design defense techniques for detecting and even blocking attacks. For example, techniques were developed that would permit an attacker to harvest credit card numbers from PCI-compliant, point-of-sale software programs. Soon, effective attacks were developed that enabled cybercriminal organizations to steal tens of millions of credit card numbers from global retail companies.

Vulnerabilities exist everywhere—in software programs, database management systems, operating systems, virtualization platforms, business processes, encryption algorithms, business processes, and personnel. As a rule, privacy and security managers should consider that every component of every type in every system has both known and unknown vulnerabilities, some of which, if exploited, could result in painful and expensive consequences for the organization. Table 3-2 contains the places where vulnerabilities may exist, together with techniques that can be used to discover at least some of them.

Third-Party Vulnerability Identification

Most organizations outsource at least a portion of their software development and IT operations to third parties. Mainly this occurs through the use of cloud-based applications and services such as software as a service (SaaS) applications and platform as a service (PaaS) and infrastructure as a service (IaaS) environments. Many organizations have the misconception that third parties take care of all security concerns in their services. Instead, organizations should thoroughly understand the security responsibility model for each outsourced service to understand which portions of security are their own responsibility and which are managed by the outsourced service.

Regardless of whether security responsibilities for any given aspect of operations are the burden of the organization or the outsourcing organization, vulnerabilities need to be identified and managed. For aspects of privacy and security that are the responsibility of the organization, the organization needs to employ normal means for identifying and managing them. For aspects of privacy and security that are the responsibility of the service provider, the provider needs to identify and manage vulnerabilities—in many cases, the service provider will make these activities available to their customers upon request.

Table 3-2	Vulnerability Context	Detection Technique
Vulnerabilities and Detection Techniques	Network device	Vulnerability scanning Penetration testing Code analysis Network architecture review
	Operating system	Vulnerability scanning Penetration testing System architecture review
	Database management system	Vulnerability scanning Penetration testing
	Software application	Vulnerability scanning Penetration testing Dynamic application scanning Static code scanning Application architecture review
	Physical security	Physical security controls review Social engineering assessments Physical penetration testing
	Business process	Process reviews Internal audits Control self-assessments
	Personnel	Social engineering assessments Competency assessments Phishing assessments (continual)

 **EXAM TIP** CDPSE candidates are expected to be familiar with the concepts of, and differences among, threat, vulnerability, probability, impact, asset value, and risk.

Risk Identification

Risk identification is the activity during a risk assessment in which various scenarios are studied for each asset. Several considerations are applied in the analysis of each risk, including these:

- **Threats** All realistic threat scenarios are examined for each asset to determine which are reasonably likely to occur.

- **Threat actors** It is important to understand the variety of threat actors and to know which ones are more motivated to target the organization and for what reasons. This further illuminates the likelihood that a given threat scenario will occur.

- **Vulnerabilities**   For each asset (both information and information systems), business process, and staff members being examined, vulnerabilities need to be identified. Then various threat scenarios are considered to determine which are made more likely because of corresponding vulnerabilities.

- **Asset value**   The value of each asset is an important factor to include in risk analysis. As described in the earlier section on asset value, assets may be valued in several ways. For instance, a customer database may have a modest recovery cost if it is damaged or destroyed; however, if that same customer database is stolen and sold on the black market, the value of the data may be much higher to cybercriminals, and the resulting costs to the organization to mitigate the harm done to customers may be higher still. Another way to examine asset value is through the revenue derived from its existence or use. The financial consequences of a ruined reputation are not included here but are a part of the impact, discussed in the next item.

- **Impact**   The risk manager examines vulnerabilities, threats (with threat actors), and asset value, and estimates the impact of the different threat scenarios. Impact is considered separately from asset value, as some threat scenarios have minimal correlation with asset value but instead are related to reputation damage. Breaches of privacy data, for example, can result in high mitigation costs and reduced business. Breaches in hospital data systems can threaten patient care. Breaches in almost any IoT or industrial control system (ICS) context can result in extensive service interruptions and life-safety issues.

Qualitative and quantitative risk analysis techniques help to distinguish higher risks from lower risks. These techniques are discussed later in this section.

Risks above a certain level are often recorded in a risk register where they will be processed through risk treatment.

Risk, Likelihood, and Impact

During risk analysis in a risk assessment, the risk manager will perform some simple calculations to stratify all of the risks that have been identified. Calculations generally resemble one or more of these:

Risk = threats × vulnerabilities
Risk = threats × vulnerabilities × asset value
Risk = threats × vulnerabilities × probabilities

ISO/IEC Guide 73, *Risk management – Vocabulary*, defines *risk* as "the combination of the probability of an event and its consequence." This is an excellent way to understand risk in simple, qualitative terms.

Likelihood

In risk assessments, likelihood is an important dimension that helps a risk manager understand several aspects related to the unfolding of a threat event. The likelihood of a serious security incident has less to do with technical details and more to do with the thought process of an adversary.

Considerations related to likelihood include the following:

- **Hygiene**   This is related to an organization's security operations practices. Organizations that do a poor job in vulnerability management, patch management, and system hardening, for example, are more likely to suffer incidents simply because they are making it easier for adversaries to gain access to their systems.

- **Data management**   Relevant in a privacy program, the quality and effectiveness of an organization's data management program will have a bearing on the probability of a privacy breach. If an organization has mature data management capabilities, it likely will be aware of anomalous behavior indicating a potential breach. On the other hand, an organization paying little attention to its data is more likely to have data compromised without the organization being aware.

- **Visibility**   This factor is related to the organization's standing: how broad and visible the organization is and how much the attacker's prestige will increase as a result of a successfully compromised target.

- **Velocity**   The timing of various threat scenarios and whether there is any warning or foreknowledge are factors. For example, an adversary who is determined to exfiltrate a large volume of data without detection is likely to do so very slowly; on the other hand, ransomware can destroy an organization's information in minutes.

- **Motivation**   It is essential to consider various types of adversaries to understand the factors that would motivate them to attack the organization. It could be about money, reputation, or rivalry.

- **Skill**   For various threat scenarios, what skill level is required to attack the organization successfully? A higher skill level does not always mean an attack is less likely; other considerations such as motivation come into play as well.

Impact

During risk assessments, impact is a key attribute of any threat scenario that a risk manager needs to understand fully. In the context of privacy, the definition of *impact* is the actual or expected result from some action, such as a breach.

Impact is perhaps the most critical attribute to understand for a threat scenario. A risk assessment can describe all types of threat scenarios, the reasons behind them, and how they can be minimized. Still, without understanding the impact of threat scenarios, a risk manager cannot determine how important one threat is from another, in terms of the urgency to mitigate the risk.

A wide range of impact scenarios is possible:

- Direct cash losses
- Reputation damage

- Loss of business—decrease in sales
- Drop in share price—less access to capital
- Reduction in market share
- Diminished operational efficiency (higher internal costs)
- Civil liability
- Legal liability
- Compliance liability (fines, censures, and so on)
- Interruption of business operations

Some of these impact scenarios are easier to analyze in qualitative terms than others, and the magnitude of most of these is difficult to quantify except in specific threat scenarios.

One of the main tools in the business continuity and disaster planning world, the business impact analysis (BIA) is highly useful for privacy and information security managers. A BIA can be conducted as part of a risk assessment or separate from it.

A BIA differs from a risk assessment. Although a risk assessment is used to identify risks and, perhaps, suggested remedies, a BIA is used to identify the most critical business processes, together with their supporting IT systems and dependencies on other processes or systems. The value that a BIA brings to a risk assessment is the understanding of which business processes and IT systems are the most important to the organization. The BIA helps the security manager better understand which processes are the most critical and, therefore, warrant the most protection, all other considerations being equal.

In qualitative risk analysis, where probability and impact are rated on simple numeric scales, a risk matrix is sometimes used to portray levels of risk based on probability and impact. Figure 3-3 shows a risk matrix.

Figure 3-3 Qualitative risk matrix

Probability		Consequences: Slightly Harmful	Consequences: Harmful	Consequences: Extremely Harmful
Probability	**Likely**	Medium Risk	High Risk	Extreme Risk
Probability	**Unlikely**	Low Risk	Medium Risk	High Risk
Probability	**Highly Unlikely**	Insignificant Risk	Low Risk	Medium Risk

Risk Analysis Techniques and Considerations

As part of a risk assessment, the risk manager examines assets, together with associated vulnerabilities and likely threat scenarios. The *risk analysis* is the detailed examination that takes place here.

Risk analysis considers many dimensions of an asset, including these:

- Asset value
- Threat scenarios
- Threat probabilities
- Relevant vulnerabilities
- Existing controls and their effectiveness
- Impact

Risk analysis can also consider business criticality if a BIA is available.

Various risk analysis techniques are discussed in the remainder of this section.

Information Gathering

A risk manager needs to gather a considerable amount of information so that the risk analysis and the risk assessment are valuable and complete. Several sources are available, including

- Interviews with process owners
- Interviews with application developers
- Interviews with privacy and security personnel
- Interviews with external privacy and security experts, including legal counsel
- Privacy and security incident records
- Analysis of incidents that occur in other organizations
- Prior risk assessments (however, caution is advised to stop the propagation of risk calculation errors from one assessment to the next)

Qualitative Risk Analysis

Most risk analysis begins with qualitative risk analysis. This technique does not seek to identify exact (or even approximate) asset value or impact or the exact probability of occurrence. Instead, these items are expressed on a scale such as high, medium, or low. The purpose of qualitative risk analysis is to understand risks relative to one another so that higher risks can be distinguished from lower risks. This is a valuable pursuit, because it gives an organization the ability to focus on more critical risks, based on impact in qualitative terms.

Semiquantitative Risk Analysis

In qualitative risk analysis, the probability of occurrence can be expressed as a numeric value, such as in the range 1 to 5 (where 5 is the highest probability). Impact can also be expressed as a numeric value, also in the range 1 to 5. Then, for each asset and each threat, risk is calculated as *probability × impact*.

For example, suppose an organization has identified two risk scenarios. The first is a risk of data theft from a customer database; the impact is scored as a 5 (highest), and probability is scored as a 4 (highly likely). The risk is scored as *5 × 4 = 20*. The second is a risk of theft of application source code; the impact is scored as a 2 (low), and probability is scored as a 2 (less likely). This risk is scored as *2 × 2 = 4*.

The risk manager understands that the data theft risk is more significant (scored as 20) as compared to the source code theft risk (scored as 4). These risk scores do not indicate that the larger risk is five times as likely to occur; neither do they mean that the larger risk is five times as expensive. They simply indicate that one risk is rated higher than the other. The scores also do not directly indicate whether the probability or the impact alone is high or low—analysis of the detailed scores is necessary to know that.

Note that some risk managers consider this a qualitative risk analysis, because the results are no more accurate in terms of costs and probabilities than the qualitative technique.

Quantitative Risk Analysis

In quantitative risk analysis, risk managers are attempting to determine actual costs and probabilities of events. This technique provides more specific information to executives about the costs they can expect to incur in various security event scenarios.

Two aspects of quantitative risk analysis prove to be a continuing challenge:

- **Event probability** It is difficult to come up with even an order-of-magnitude estimate on the probability of nearly every event scenario. Even with better information coming from industry sources, the probability of high-impact incidents depends on many factors, some of which are difficult to identify or even quantify.

- **Event cost** It is difficult to put an exact cost on any given privacy or security incident scenario. Privacy and security incidents are complex events that involve many parties and have unpredictable short- and long-term outcomes. Despite ever-improving information from research organizations on the cost of breaches, these are still rough estimates and may not take all aspects of cost into account.

Because of these challenges, quantitative risk analysis should be regarded as an effort to develop estimates, not exact figures. Partly this is because risk analysis is a measure of events that *may* occur, not a measure of events that *do* occur.

Standard quantitative risk analysis involves the development of several figures:

- **Asset value (AV)** This is the value of the asset, which is usually (but not necessarily) the asset's replacement value. Depending on the type of asset, different values may need to be considered.

- **Exposure factor (EF)**   This is the financial loss that results from the realization of a threat, expressed as a percentage of the asset's total value. Most threats do not eliminate the asset's value; instead, they reduce its value. For example, if an organization's $120,000 server is rendered unbootable because of malware, it will still have salvage value, even if that is only 10 percent of the asset's total value. In this case, the EF would be 90 percent. Note that different threats will have various impacts on EF because the realization of different threats will cause varying amounts of damage to assets.

- **Single loss expectancy (SLE)**   This value represents the financial loss when a threat scenario occurs one time. SLE is defined as $AV \times EF$. Note that different threats have a varied impact on EF, so those threats will have the same multiplicative effect on SLE.

- **Annualized rate of occurrence (ARO)**   This is an estimate of the number of times that a threat will occur per year. If the probability of the threat is 1 in 50 (one occurrence every 50 years), then ARO is expressed as 0.02. However, if the threat is estimated to occur four times per year, then ARO is 4.0. Like EF and SLE, ARO will vary by threat.

- **Annualized loss expectancy (ALE)**   This is the expected annualized loss of asset value due to threat realization. ALE is defined as $SLE \times ARO$.

ALE is based upon the verifiable values AV, EF, and SLE, but because ARO is only an estimate, ALE is only as good as the ARO. Depending upon the asset's value, the risk manager may need to take extra care to develop the best possible estimate for ARO, based upon whatever data is available. Sources for estimates include the following:

- History of event losses in the organization
- History of similar losses in other organizations
- History of dissimilar losses
- Best estimates based on available data

When performing a quantitative risk analysis for a given asset, the ALE for all threats can be added together. The sum of all ALEs is the annualized loss expectancy for the complete array of threats. An unusually high sum of ALEs would mean that a given asset is confronted with a lot of significant threats that are more likely to occur. But in terms of risk treatment, ALEs are better off left as separate and associated with their respective threats.

OCTAVE

Operationally Critical Threat, Asset, and Vulnerability Evaluation (OCTAVE) is a risk analysis approach developed by Carnegie Mellon University. The latest version is known as OCTAVE Allegro and is used to assess privacy and security risks so that an organization can obtain meaningful results from a risk assessment.

The OCTAVE Allegro methodology uses eight steps:

- **Step 1: Establish risk measurement criteria**   The organization identifies the most critical impact areas, which in the model include *reputation/customer confidence, financial, productivity, safety and health, fines/legal penalties,* and *other*. For example, reputation may be the most critical impact area for one organization, while privacy or safety may be most important for others.

- **Step 2: Develop an information asset profile**   The organization identifies its in-scope information assets and develops a profile for these assets that describe its features, qualities, characteristics, and value. Noting whether regulated personal information is included is of particular use for an organization's privacy program.

- **Step 3: Identify information asset containers**   The organization identifies all the internal and external information systems that store, process, and transmit in-scope assets, especially personal information. Note that many of these systems may be operated by third-party organizations.

- **Step 4: Identify areas of concern**   This is the start of identifying threats that, if realized, could cause harm to information assets. Typically, this is identified in a brainstorming activity.

- **Step 5: Identify threat scenarios**   This is a continuation of step 4, where threat scenarios are expanded upon (and unlikely ones are eliminated). A threat tree may be developed that first identifies actors and basic scenarios and is then expanded to include more details.

- **Step 6: Identify risks**   A continuation of step 5, the consequences of each threat scenario are identified.

- **Step 7: Analyze risks**   This simple quantitative measure is used to score each threat scenario based on risk criteria developed in step 1. The output is a ranked list of risks.

- **Step 8: Select mitigation approach**   A continuation of step 7, the risks with higher scores are analyzed to determine methods available for risk reduction.

The OCTAVE Allegro methodology includes worksheets for each of the steps described here, making it easy for a person or team to perform a risk analysis based on this technique.

 **NOTE**   Further information about OCTAVE Allegro is available at www.cert .org/resilience/products-services/octave/.

Other Risk Analysis Methodologies

Additional risk analysis methodologies provide more complex approaches that may have usefulness for certain organizations or in selected risk situations.

- **Delphi method**   In this method, questionnaires are distributed to a panel of experts in two or more rounds. A facilitator will anonymize the responses and distribute them to the experts. The objective is for the experts to converge on the most critical risks and mitigation strategies.

- **Event tree analysis (ETA)**   Derived from the fault tree analysis method (described next), ETA is a logic modeling technique for analysis of success and failure outcomes given a specific event scenario—in this case, a threat scenario.

- **Fault tree analysis (FTA)**   This logical modeling technique is used to diagram all the consequences for a given event scenario. FTA begins with a specific scenario and proceeds forward in time with all possible outcomes. A large "tree" diagram can result, which depicts many different chains of events.

- **Monte Carlo analysis**   Derived from Monte Carlo computational algorithms, this analysis begins with a given system with inputs, where the inputs are constrained to a minimum, likely, and maximum values. Running the simulation provides some insight into actual likely scenarios.

 **EXAM TIP**   CDPSE candidates are not expected to memorize risk analysis techniques, but you should be familiar with general risk analysis concepts.

Risk Evaluation and Ranking

Upon completion of a risk assessment, when all risks have been identified and scored, the risk manager, together with others in the organization, will analyze the results and begin to develop a strategy for going forward. Risks can be evaluated singly, but the organization will better benefit from analysis of all the risks together. This is because many risks are interrelated, and the right combination of mitigation strategies can result in many risks being adequately treated.

The results of a risk assessment should be analyzed in several different ways, including the following:

- Looking at all risks by business unit or service line

- Looking at all risks by asset type (in particular, personal information)

- Looking at all risks by activity type

- Looking at all risks by type of consequence

Because no two organizations (or their risk assessment results) are alike, this type of analysis is likely to identify risk treatment *themes* that may have broad implications across many risks. For example, an organization may identify several tactical risks all associated with access management and vulnerability management. Rather than treating individual tactical risks, a better approach may be to improve or reorganize the access or vulnerability management programs from the top down, resulting in many identified risks being mitigated programmatically. Organizations need to consider not just the details in a risk assessment, but the big picture.

Another type of risk to look for is one with a low probability of occurrence and high impact, which is typically the type of risk treated by transfer. *Risk transfer* most often comes in the form of cyber insurance, but it is also relevant to privacy and security monitoring when it includes indemnification.

Risk Ownership

When considering the results of a risk assessment, the organization needs to assign individual risk management tasks to individual people, typically middle- to upper-management leaders. These leaders, who should also have ownership of controls that operate within their span of oversight, use a budget, staff, and other resources in daily business operations. These are the risk owners, and, to the extent that there is a formal policy or statement in place on risk tolerance or risk appetite, they should be the people making risk-treatment decisions for risks in their domain. To the extent that these individuals are accountable for operations in their part of the organization, they should also be responsible for risk decisions, including risk treatment, in their operational areas. A simple concept to approach risk ownership is that if nobody owns the risk, then nobody is accountable for managing the risk, which will lead to a higher probability of the risk becoming an ongoing, unresolved issue with negative impacts on the business, along with the possible identification of a scapegoat who will be blamed if an event occurs.

Risk Treatment

To determine the best risk treatment plan, management can view reports and other information that lists risks that have been identified, assessed, and analyzed. At this stage, the organization has completed identifying risks and begins to determine what should be done about those risks. Risk treatment comprises the decisions and the activities that follow. A key element in deciding the appropriate risk treatment is ensuring that the right people at the right level of the organization are actively involved in deciding upon the risk treatment. This is achieved by having a formalized risk management program that includes all the key elements of an effective program outlined in this chapter.

In a general sense, risk treatment represents the actions that take place that the organization undertakes to reduce risk to an acceptable level. More specifically, for each risk identified in a risk assessment, an organization can consider four responses, or actions:

- Risk acceptance
- Risk mitigation
- Risk avoidance
- Risk transfer

These four actions are explained in more detail in the following sections.

There is a fifth potential action—or, rather, *inaction*—related to risk treatment, known colloquially as *ignoring the risk*. A potentially dangerous undertaking, ignoring a risk amounts to an organization pretending that the risk does not exist. In this case, the organization has unofficially accepted the risk and is responsible if the risk becomes an issue later on. By unofficially accepting the risk and not assigning a risk owner, the organization is possibly increasing the impact and likelihood of the risk evolving into an incident.

Ignoring a *known risk* is different from an organization's ignoring an *unknown risk*. This is usually a result of a risk assessment or risk analysis that is not sufficiently thorough in identifying all relevant risks. The best solution for these "unknown unknowns" is to have an external, competent firm perform an organization's risk assessment every few years or for an organization to examine an organization's risk assessment thoroughly to discover opportunities for improvement, including expanding the span of threats, threat actors, and vulnerabilities so that there are fewer or no unknown risks.

Risk Acceptance   In deciding to accept a risk, the organization determines that the risk requires no reduction or mitigation.

If only risk acceptance were this simple! Further analysis of risk acceptance shows that there are conditions under which an organization will elect to accept risk:

- The cost of risk mitigation is higher than the value of the asset being protected.
- The impact of compromise is low, or the value or classification of the asset is low.

Organizations may elect to establish a framework for risk acceptance, like the one shown in Table 3-3.

When an organization accepts a risk, instead of closing the matter for perpetuity, the organization should review it at least annually for the following reasons:

- The value of the asset may have changed during the year.
- The way that the asset is used may have changed during the year.
- The value of the business activity related to the asset may have changed during the year.
- The potency of threats may have changed during the year, potentially leading to a higher risk rating.
- The cost of mitigation may have changed during the year, potentially leading to greater feasibility for risk mitigation or transfer.

As with other risk treatment activities, detailed recordkeeping helps the risk manager better track matters such as risk assessment review.

Risk Level	Level Required to Accept
Low	Business department leader, plus chief information officer (CIO) or manager of information security
Medium	Business unit leader, plus CISO or director of information security
High	Chief executive officer (CEO), chief operating officer (COO), or organization president
Severe	Board of directors

Table 3-3   Framework for Risk Acceptance

Risk Mitigation   In risk mitigation, the organization decides to reduce the risk through some means, such as by changing a process or procedure, by improving a privacy or security control, or by adding a privacy or security control.

Risk mitigation is generally chosen when management understands that performing risk mitigation costs less than the value of the asset being protected. Sometimes, however, an asset's value is difficult to measure, or there may be a high degree of goodwill associated with the asset. For example, the value of a customer database that contains personal information including bank account or credit card information may itself be low; however, the impact of a breach of this database may be higher than its book value because of the fines, loss of business, or adverse publicity that may result.

Risk mitigation may result in a task that can be carried out in a relatively short time. However, risk mitigation may also involve one or more major projects that start in the future, perhaps in the next budget year or many months or quarters in the future. Further, such a project may be delayed, its scope may change, or it may be canceled altogether. Thus, the risk manager needs to monitor risk mitigation activities carefully to ensure that they are completed as originally agreed so that the risk mitigation is not forgotten or set aside.

Risk Avoidance   In risk avoidance, the organization decides to discontinue an activity that precipitates the risk. Often, risk avoidance is selected in response to an activity that was not formally approved in the first place. For example, a risk assessment may have identified a department's use of an external service provider that represented a measurable risk to the organization. The service provider may or may not have been formally vetted in the first place. Regardless, after the risk is identified in a risk assessment (or by other means), the organization may choose to cease activities with that service provider to avoid the risk.

Risk Transfer   After deciding to transfer a risk, the organization will employ an external organization to accept the risk. In this case, the organization does not have the operational or financial capacity to accept the risk, and risk mitigation is not the best choice. In risk transfer, an organization may have identified a significant financial risk related to a breach of its stores of personal information, for example. The risk transfer decision, in this case, may involve the purchase of cyber insurance that would offset the costs associated with such a breach. A risk transfer decision may also include the purchase of an incident response retainer, which is essentially a prepurchase of incident response services in the event of a breach.

A risk assessment may reveal the absence of security monitoring of a critical system. Another form of risk transfer involves using an external security services provider to monitor the critical system.

Residual Risk   When an organization undergoes risk treatment for risks that it identifies, in most cases, the risk treatment does not eliminate the risk but reduces it to some degree. *Residual risk* is the risk that remains after risk treatment is applied.

Some organizations approach risk treatment and residual risk improperly. They identify a risk, employ some risk treatment, and then fail to understand the residual risk and close the risk matter. A better way to approach residual risk is to analyze it as though it were a new risk and apply risk treatment to the residual risk. This iterative process provides organizations with an opportunity to revisit residual risk and make new risk treatment decisions. Ultimately, after one or more iterations, the residual risk will be accepted, and then the matter can be closed.

For instance, suppose a privacy manager identifies a risk in the organization's access management system where multifactor authentication is not used. This is considered high risk, and the IT department implements a multifactor authentication solution. When the privacy manager reassesses the access management system, she finds that multifactor authentication is required in some circumstances but not in others. A new risk is identified, at perhaps a lower level of risk than the original risk. But the organization once again has an opportunity to examine the risk and make a decision about it. It may improve the access management system further by requiring multifactor authentication in more cases than before, further reducing risk, which should be examined again for further risk treatment opportunities. Finally, the risk will be accepted as-is when the organization is satisfied that the risk has been sufficiently reduced.

In addition to the risk treatment life cycle, subsequent risk assessments and other activities will identify risks that represent residual risk from earlier risk treatment activities. And over time, the nature of residual risk may change, based on changing threats, vulnerabilities, or business practices, resulting in an initially acceptable residual risk that is no longer acceptable.

Controls

A common outcome of risk treatment, when mitigation is chosen, is the enactment of controls. Put another way, when an organization identifies a risk in a risk assessment, the organization may decide to develop (or improve) a control that will mitigate the risk that was found.

Suppose an organization determined that its procedures for terminating access for departing employees were resulting in many user accounts *not* being deactivated. The existing control was a simple, open-loop procedure, in which analysts were instructed to deactivate user accounts. Often, they were deactivating user accounts late or not at all. To reduce this risk, the organization modified the procedure (updated the control) by introducing a step where a second person would verify all account terminations daily.

Controls are measures put in place to ensure desired outcomes. Controls can come in the form of procedures, or they can be implemented directly in a system. There are many categories and types of controls, as well as standard control frameworks. You can find a thorough discussion of controls in the book *CISM Certified Information Security Manager All-In-One Exam Guide*, in Chapter 4.

Costs and Benefits

As organizations ponder options for risk treatment (and in particular, risk mitigation), they generally will consider the costs of the mitigating steps and the expected benefits they may receive. When an organization understands the costs and benefits of risk mitigation, this helps them develop strategies that are more cost effective or that result in greater cost avoidance.

When weighing mitigation options, an organization needs to understand several cost- and benefit-related considerations, including these:

- **Change in threat probability**   Organizations need to understand how a mitigating control changes the probability of threat occurrence and what that means in terms of cost reduction and avoidance.

- **Change in threat impact**   Organizations need to understand the change in the impact of a mitigated threat in terms of an incident's reduced costs and avoided costs versus the cost of the mitigation.

- **Change in operational efficiency**   Aside from the direct cost of the mitigating control, organizations need to understand the impact on the mitigating control on other operations. For instance, adding code review steps to a software development process may mean that the development organization may complete fewer fixes and enhancements in a given time period.

- **Total cost of ownership (TCO)**   When an organization considers a mitigation plan, the best approach is to understand its TCO, including the following costs:

  - Acquisition

  - Deployment and implementation

  - Recurring maintenance

  - Testing and assessment

  - Compliance monitoring and enforcement

  - Reduced throughput of controlled processes

  - Training

  - End-of-life decommissioning

- **Compliance-related fines and penalties**   The matter of compliance with privacy and security laws and regulations often involves fines and penalties when there are findings of noncompliance. The potential for fines and penalties needs to be considered as a potential consequence for failing to mitigate some risks.

While weighing costs and benefits, organizations need to keep in mind several things:

- Estimating the probability of a specific threat or event is difficult—particularly infrequent, high-impact events such as large-scale data thefts.

- Estimating the impact of any particular threat is difficult, especially those rare, high-impact events.

Thus, the precision of cost-benefit analysis is no better than estimates of event probability and impact.

An old adage in information security states that an organization would not spend $20,000 to protect a $10,000 asset. Although that may be true in some cases, there is more to consider than just the asset's replacement (or depreciated) value. For example, loss of the asset could result in an embarrassing and costly public relations debacle, or the asset may play a key role in the organization's earning hundreds of thousands of dollars in revenue each month.

Still, the principle of proportionality is valid and is often a good starting point for making cost-conscious decisions on risk mitigation. The principle of proportionality is described in NIST's generally accepted security systems principles (GASSP) and section 2.5 of its generally accepted information security principles (GAISP).

Privacy Impact Assessments

A privacy impact assessment (PIA), sometimes confused with a data protection impact assessment (DPIA, as coined in the General Data Protection Regulation [GDPR]), is a targeted risk assessment undertaken to identify impacts to individual privacy and impacts to an organization's ability to protect information resulting from a proposed change to a business process or information system. PIAs are not a new concept. They have been required since 2002 for US government electronic services and processes under the E-Government Act of 2002, and the European Union Article 29 Working Party endorsed the requirement to conduct PIAs for radio frequency identification applications in 2011. However, the GDPR raised the visibility of this process because of the applicability to all businesses that process personal data. Generally, a PIA is conducted for new processes or systems that will collect, store, or transmit PII or when there is a significant modification to a process or system that may create a new privacy risk. The purpose of a PIA is to ensure that personal information collected is used only for the intended purpose, and it identifies the impact(s) that any process or system change has on the organization's compliance with its privacy policy and applicable privacy laws and regulations.

One could also say that the purpose of a PIA is to validate the proposed change from a privacy perspective. By *validate*, I imply that the proposed change has been well-designed (presumably with privacy and security by design) and that the impact on privacy is neutral or better.

 **EXAM TIP** CDPSE candidates are not expected to memorize PIA procedures. You should, however, be familiar with the concepts, purposes, and approaches for performing PIAs.

Two excellent resources for PIAs and how PIAs achieve these objectives are

- https://www.gsa.gov/reference/gsa-privacy-program/privacy-impact-assessments-pia
- https://www.ftc.gov/site-information/privacy-policy/privacy-impact-assessments

PIA Procedure

Following is the procedure for conducting a PIA:

1. Obtain a description of the project or proposed change, including the purpose of PII data collection and relevant details on business processes or information systems.

2. Identify all changes to data collection, data flows, and storage of personal information.

3. Determine whether the proposed change violates any terms of the organization's external privacy policy, internal privacy policy, or security policy. Identify and describe all such violations. Identify any compensating controls or changes that would reduce or eliminate the violations. This determination must include identifying the original purpose of the collection and/or use of personal information, and whether the proposed change violates or exceeds the purpose.

4. Determine whether the proposed change violates any terms of privacy or security laws, regulations, or related guidelines or codes of conduct.

5. Determine whether the proposed change introduces any new security risks and, if so, what potential alterations to the proposed change may reduce or eliminate those risks.

6. Determine whether the proposed change alters any previously known security risks and, if so, what potential alterations to the proposed change may reduce or eliminate those risks.

7. Develop a list of all such impacts (and possible countermeasures) identified in the preceding steps.

8. Write a formal report describing all of the above.

 **NOTE** A PIA is nothing more than a risk assessment whose focus is the privacy and security of subject data in some specific context.

Engaging Data Subjects in a PIA

The GDPR, in Article 35, Paragraph 9, suggests that the organization consult with data subjects or their representatives to obtain their opinion of proposed changes to a PIA. Such engagement could take the form of

- Surveys
- Focus groups
- Announcement of the proposed change with a request for comments

 **NOTE** The EU GDPR describes the PIA procedure in Articles 35 and 36.

The Necessity of a PIA

Not all regulations explicitly require an analysis of proposed changes to stated or implied data subject rights or the security of their personal information. The absence of a specific requirement for a risk assessment does not absolve an organization from performing one, however. It is a standard business practice to analyze various forms of impact

PART I

of a proposed change to a business process or supporting information system. A failure to assess the impact of a proposed change could even be seen as negligence: a reasonable person would find such an organization at fault for not seeking to understand the potential impacts of a proposed change upon the security, privacy, or proper use of personal information.

Integrating into Existing Processes

Organizations building their privacy programs need to place "hooks" into three types of existing business processes:

- **Product development**   The PIA is a tool designed to promote "privacy by design" by ensuring that privacy is considered in technical, organizational, and security measures at the beginning stage of product development and throughout the product life cycle.

- **IT change control**   This process needs to include a security risk assessment and a privacy impact assessment. The data privacy officer must be informed of all changes and counted among the approvers for changes that potentially impact privacy.

- **Business process change control**   This process must include a PIA. The data privacy officer needs to be informed of all changes and counted among the approvers for changes that potentially impact privacy.

CAUTION   Organizations lacking IT change control or business process change control need to implement these processes and ensure that all relevant personnel are aware of, and will comply with, the terms of these processes.

Recordkeeping and Reporting

All PIAs that are performed must be preserved as a part of the organization's recordkeeping. The DPO is typically assigned this responsibility. The DPO may elect to include statistics about PIAs as a part of regular privacy metrics reported to executive management and the board of directors. Aspects of this reporting may include

- The number of PIAs performed and the level of effort expended

- The projects associated with PIAs that were performed

- The number of exceptions noted where processes or systems required remediation

- The regulations in scope for PIAs (applicable for organizations subject to multiple privacy laws)

- If findings for PIAs are placed into a risk register, statistics on the contents of the risk register, including the number of items, aging, time to remediation, and context

Risks Specific to Privacy

To perform an effective PIA, the privacy specialist needs to be aware of privacy-specific threats, vulnerabilities, and attacks. A typical risk assessment considers threats and vulnerabilities in the context of some particular asset. Being familiar with specific threats and vulnerabilities will result in a better PIA.

Privacy Vulnerabilities

During a PIA, the privacy specialist must identify vulnerabilities in an information system, business process, or whatever the PIA is focused upon. A good definition of the term *vulnerability* is "a weakness that may be present in a system that makes the probability of one or more threats more likely."

Weaknesses in processes or systems represent opportunities for business processes to deviate from what is expected. The nature of the deviation may be a skipped step in a manual procedure, a software program that behaves unexpectedly when presented with unexpected input, or the ability for a worker to deliberately perform something incorrectly without being detected.

Identifying Vulnerabilities   When examining business processes, the following may indicate the presence of vulnerabilities:

- Manual steps that rely upon human decision-making
- Steps that require workers to be proactive (such as checking a mailbox for incoming requests)
- Steps that involve data entry (the possibility of miskeying data)
- Steps performed that do not include recordkeeping
- The absence of reconciliation procedures (such as matching the number of incoming requests to the number of outgoing replies)
- Lack of written documentation describing the steps in processes and procedures
- Lack of training for personnel who perform processes and procedures
- Lack of oversight for personnel who perform processes and procedures

When examining information systems, the techniques performed in typical system vulnerability assessments apply and include the following:

- Misconfiguration
- Missing security patches
- Software vulnerabilities that permit an attacker to cause software to behave in unintended ways (such as script injection, SQL injection, denial of service)
- Poor user access management controls (including easily guessed passwords, failure to remove user accounts for terminated users, password settings that invite brute-force attacks)

Automated tools such as port scanners, vulnerability scanners, and code scanning tools are often used to perform these activities.

Vulnerability Severity

When identifying vulnerabilities, you should also rate the severity of each vulnerability. The severity can be expressed on a high–medium–low scale or a numeric scale such as 1–5 or 1–10. A severity rating is generally associated with the ease of exploiting the vulnerability, the skill level required to exploit it, and the result of exploiting it.

A standard vulnerability rating scale, the Common Vulnerability Scoring System (CVSS), is used in the information security world. The severity of vulnerabilities identified in information systems using CVSS is a 0–10 range, where 10 is the highest severity. A CVSS score is calculated using several inputs, including attack complexity, whether authentication is required, whether the attacker must be in the physical proximity of the target system, and the impact of an exploitation upon the confidentiality, integrity, and availability of information in the system.

Data Storage and Data Flow Vulnerabilities   It's also necessary to understand the flow of personal information in the context of the PIA. The examiner needs to identify all instances of data storage, and then examine access control processes and the security of systems associated with stored data.

In addition, the examiner needs to identify all instances of data movement within the organization as well as data leaving and entering the organization. For each case, the measures taken to protect the confidentiality and integrity of personal information in transit must be identified.

Application Security Vulnerabilities   If an application is new or is undergoing significant changes, it should be subjected to a variety of tests to ensure that it is free of vulnerabilities that could be exploited by an attacker to illicitly obtain personal information or cause a malfunction of the application. Testing includes penetration tests, Static Application Security Testing (SAST) code reviews, and Dynamic Application Security Testing (DAST) code reviews.

Privacy Threats

The performance of a PIA should consider reasonable and likely threats that may occur within the business process or system being examined. As mentioned, a PIA is a risk assessment focused on the potential impact on privacy compliance and security, targeted to a process or system that will be undergoing changes. A good definition of the term *threat* is "an event that, if realized, would bring harm to an asset."

In this case, of course, the asset is personal information, but a threat can also include the business process being examined, the underlying information systems that facilitate the operation of the business process, the persons who perform the process, and those who operate and manage the systems. All of these are a part of the *attack surface* that must be considered.

When analyzing a variety of threats, the privacy specialist considers each threat and determines

- Whether the threat is relevant
- How the threat may be carried out (including consideration for any corresponding vulnerabilities that have been identified)

- The likelihood that the threat will be carried out
- The impact on the organization (and the data subject) if the threat were carried out

In a typical PIA, the analyst will create a chart listing all reasonable threats. Each threat is scored in terms of relevance, likelihood of occurrence, and impact of occurrence. The scoring may be in the form of qualitative values such as high–medium–low or on a numeric scale such as 1–5 or 1–10 (where the highest number in the scale represents the highest probability and impact).

 **NOTE** The lists of threats found in Appendix E of NIST SP 800-30 and Appendix C of ISO/IEC 27005 represent good starting points for threat analysis.

Privacy Countermeasures

After the privacy analyst has completed the vulnerability and threat analysis of the process or system being examined, she may conclude that one or more threats or vulnerabilities represent unacceptable conditions. She may then suggest that one or more countermeasures be enacted to reduce risks considered unacceptable.

Here are some example countermeasures:

- In a data subject request (DSR) process, have a second employee check the contents of the response to be sent back to the data subject.
- Implement a web application firewall (WAF) to offer further protection for a web application that collects and manages personal information.
- Implement automated search tools to ensure more accurate (and timely) results in a DSR.

PIA Case Study

Let's take a look at a case study that represents a likely scenario. An organization in the retail industry has conducted most of its business through a web application. The organization has written a mobile app for Apple and Android to make it more convenient for its customers to place orders and check on the status of their orders. The privacy officer was informed of the mobile app late in its development.

As is typical in the organization, a security firm was commissioned to perform a vulnerability analysis on the mobile app. Several vulnerabilities were identified and later remediated by the organization's developers. A retest confirmed that the vulnerabilities were remediated.

The privacy officer also determined that the mobile app uses simple user ID and password authentication with no other options. As a result, the privacy officer's PIA made the following recommendations:

- Make multifactor authentication available to mobile app users who prefer to use it.

- Change the systems development life-cycle process so that a PIA can be completed on the design of a new application or system rather than an analysis of the nearly finished product.

- Include privacy requirements in future systems development projects.

Chapter Review

Privacy programs can succeed if they include regular risk assessments as part of a formal risk management program. Risk management is a core tenet of cybersecurity programs, often referred to as information security management systems (ISMSs).

Risk management is a cyclical, iterative activity used to acquire, analyze, and treat risks. This book focuses on privacy risk, but overall, the life cycle for privacy risk is functionally similar to that for information risk or even business risk: a new risk is introduced into the process, the risk is studied, and a decision is made about how to deal with it.

The risk management life cycle consists of risk assessment, risk identification, risk analysis, and risk treatment. The life cycle is bounded by the scope of the risk management program and the organization's risk appetite.

After a risk assessment's scope has been determined, the initial step in a risk assessment is to identify assets and determine each asset's value. In a typical information risk assessment, assets will consist of various types of information (including intellectual property, internal operations, and personal information) and the information systems that support and protect those information assets.

In asset classification, an organization assigns an asset to a category representing usage or risk. The purpose of asset classification is to determine, for each asset, its level of criticality to the organization. In an organization with a formal privacy program, asset classification will include one or more classifications for assets related to personal information.

A crucial part of a risk assessment is the identification of the value of an asset. In the absence of an asset's value, it is more difficult to calculate risks associated with an asset, even when qualitative risk valuation is employed. Without a known valuation, the impact of loss is more difficult to determine.

The identification of threats is a key step in a risk assessment. A threat is defined as an event that, if realized, would bring harm to an asset and, thus, to the organization.

The identification of vulnerabilities is an essential part of any risk assessment. A vulnerability is any weakness in a process or system that permits an attacker to compromise a target process or system successfully.

Risk identification is performed during a risk assessment, and various scenarios are studied for each asset. Considerations include threats, threat actors, vulnerabilities, asset value, and impact of a threat event.

In a risk assessment, the risk manager examines assets, together with associated vulnerabilities and likely threat scenarios. The risk analysis is the detailed examination of each risk.

Qualitative risk analysis uses rankings such as high–medium–low or simple numeric scales. In contrast, quantitative risk analysis expresses risks in financial terms, including asset value (AV), exposure factor (EF), single loss expectancy (SLE), annualized rate of occurrence (ARO), and annualized loss expectancy (ALE).

Operationally Critical Threat Asset and Vulnerability Evaluation (OCTAVE) is a risk analysis approach developed by Carnegie Mellon University. The latest version, OCTAVE Allegro, is used to assess privacy and security risks so that an organization can obtain meaningful results from a risk assessment.

Other risk analysis techniques include Delphi, event tree analysis (ETA), fault tree analysis (FTA), and Monte Carlo analysis.

In a general sense, risk treatment represents the actions that an organization undertakes to reduce risk to an acceptable level. The four possible actions are risk acceptance, risk mitigation, risk avoidance, and risk transfer.

A privacy impact assessment (PIA), sometimes known as a data protection impact assessment (DPIA, as coined in the GDPR), is a targeted risk assessment undertaken to identify impacts to individual privacy and impacts to an organization's ability to protect information resulting from a proposed change to a business process or information system. The purpose of a PIA is to ensure that personal information collected is used only for the intended purpose, and it identifies the impact(s) that any process or system change has on the organization's compliance with its privacy policy and applicable privacy laws and regulations.

A PIA considers vulnerabilities, threats, data storage, and data flows to identify potential impacts on proposed changes in business processes and information systems.

After a privacy analyst has completed the vulnerability and threat analysis of the process or system being examined in a PIA, they may conclude that one or more threats or vulnerabilities represent unacceptable conditions and suggest countermeasures be enacted to reduce the risk.

Quick Review

- Because data protection is a part of privacy, information security cannot be entirely separated from privacy. Thus, many processes, including risk management, work better when they address security *and* privacy.

- Important risk assessment standards include NIST SP 800-30, *Guide for Conducting Risk Assessments*; NIST SP 800-39, *Managing Information Security Risk: Organization, Mission, and Information, System View*; and Factor Analysis of Information Risk (FAIR).

- Important risk management standards include ISO/IEC 27005, *Information technology – Security techniques – Information security risk management*; ISO/IEC 27701, *Security techniques – Extension to ISO/IEC 27001 and ISO/IEC 27002 for privacy information management – Requirements and guidelines*; and the NIST Privacy Framework.

- It is difficult to put a valuation on the personal information stored or processed by an organization for risk-management purposes. Typical evaluation methods include the cost of recovering, reconstituting, or reacquiring personal information.

- The terms *threat* and *vulnerability* are often misused. These terms are distinguished from one another in this way: a vulnerability is the weaknesses in a system that could permit an attack to occur. A vulnerability is not the attack vector or technique—this is known as a threat.

- ISO/IEC Guide 73, *Risk management – Vocabulary*, defines *risk* as "the combination of the probability of an event and its consequence."

- Although the impact of a threat can be calculated, it is generally far more difficult to know a threat's probability of occurrence.

- The failure of an organization to address a risk constitutes tacit risk acceptance.

- A PIA is nothing more than a risk assessment whose focus is the privacy and security of subject data in some specific context.

- A privacy offer may decide to involve to survey data subjects during a PIA to understand their opinions regarding the focus of the assessment.

Questions

1. A risk manager is planning a first-ever risk assessment in an organization. What is the best approach for ensuring success?

   A. Interview personnel separately so that their responses can be compared.

   B. Select a framework that matches the organization's control framework.

   C. Work with executive management to determine the correct scope.

   D. Do not inform executive management until the risk assessment has been completed.

2. As a part of a privacy impact assessment (PIA), a security manager has completed a vulnerability scan and has identified numerous vulnerabilities in production servers that could result in the exposure of personal information. What is the best course of action?

   A. Recommend that vulnerabilities be remediated.

   B. Notify regulators.

   C. Notify system owners.

   D. Add individual vulnerability entries to the risk register.

3. The concept of privacy and security tasks in the context of a SaaS or an IaaS environment is depicted in a:

   A. Discretionary control model

   B. Mandatory control model

   C. Monte Carlo risk model

   D. Shared responsibility model

4. What are the categories of risk treatment?

   A. Risk avoidance, risk transfer, risk mitigation, and risk acceptance

   B. Risk avoidance, risk transfer, and risk mitigation

   C. Risk avoidance, risk reduction, risk transfer, risk mitigation, and risk acceptance

   D. Risk avoidance, risk treatment, risk mitigation, and risk acceptance

5. An organization is contemplating significant changes to a business process that involves the management of personal information. When should a PIA be performed?

   A. After requirements have been developed

   B. Before requirements have been developed

   C. After the process has been changed

   D. After the process design changes have been completed

6. When would it make sense to spend $50,000 to protect an asset worth $10,000?

   A. If the protective measure reduced threat impact by more than 90 percent.

   B. It would never make sense to spend $50,000 to protect an asset worth $10,000.

   C. If the asset was required for realization of $500,000 monthly revenue.

   D. If the protective measure reduced threat probability by more than 90 percent.

7. Which of the following statements is true about compliance risk?

   A. Compliance risk can be tolerated when fines cost less than controls.

   B. Compliance risk is just another risk that needs to be measured.

   C. Compliance risk can never be tolerated.

   D. Compliance risk can be tolerated when it is optional.

8. A privacy and security steering committee empowered to make risk-treatment decisions has chosen to accept a specific risk. What is the best course of action?

   A. Refer the risk to a qualified external security audit firm.

   B. Perform additional risk analysis to identify residual risk.

   C. Reopen the risk item for reconsideration after one year.

   D. Mark the risk item as permanently closed.

9. A privacy steering committee has voted to mitigate a specific risk. Some residual risk remains. What is the best course of action regarding the residual risk?

   A. Accept the residual risk and close the risk ledger item.

   B. Continue cycles of risk treatment until the residual risk reaches an acceptable level.

   C. Continue cycles of risk treatment until the residual risk reaches zero.

   D. Accept the residual risk and keep the risk ledger item open.

10. A privacy manager has been directed by executive management *not* to document a specific risk in the risk register. This course of action is known as:

   A. Burying the risk

   B. Transferring the risk

   C. Accepting the risk

   D. Ignoring the risk

11. A security manager is performing a risk assessment on a business application. The security manager has determined that security patches have not been installed for more than a year. This finding is known as a:

   A. Probability

   B. Threat

   C. Vulnerability

   D. Risk

12. A security manager is performing a risk assessment on a data center. The security manager has determined that unauthorized personnel can enter the data center through the loading dock door and shut off utility power to the building. This finding is known as a:

   A. Probability

   B. Threat

   C. Vulnerability

   D. Risk

13. A privacy manager has developed a scheme that prescribes required methods to protect information at rest, in motion, and in transit. This is known as a(n):

   A. Data classification policy

   B. Asset classification policy

   C. Data loss prevention plan

   D. Asset loss prevention plan

14. A security manager is developing a strategy for making improvements to the organization's incident management process. Why would the organization's privacy officer be requesting that a PIA be performed regarding the planned changes?

    A. To reduce the impact of privacy incidents

    B. To reduce the probability of privacy incidents

    C. To ensure that privacy incidents do not occur

    D. To ensure that a privacy incident is properly managed

15. What is usually the primary objective of risk management?

    A. Fewer and less severe privacy and security incidents

    B. No privacy or security incidents

    C. Improved compliance

    D. Fewer audit findings

Answers

1. **C.** The best approach for success in an organization's risk management program, and during risk assessments, is to have support from executive management. Executives need to define the scope of the risk management program, whether by business unit, geography, or other means.

2. **A.** In this instance, a PIA is being performed that includes a vulnerability assessment of a system that stores or processes personal information. The results of the PIA will include a recommendation that the identified vulnerabilities be remediated. Notifying system owners will be a part of the recommended remediation after the completion of the PIA.

3. **D.** The shared responsibility model, sometimes known as a shared responsibility matrix, depicts the operational model for SaaS and IaaS providers where client organizations have some privacy and security responsibilities (such as end user access control) and service provider organizations have some privacy and security responsibilities (such as physical access control).

4. **A.** The four categories of risk treatment are risk mitigation (risks are reduced through a control or process change), risk transfer (risks are transferred to an external party such as an insurance company or managed services provider), risk avoidance (the risk-producing activity is discontinued), and risk acceptance (management chooses to accept the risk).

5. **D.** The best time to perform a PIA is after process design changes have been completed, and before any implementation is started. If the PIA is performed before the design's completion, the PIA may miss details present only in the design itself. If the PIA is performed after the process has been changed, there may be reluctance by the organization to make further changes to the process. Ideally, privacy and security are included in requirements development, so that the proposed changes are private and secure by design.

6. **C.** Ordinarily, it would not make sense to spend $50,000 to protect an asset worth $10,000. Sometimes, however, other considerations, such as revenue realization or reputation damage, can be difficult to quantify and would be worth spending more to protect.

7. **B.** In most cases, compliance risk is just another risk that needs to be understood. This includes the understanding of potential fines and other sanctions concerning the costs required to reach a state of compliance. In some cases, however, being out of compliance can also result in reputation damage, as well as larger sanctions if the organization suffers from a security breach because of the noncompliant state.

8. **C.** A risk register item that has been accepted should be shelved and considered after some time, perhaps one year. This is a better option than closing the item permanently; in a year's time, changes in business conditions, security threats, and other considerations may compel the organization to take different actions.

9. **B.** After risk reduction through risk mitigation, the residual risk should be treated like any new risk: it should be reexamined, and a new risk treatment decision should be made. This should continue until the final remaining residual risk is accepted.

10. **D.** An organization that refuses to formally consider a risk is ignoring the risk. This is not a formal method of risk treatment because of the absence of deliberation and decision-making. It is not a wise business practice to keep some risk matters "off the books."

11. **C.** The absence of security patches on a system is considered a vulnerability, defined as a weakness in a system that could permit an attack to occur.

12. **B.** Any undesired action that could harm an asset is considered a threat.

13. **A.** A data classification policy is a statement that defines two or more classification levels for data, together with procedures and standards for the protection of data at each classification for various use cases such as storage in a database, storage on a laptop computer, transmissions via e-mail, and storage on backup media.

14. **D.** Requesting a PIA is a reasonable request to ensure that changes to an organization's security incident response process do not negatively impact the organization's privacy obligations. Changes to an incident response process generally will not result in any changes to the impact or probability of incidents, although the changes could impact the organization's response to incidents.

15. **A.** The most common objective of a risk management program is to reduce the number and severity of privacy and security incidents.

PART II

Privacy Architecture

Infrastructure

In this chapter, you will learn about
- Technology stacks
- Cloud services
- Endpoints
- Remote access
- System hardening techniques

This chapter covers Certified Data Privacy Solutions Engineer job practice 2, "Privacy Architecture," part A, "Infrastructure." The entire Privacy Architecture domain represents 36 percent of the CDPSE examination.

Technology infrastructure relates to privacy in this way: part of privacy involves governing the proper handling of stored personal information, and part involves the protection of stored personal information—this is the discipline of information security. For a professional to be proficient in information security, he or she must understand how the underlying technology infrastructure works. It is often said that if you don't understand how the technology works, you cannot possibly know how to protect it.

Technology Stacks

Effective privacy requires effective security. Effective security needs an IT architecture with integrity and protective controls. One common way that IT architecture is discussed is in the form of a *technology stack*, often referred to as a *service stack*. The technology stack is a set of technologies that make up a system, often expressed in bottom-up notation, such as Linux-Apache-MySQL-Perl, or Windows-IIS-SQL-ASP.Net. These are shorthand ways to describe the major building blocks of a system. There's an even shorter way to express technology stacks. For example, *LAMP* is Linux-Apache-MySQL-PHP/Perl/Python, and *WISA* is Windows-IIS-SQL-ASP.NET. More than a dozen such stack names are in use—IT people love their acronyms. Editorially, LAMP is exceedingly popular, in part because all of its components are open source. Thus, the industry has probably tens of thousands of able professionals who are familiar with the design and operation of dynamic web servers (of which the world has billions).

In this section, I discuss the primary components of technology stacks: operating systems, database management systems, web servers, and software languages. However, because of the structure of the CDPSE job practice, software development is discussed in Chapter 5.

To round out the discussion of technology stacks, a similar term is used to describe the components in a software development environment: *development stacks*. For example, Eclipse-Subversion-Jenkins describes the components in a development stack. In this example, Eclipse is the desktop tool that developers use to write and debug code, Subversion is the source code repository, and Jenkins is the automated build and deploy environment. Pick up a copy of McGraw Hill's *CSSLP Certification All-In-One Exam Guide* for more insight into secure software development.

 EXAM TIP The CDPSE job practice touches on the secure development life cycle and its concepts, but there is no further discussion on development stacks in the job practice or in this book.

Hardware

Hardware is the physical machinery used in IT computing and communications. Hardware comes in many forms, including mainframes, servers, network devices such as routers and firewalls, laptop computers, and smartphones. The fundamental concepts of hardware are discussed in this section.

As mentioned, security professionals can protect organizations only to the extent that they understand all of the underlying technology in use. This is an accurate assertion, because security professionals (and today, privacy professionals) lacking this knowledge will fail to recognize unsafe practices that can lead to the compromise of systems and the loss or exposure of sensitive and personal information.

 **NOTE** This book will cover the basics of hardware; lengthier explanations may be found in McGraw Hill's *All-In-One Exam Guides* for the CISA, CISM, and CISSP certifications.

Computers

Computers are general-purpose machines that run operating system software, which in turn runs subsystem or application software. Computers consist of the following components:

- **Central processing unit (CPU)** This is the component in which software instructions are decoded and executed. Many computers consist of multiple CPUs, or *cores*, that facilitate improved performance.

- **Main storage** Commonly known as *RAM* (random access memory), this is the high-speed memory used by the CPU to store the contents of programs currently being run. The computer's operating system will also occupy some of a computer's RAM. Main memory is generally volatile—that is, when power is removed from the computer, the contents of main memory are lost.

- **Secondary storage**   This is a computer's permanent storage, traditionally occupying hard disk drives (HDDs), but more commonly, solid-state drives (SSDs) are used because of their superior performance. Secondary memory is used to store the operating system, programs, tools, and data within organized structures known as *file systems*. Unlike main storage memory, which is volatile, secondary storage memory is *persistent* and remains in place even if power is removed from the computer.

- **Bus**   This is the means for communication among various components of the computer, including its CPU, main storage, secondary storage, and adaptors. Modern computers often have more than one bus—one or more for internal communications, and one or more for external communications via adaptors. Bus standards in use include Universal Serial Bus (USB).

- **Adaptors**   These plug-in devices are used to connect other components (sometimes called *peripheral devices*, or peripherals) to a computer, such as additional storage, networks, printers, monitors, and keyboards. In larger (desktop, server, mainframe) computers, adaptors are separate components that plug into special connectors. In laptop computers, tablets, and smartphones, adaptors are integrated into the computer's main logic board and are not removable.

Storage

Aside from storage within computers themselves, server and mainframe computers often employ separate storage systems for storing large amounts of data in database management systems as well as unstructured data. Storage systems are configured by organizing one or more *volumes*, within which file systems are created, and data of some kind is stored within the file systems. Volumes can vary in size, and often their size can be adjusted dynamically.

Types of storage systems used include

- **Storage area network (SAN)**   This stand-alone storage system can be configured to contain several virtual volumes and can be connected to many servers through fiber-optic cables.

- **Network-attached storage (NAS)**   This stand-alone storage system can be configured to contain several virtual volumes and can be connected to servers via a local area network (LAN).

- **Cloud-based storage**   In this subscription service, a storage service provider employs large storage systems that organizations access over the Internet or dedicated communications connections.

Storage systems are required to be reliable. A common standard in place to ensure reliability is a redundant array of independent disks (RAID). RAID ensures that data will be intact and available even in the event of the failure of one or more individual hard drives within the storage system.

Networks

Networks are the means through which computers communicate with one another. Whether networked computers are in a single room or spread across the globe, and whether or not a telecommunications company is involved, networks facilitate all computer communications, both wired and wireless. Network technology is discussed in detail in Chapter 6.

 **EXAM TIP** CDPSE requires that you have only a cursory understanding of computing and networking hardware.

Operating Systems

Computer operating systems (OSs) are large, general-purpose programs that control computer hardware and facilitate the use of software applications. Operating systems perform the following functions:

- **Access to peripheral devices** The operating system controls and manages access to all devices and adaptors that are connected to the computer. This includes storage devices, display devices, and communications adaptors.

- **Storage management** The operating system provides for the orderly storage of information on storage hardware. For example, operating systems provide file system management for the storage of files and directories on SSDs or hard drives.

- **Process management** Operating systems facilitate the existence of multiple processes, some of which will be computer applications and tools. Operating systems ensure that each process has private memory space and is protected from interference and eavesdropping by other processes. These safeguards are collectively known as *process isolation*.

- **Resource allocation** Operating systems facilitate the sharing of resources on a computer such as memory, communications, and display devices.

- **Communication** Operating systems facilitate communications with users via peripheral devices and also with other computers through networking. Operating systems typically have drivers and tools to facilitate network communications.

- **Security** Operating systems restrict access to protected resources through process, user, and device authentication.

Examples of popular operating systems include Linux, Solaris, macOS, Android, iOS, Chrome OS, and Microsoft Windows.

The traditional context of the relationship between operating systems and computer hardware is this: one copy of a computer operating system runs on a computer at any given time. Virtualization, however, has changed all of that. Virtualization is discussed later in this section.

Server Clustering

Using special software, a group of two or more computers can be configured to operate as a *cluster*. This means that the group of computers will *appear* as a single computer for the purpose of providing services. Within the cluster, one computer will be active and the other computer(s) will be in passive mode; if the active computer should experience a hardware or software failure and crash, the passive computer(s) will transition to active mode and continue to provide service. This is known as *active-passive* mode. The transition is called a *failover*.

Clusters can also operate in *active-active* mode, where all computers in the cluster provide service; in the event of the failure of one computer in the cluster, the remaining computer(s) will continue providing service.

Grid Computing

Grid computing is a technique used to distribute a problem or task to several computers at the same time, taking advantage of the processing power of each to solve the problem or complete the task in less time. Grid computing is a form of distributed computing, but in grid computing, the computers are coupled more loosely and the number of computers participating in the solution of a problem can be dynamically expanded or contracted at will.

Virtualization

Virtualization refers to the set of technologies that enables two or more running operating systems (of the same type or different types) to reside on a single physical computer. Virtualization technology enables organizations to use computing resources more efficiently.

Before I explain the benefits of virtualization, I should first state one of the principles of computer infrastructure management: It is a sound practice to use a server for one single purpose. Using a single server for multiple purposes can introduce a number of problems, such as these:

- Tools or applications that reside on a single computer may interfere with one another.

- Tools or applications that reside on a single computer may interact with each other or compete for common resources.

- A tool or application on a server could, although rarely, cause the entire server to stop running; on a server with multiple tools or applications, it could cause the other tools and applications to stop functioning.

Prior to virtualization, the most stable configuration for running many applications and tools was to run each on a separate server. This would, however, result in highly inefficient use of computers and capital, as most computers with a single operating system spend much of their time in an idle state.

Virtualization permits IT departments to run many applications or tools on a single physical server, each within its own respective operating system, thereby making more efficient use of computers (not to mention electric power and data center space).

Virtualization software emulates computer hardware so that an operating system running in a virtualized environment does not know that it is running on a *virtual machine*. Virtualization software, known as a *hypervisor*, includes resource allocation configuration settings so that each *guest* (a running operating system) will have a specific amount of memory, hard disk space, and other peripherals available for its use. Virtualization also facilitates the sharing of peripheral devices such as network connectors so that many guests can use an individual network connector. However, each will have its own unique IP address.

Virtualization is the basis of cloud-based infrastructure as a service (IaaS) services such as Amazon AWS, Google Cloud, and Microsoft Azure.

Virtualization software provides security by isolating each running operating system and preventing it from accessing or interfering with others. This is similar to the concept of *process isolation* within a running operating system, where a process is not permitted to access resources used by other processes.

A server with running virtual machines is depicted in Figure 4-1.

Many security issues need to be considered in a virtualization environment, including

- **Access control**   Access to virtualization management and monitoring functions should be restricted to those personnel who require it.

- **Resource allocation**   A virtualization environment needs to be carefully configured so that each virtual machine is given the resources it requires to function correctly and perform adequately.

- **Logging and monitoring**   Virtual environments need to be carefully monitored so that any sign of security compromise will be quickly recognized and acted on.

- **Hardening**   Virtual environments need to be configured so that only necessary services and features are enabled, and all unnecessary services and features are either disabled or removed.

- **Vulnerability management**   Virtualization environments need to be monitored as closely as operating systems and other software so that the IT organization is aware of newly discovered security vulnerabilities and available patches.

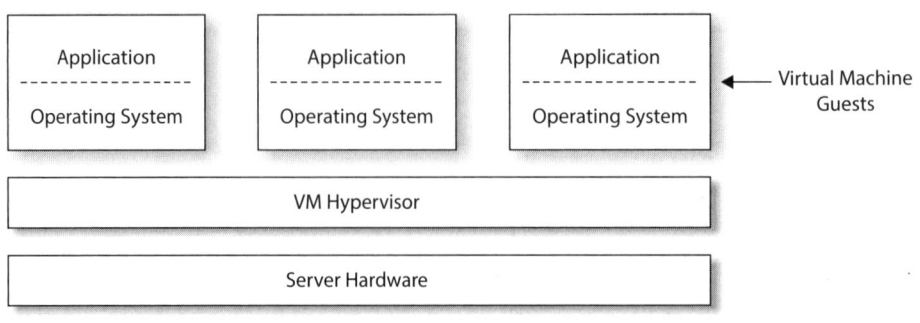

Figure 4-1   Virtualization

Cloud Computing

Cloud computing refers to dynamically scalable and usually virtualized computing resources that are used internally or provided as a service. Cloud computing services may be rented or leased so that an organization can have scalable application environments without the need for supporting hardware or a data center. Or cloud computing may include networking, computing, and even application services in an IaaS, software as a service (SaaS), or platform as a service (PaaS) model. Cloud computing is discussed in more detail later in this chapter.

Containerization

Containerization is a form of virtualization whereby multiple applications may inhabit a single running operating system. Each such application is completely isolated from every other and can access only the resources allocated to the container.

Containerization is a different approach for providing isolated resources to applications in large environments. Suppose, for example, that an engineer wants to run ten isolated applications on a physical server. With virtualization, the engineer would create ten virtual machines, each running a full operating system with one application each. In containerization, the engineer would run a single OS instance with ten containers, one for each application.

The primary advantage of containerization over virtualization is the more efficient use of hardware resources. One disadvantage of containerization is that all containers are running under a single version of the operating system, whereas in virtualization, many different operating systems can run in virtual machines.

File Systems

A *file system* is a logical structure that facilitates the storage of data on a digital storage medium such as a hard drive, SSD, optical disc, or flash memory device. The structure of the file system facilitates the creation, modification, expansion and contraction, and deletion of data files. A file system may also be used to enforce access controls to control which users or processes are permitted to access, alter, or create files in a file system.

It can also be said that a file system is a special-purpose database designed for the storage and management of files.

Modern file systems employ a storage hierarchy that consists of two main elements:

- **Directories**   A *directory* is a structure that is used to store files. A file system may contain one or more directories, each of which may contain files and subdirectories. The topmost directory in a file system is usually called the "root" directory. A file system may exist as a hierarchy of information, in the same way that a building can contain several file rooms, each of which contains several file cabinets, which contain drawers that contain dividers, folders, and documents. Directories are called *folders* in some computing environments.

- **Files**   A *file* is a sequence of zero or more characters that are stored as a logical whole. A file may be a document, a spreadsheet, an image, a sound file, a computer program, or data that is used by a program. A file can be small as zero characters in length (an empty file) or as large as many gigabytes (trillions of characters).

A file occupies units of storage on storage media (which could be a hard disk, SSD, or flash memory device, for example) that may be called blocks or sectors; however, the file system hides these underlying details from the user so that the file may be known simply by its name, the directory in which it resides, and its contents.

Well-known file systems in use today include the following:

- **FAT (file allocation table)**   This file system has been used in MS-DOS and early versions of Microsoft Windows, and it is often used as the file system on portable media devices such as flash drives. Versions of FAT include FAT12, FAT16, and FAT32. FAT does not support security access controls, including specifying access permissions to files and directories. FAT also does not include any journaling (the process of recording changes made to a file system, which aids in file system recovery) features, making it more vulnerable to corruption if power is removed during write operations.

- **NTFS (NT File System)**   This is used in newer versions of Windows, including desktop and server editions. NTFS supports file- and directory-based access control and file system journaling.

- **EXT3**   This journaled file system is used by the Linux operating system.

- **HFS (hierarchical file system)**   This file system is used on computers running the Apple macOS operating system.

- **APFS (Apple File System)**   This file system is used on computers running the Apple macOS operating system.

- **ReFS (Resilient File System)**   This file system is used on Windows Server 2012 and later versions and is intended to become the replacement for NTFS.

- **ISO/IEC 9660**   This file system is used by CD-ROM and DVD-ROM media.

- **UDF (Universal Disk Format)**   This optical media file system is considered a replacement for ISO/IEC 9660. UDF is widely used on rewritable optical media.

Database Management Systems

A *database management system*, or DBMS, is a software program or collection of programs that facilitates the storage and retrieval of potentially large amounts of structured information. A DBMS contains methods for inserting, updating, and removing data; computer programs and software applications can use these functions to manipulate data in the database. A DBMS also usually contains authentication and access control, thereby permitting control over which users and programs may access what data.

DBMS Organization

Most DBMSs employ a data definition language (DDL) that is used to define the structure of the data contained in a database. The DDL defines the types of data stored in the database as well as relationships between different portions of that data.

DBMSs employ some sort of a data dictionary (DD) or directory system (DS) that is used to store information about the internal structure of databases stored in the DBMS. To understand how they relate to each other, you can think of the DDL as the instructions for building a database's structure and data relationships; the DD or DS is where the database's structure and relationships are stored and used by the DBMS.

DBMSs also employ a data manipulation language (DML) that is used to insert, delete, and update data in a database. SQL is a popular DML that is used in the Oracle and SQL Server DBMSs.

DBMS Structure

Three principal types of DBMSs are in use today: relational, object, and hierarchical. Each is described in this section.

Relational Database Management Systems
The relational database management system (RDBMS) represents the most popular model used for DBMSs. A relational database permits the design of a structured, logical representation of information.

Many relational databases are accessed and updated through the SQL (Structured Query Language) computer language. Standardized in ISO/IEC and ANSI standards, SQL is used in many popular relational DBMS products, including Oracle Database, Microsoft SQL Server, MySQL, and IBM DB2.

RDBMS Basic Concepts
A relational database consists of one or more *tables*. A table can be thought of as a simple list of records, such as lines in a data file. The records in a table are often called *rows*, and the different data items that appear in each row are usually called *fields*.

A table often has a *primary key*, which is simply one of the table's fields that is populated with values that are unique in the table. For example, a table of healthcare patient names can include each patient's identification number, which can be made the primary key for the table.

One or more *indexes* can be built for a table. An index facilitates rapid searching for specific records in a table based upon the value of one of the fields other than the primary key. For instance, a table that contains a list of assets and their serial numbers can have an index of the table's serial numbers. This will permit a rapid search for a record containing a specific serial number; without the index, RDBMS software would have to examine every record in the table sequentially until the desired records were found.

One of the most powerful features of a relational database is the use of *foreign keys*. A foreign key is a field in a record in one table that can reference a primary key in another table. For example, a table that lists sales orders includes fields that are foreign keys, each of which references records in other tables. This is shown in Figure 4-2.

Relational databases enforce *referential integrity*. This means that the database will not permit a program (or user) to delete a row from a table if there are records in other tables whose foreign keys reference the row to be deleted. The database instead will return an error code that will signal to the requesting program that rows in other tables would be "stranded" if the row were deleted. Using the example in Figure 4-2, a relational database will not permit a program to delete salesperson #2 or #4 from the Salesperson table, because records in the Orders table reference those rows.

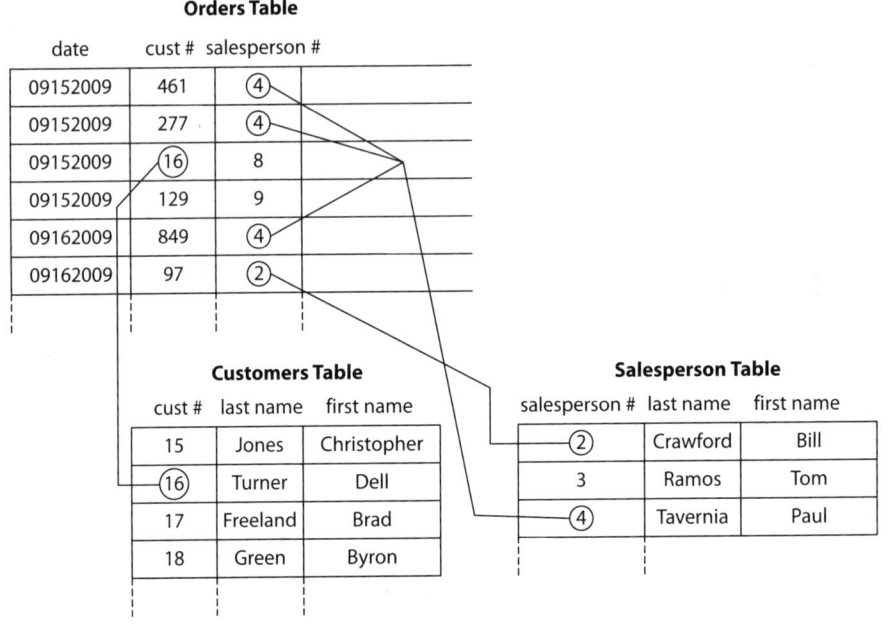

Figure 4-2 Fields in a sales orders table point to records in other tables.

The power of relational databases comes from their design and from SQL. Queries are used to find one or more records from a table using the SELECT statement. An example statement is

```
SELECT * FROM Orders WHERE Price > 100 ORDER BY Customer
```

One powerful feature in relational databases is a special query called a *join*, where records from two or more tables are searched in a single query. An example join query is

```
SELECT Salesperson.Name, count(*) AS Orders FROM Salesperson JOIN Salesperson_
Number ON Salesperson.Number = Orders.Salesperson GROUP BY Salesperson.Name
```

This query will produce a list of salespersons and the number of orders they have sold.

Relational Database Security   Relational databases in commercial applications need to have some security features, including these three primary security features:

- **Access controls**   Most relational databases have access controls at the table and field levels, so that a database can permit or deny a user the ability to read data from or write data to a specific table or even a specific field. To enforce access controls, the database needs to authenticate users so that it knows the identity of each user making access requests. DBMSs employ a data control language (DCL) to control access to data in a database.

- **Encryption**   Sensitive data, such as financial or medical records, may need to be encrypted. Some relational databases provide field-level database encryption that enables a user or application to specify certain fields that should be encrypted. Encryption protects the data by making it difficult to read if an intruder is able to obtain the contents of the database by some illicit means.

- **Audit logging**   DBMSs provide audit logging features that enable an administrator or auditor to view some or all activities that take place in a database. Audit logging can show precisely the activities that take place, including details of database changes and the users who made those changes. The audit logs themselves can be protected so that they resist tampering, which can make it difficult for someone to make changes to data and erase their tracks.

Database administrators can also create *views*, which are virtual tables created via stored queries. Views can simplify viewing data by aggregating or filtering data. They can improve security by exposing only certain records or fields to users.

NoSQL   *NoSQL* DBMSs are nonrelational and designed to support large, sometimes disparate, data sets across multiple systems. Several types of NoSQL databases are in use, including Column, Document, Key-Value, and Graph. The motivation for the use of NoSQL databases is primarily applicability and usefulness: relational databases are not always the best choice for a DBMS in every application.

Application Servers

An application server is a system that runs one or more business software applications that are designed to support one or more business processes. The application(s) run under the control of a server operating system (such as Linux, Windows, Solaris, AIX, IBM I, or z/OS), which in turn either runs directly on server hardware, under the control of a hypervisor as a virtual machine, or in a container.

Modern business applications communicate with a DBMS where business records are stored and processed. These DBMSs may reside on the same operating system or, more typically, on a separate server.

Business applications that interact with humans typically communicate with a web server (which may be on the same operating system) or through APIs to companion mobile apps that run on mobile devices such as tablets and smartphones.

More details about applications are included in Chapter 5.

Cloud Services

The term "cloud" is an abstraction that may refer to one or more of these:

- Computers made available to customers as a service

- A collection of virtual machines located on premises or at a remote location

- Software applications (including those that emulate hardware devices such as firewalls) made available to customers as a service

- Data storage made available to customers as a service
- Network connectivity made available to customers as a service

Fairly standard models describing types of cloud services have been developed, including IaaS, PaaS, and SaaS, which are described in the remainder of this section.

Infrastructure as a Service

Infrastructure as a service, or IaaS, refers to servers where computing, network, or storage infrastructure is offered as a service to customers. IaaS solutions are generally accessed via the Internet, but they can also be accessed via dedicated network connections. All forms of IaaS employ virtualization technology to make services accessible to customers. (Virtualization was discussed earlier in this chapter.)

Often, IaaS is consumed in the form of customer-managed operating systems that customer organizations build, configure, and maintain. This represents a direct replacement for traditional on-premises servers. Customers manage their IaaS environments through a management interface that enables them to build, start, stop, or destroy individual server operating systems. IaaS environments also enable customers to arrange their servers in a designed virtual network, which may include firewalls and other security devices, resulting in a network architecture not unlike that of an on-premises or colocated environment.

Elasticity

A favorite feature of IaaS environments is *elasticity*, where the cloud service provider will automatically instantiate additional copies of running operating systems (with their respective business applications) to meet demand levels. The rules, parameters, and limits regarding elasticity are configurable by customers. For instance, a customer that is an e-commerce organization may employ 20 web servers and 20 database servers to manage demand. The customer can configure the cloud to add as many as 10 additional web servers and database servers during peak periods of demand. Since cloud service providers charge their customers by the number of servers used and the work they perform, these limits ensure that excessively high demand will not result in excessive charges.

Types of IaaS services

IaaS services are designed and delivered in various ways. Common terms have been adopted for these services, including the following:

- **Public cloud**   The use of infrastructure that is managed and accessed over the public Internet
- **Private cloud**   The use of IaaS with its management and virtualization capabilities, dedicated to a single customer and hosted by the IaaS provider, a colocation provider, or onsite
- **Hybrid cloud**   The use of a combination of public cloud and private cloud

> **NOTE** Organizations typically employ private cloud to meet regulatory or contractual requirements that forbid the use of public cloud.

Notable IaaS providers include Amazon Web Services (AWS), Microsoft Azure, Google Cloud, and Rackspace Cloud.

Platform as a Service

Platform as a service (PaaS) is a service platform hosted by a service provider and available over the Internet. Service platforms typically have a core service that resembles software as a service (SaaS), but with the ability for other organizations to attach or associate their own software programs and/or databases that are integrated into the core platform. Examples of PaaS services include SalesForce, SAP Concur, and Heroku.

Software as a Service

Software as a service (SaaS) is the model through which software vendors make their applications available to customers. Rather than the software being available to customers to install on their own servers, software vendors run the software on their computers and make it available to customers in a leasing arrangement.

SaaS vendors typically run very large instances of their applications in a multitenant architecture, whereby multiple customers use the application concurrently. SaaS vendors implement logical controls that logically partition customer data so that customers can access only their own data and not that of other customers.

Application vendors have found that supporting only their own running instances of their software is far more efficient and effective than supporting customers running the software on their own servers. Further, the growth of IaaS means that fewer organizations have their own server environments, making SaaS a more attractive prospect for many customers.

SaaS reduces the complexity of application management for organizations. Instead of having to build and maintain server operating systems and database management systems, as well as installing application patches, organizations simply log in to SaaS applications that are hosted by the software companies themselves, being the true users of business applications without having to do any of the work supporting their infrastructure.

Serverless Computing

Serverless computing is a cloud service model in which a customer organization deploys its software applications to a cloud service provider that manages the underlying operating systems. The term "serverless computing" does not imply the absence of a server, but rather the absence of the need for the customer organization to manage it. Examples of serverless computing services include AWS Lambda, Google Cloud Functions, Azure Stack, and IBM Cloud Functions.

Mobile Backend as a Service

A relative newcomer, *mobile backend as a service* (MBaaS) is a cloud-based service platform that facilitates the use of data-backed mobile apps. These platforms commonly include user management, APIs, and notifications. Some MBaaS services feature integration with social media platforms such as Facebook, Twitter, and Instagram.

Cloud Responsibility Models

Many organizations adopt cloud-based services with the belief that cloud service providers take care of all aspects of information security. This errant thinking has led to many breaches over the years. A cloud responsibility model defines the parties that are responsible for various aspects of information security. Having a clear understanding of your responsibilities and capabilities in managing the cloud environment is critical to your ability to uphold privacy and security requirements. For example, in a SaaS model, the SaaS provider manages network and operating system security, while the SaaS customer is responsible for user access management. In an IaaS model, the IaaS provider will take care of physical security, but the IaaS customer is responsible for the configuration and patching of operating systems. The following illustrates a typical shared responsibility model.

Responsibility	SaaS	PaaS	IaaS	On-prem
Data governance and rights management	Customer	Customer	Customer	Customer
Client endpoints	Customer	Customer	Customer	Customer
Account and access management	Customer	Customer	Customer	Customer
Identity and directory infrastructure	Both	Both	Customer	Customer
Application software	Vendor	Both	Customer	Customer
Network controls	Vendor	Both	Customer	Customer
Operating system	Vendor	Vendor	Customer	Customer
Physical hosts	Vendor	Vendor	Vendor	Customer
Physical networks	Vendor	Vendor	Vendor	Customer
Physical data center	Vendor	Vendor	Vendor	Customer

Shadow IT and Citizen IT

The advent of cloud computing and SaaS made it all too easy for organization departments to procure their own applications directly. Often this occurred without consulting with, or even informing, the IT department. Most often, corporate departments are simply trying to "do the right thing" by streamlining business processes to reduce cost or increase competitiveness. However, without IT, security, or privacy professionals being involved, this *shadow IT* (sometimes called citizen IT) is wrought with security and privacy risks, as there may be few or no persons in these departments that are aware of these risks and the techniques used to mitigate them.

Often, shadow IT occurs when an organization's IT department is out of step with or not aligned with its internal users. Frustrated that their IT departments are not supporting them, they go around IT to solve their information processing problems.

Shadow IT can be difficult to detect, but tools such as a cloud access security broker (CASB), data loss prevention (DLP), and web content filtering systems can detect shadow IT activity.

 **NOTE**   As of this writing, the problems with shadow IT have been exacerbated by the COVID pandemic, as many organizations are permitting their remote workers to use personally owned computers (bring-your-own-device, or BYOD, discussed later in this chapter), which without the proper planning and controls, can make users blind to the storage and use of sensitive and personal information.

Endpoints

An *endpoint* is any of several types of end-user devices, including desktop computers, laptop computers, tablet computers, and smartphones. The term may also refer to connected devices, which are discussed later in this chapter.

In some organizations, endpoints include connected devices in the categories of IoT (Internet of things), industrial control systems (ICS), and supervisory control and data acquisition (SCADA). Other organizations use the terms IoT, ICS, SCADA, and so on, to distinguish these endpoints from those used by end users. Industry lexicons are still developing in this area, and there is not yet a consensus on the precise definitions of these terms.

Laptop and Desktop Computers

Laptop and desktop computers are the mainstays of end-user computing in most organizations. For the most part, IT departments issue laptop or desktop computers to employees, and these IT departments utilize a service desk function to support the end users when they have questions about or problems with their computers. Microsoft Windows remains the dominant laptop and desktop operating system, with Apple macOS also in popular use. Some organizations also support ChromeOS on Chromebooks as well as the Linux operating system.

Configuration Management

For improved efficiency and security, most organizations employ standard computer *images*, which are preconfigured operating systems with all of the configuration settings, security settings, agents, and tools preinstalled. The use of images saves considerable time over the alternative: installing all of the tools and applications manually, which typically takes several hours.

IT departments generally use management tools to centrally control the configuration of the laptops and desktops in the organization. With a mouse click, an IT administrator can change a configuration setting on many, or all, of the laptops and/or desktops in the organization.

Security and Privacy

Security configurations and tools on endpoints are critical. Without robust security architectures and effective security operations, endpoints would be quickly compromised by attackers, resulting in malware that will cause a variety of harmful issues, including theft of login credentials, theft of stored data, or use of ransomware.

Organizations generally implement a number of essential security capabilities on endpoints:

- **Antimalware**   Whether they are traditional antivirus software or any of the more advanced antimalware programs, these tools detect, block, and remove most malware attacks when they occur.

- **Firewall**   These are especially important for laptop computers, which are often used away from the corporate network and its firewall(s) protecting all endpoints.

In addition, many organizations also employ one or more of these security tools on endpoints:

- **Intrusion prevention system (IPS)**   Similar to enterprise network IPS, this detects and blocks attacks.

- **Data loss prevention (DLP)**   These tools have a variety of capabilities, including monitoring and controlling the handling of sensitive information. DLP is an important tool that gives organizations specific visibility and control of the handling of data containing personal information.

- **Application whitelisting**   This is used to restrict what applications are permitted to run on an endpoint.

- **Web content filtering**   This is used to protect endpoints from malicious web sites in watering hole attacks, as well as to block access to web sites based on their subject matter.

Further, in many organizations, end users are not local administrators on their laptops and desktops. This helps to reduce the number and impact of end-user support issues and the impact of successful malware attacks.

Support

In all but the smallest organizations, IT departments use tools that offer service desk personnel the ability to access end-user laptop and desktop computers remotely. This kind of administrative access takes two forms: First, an end user can "share his or her screen" with an administrator who can temporarily take control of the computer to troubleshoot an issue or show a user how to operate a program. Second, an administrator can access a user's computer *without* the user's knowledge or consent. The latter method may be used as a part of an investigation or other circumstance where IT must better understand either the contents of the computer or actions taking place on it, without the user's knowledge.

Virtual Desktop Infrastructure

While the desktop computing paradigm has been popular for decades, *virtual desktop infrastructure* (VDI) technology is gaining in popularity. In VDI, the programs and storage are on centralized servers that use a remote display protocol to give end users the appearance of local computing. VDI offers operational, security, and privacy advantages:

- Central control of desktop operating systems
- Less expensive endpoint computers, since they are not storing any data or running applications
- Improved security through network access restriction—endpoints can be permitted to communicate only with their VDI servers
- Better control of sensitive and personal information, since it is never stored on endpoints

The primary disadvantage of VDI is that a network connection is always required in order to run programs and access data. However, hybrid environments can also be engineered, allowing for occasional local storage of data.

Mobile Devices

The power and ubiquity of mobile devices makes them excellent candidates for conducting business while away from the workplace or home office. Virtually every office worker owns a smartphone capable of connecting to company e-mail, running browsers and applications, and storing copious amounts of data.

Permitting workers to use personally owned mobile devices for accessing company e-mail and apps can make workers far more productive. Workers often check on their e-mail before and after work, on weekends, and even on vacation. The flexibility of always being connected means workers are no longer tethered to workplaces or home offices.

To counter risks, including data leakage and credential theft, many organizations actively manage mobile devices with *mobile device management* (MDM) systems that are used to enforce security configurations and further protect sensitive and personal information. Some MDM products have the ability to prevent end users from inadvertently leaking company information through restrictions on local file storage and handling.

The problem with personally owned mobile devices, however, is that they are not the property of the organization, meaning the organization does not have complete control over these devices. This poses a problem for organizations that need visibility and control of data usage and data movement.

Bring-Your-Own _____

The phenomenon of *bring-your-own-device* (BYOD) began in the early 1990s when organizations first created "remote access." BYOD accelerated in the 2000s with the Apple iPhone and its native ability to connect directly with organizations' Outlook Web Access e-mail servers, thereby bringing business e-mail to personally owned devices without the consent or assistance from company IT departments. Although organizations slowly responded by restricting such access, they had little choice but to accept the trend and manage it accordingly.

Another significant hotbed of BYOD lies in the remote access capability. Many organizations' VPN services do not check to see whether the computer requesting remote access is a company-owned and company-managed machine. Consequently, many workers accessing company networks and systems remotely from home are doing so with their personally owned computers instead of their corporate-issued laptops. This, too, represents a significant data leakage risk as organizations will be completely blind to whatever its workers do with sensitive data once it leaves the control of the organization.

BYOD is similar to *bring-your-own-apps* (BYOA), which is when an organization's employees load personally owned applications onto their company-issued computers, or even use free or nominal-cost cloud apps to store or process sensitive and personal information.

To counter BYOD and BYOA, IT departments must implement additional controls, including the following:

- **Network Access Control (NAC)**   This protocol controls which devices are permitted to connect to a wired or wireless network, as well as by remote access. MAC address filtering can also be used, but this technique does not scale well to larger organizations.

- **Application whitelisting**   This tool permits only approved programs to execute on a computer.

- **Cloud access security broker (CASB)**   This tool manages and controls users' access to cloud servers to prevent interaction with cloud services not permitted by the organization. Web content filtering can also perform this function in many cases.

- **Data loss prevention (DLP)**   These tools observe and restrict the movement of files containing sensitive and personal information. This can help to prevent data from being exported to cloud-based services not permitted by the organization.

- **User behavior analytics (UBA)**   These tools observe network and system patterns over time and generate alerts when anomalous events occur.

Zero Trust Architecture

The concept of *zero trust* represents a new way of thinking about information security. Traditionally, information security was concerned with the protection and control of an entire computing and network ecosystem, including servers and data, endpoints, and the networks connecting them all together. The perimeter was vast and included everything in the organization's control.

Zero trust changes all of that. First, it changes the definition of the perimeter, from that of the entire enterprise, to microperimeters close to the actual information being protected. These concepts are part of zero trust:

- **Untrusted endpoints**   With regard to the protection of sensitive and personal information, end-user computing is considered untrusted, as though the organization's users were users on the Internet with unmanaged and untrusted endpoints.

- **Untrusted data center systems**   With regard to the protection of data in DBMSs, database servers no longer blindly trust all systems in the data center, but instead require any system (such as an application server) to authenticate itself.

- **Untrusted networks**   With regard to the security of application servers, these servers no longer accept incoming connections from endpoints on enterprise networks without first requiring strong authentication. Also, from the perspective of data center networks, all other networks, including end-user networks and the Internet, are considered untrusted and hazardous.

Connected Devices and Operational Technology

If the endpoints, mobile devices, servers, and network devices in the IT ecosystem are the "visible part of an iceberg," then the world of connected devices, ICSs, and SCADA systems are located "below the surface." In many industries, including energy, public utilities, manufacturing, air traffic control, medical research, and more, legions of devices of all kinds are communicating with one another on data networks. And many of these devices have been in use for decades.

There has been an explosion of connected devices in the consumer world as well. Doorbells, televisions, thermostats, kitchen appliances, barbecue grills, security cameras, medical devices, and automobiles—nearly everything is connected to networks, and much is connected to the Internet itself.

Privacy and security are often overlooked in the design and implementation of many of these so-called smart devices, but privacy and security issues abound, and they are not easily managed or mitigated. Consider the following:

- Poorly designed security, or no security at all

- Lack of security update capability—no way to patch devices already in use

- Devices running outdated and unsupported versions of operating systems and subsystems

- Troves of backend data containing biometric information, images, location data, financial data, medical data—all waiting to be compromised and stolen

- Poorly designed security controls protecting critical infrastructure, including power generation and distribution, water supply and treatment, vital manufacturing facilities, air traffic control, and more

- Increased connectivity between these control networks with office worker networks and the Internet itself exposing these control networks to attacks

The heightened importance and the inherent weaknesses in connected devices compel organizations to enact additional protective controls, including network segmentation, to isolate these devices from the rest of the organization. Tight access controls restrict access into and out of these networks, and studious monitoring of communications in these networks is used to detect anomalies that may be signs of tampering or intrusion.

The other side of the problem of connected devices is the information that they gather, some of which is related to individual persons. This information includes

- Location history—detailed accounts of the individual's whereabouts
- Facial recognition
- Financial transactions
- Medical information, including vital signs and health history

Arguably, numerous benefits are derived from the collection of this information, which amounts to improvements in the quality of life of the people using the devices. But the accumulation of this information also increases the risk of major breaches should this information not be fully protected from all threats.

The other dimension of the problem of these vast information stores with myriad details about the personal lives of individuals is the potential misuse of this information. Without proper governance, organizations tend to "push the envelope" regarding the ways in which this personal information may be used.

Finally, there are abuses of government agencies that, through various means (legal and otherwise), are able to obtain all or parts of these information stores as a part of the "surveillance state."

Smart Devices: Back to the Security Stone Age

Just as major vendors were making security important in their products, and many organizations were getting their vulnerability management acts together, along came so-called smart devices. From a security perspective, most of them are simply awful: they use outdated and vulnerable libraries and operating systems, they are configured poorly, and they cannot be patched in the field. These devices are everywhere—in homes and in businesses—and they are turning the world of infosec upside-down because of their poor design and utter lack of support. They are a menace on networks because of their poor securability. Better organizations are taking matters into their own hands and either blocking them outright or placing them on well-defended isolated networks, where their weaknesses can't hurt the rest of the organization.

In homes, these devices are spying on users, eavesdropping on conversations, leaking personal data, and joining botnets. Unlike better organizations, most home users don't know or don't fully appreciate the hazards presented by these devices.

Remote Access

Remote access is defined as the means of providing remote connectivity to an internal corporate network through a data link. Remote access is provided by many organizations so that employees who are temporarily or permanently offsite can access internal network resources from their remote locations.

Remote access was initially provided using dial-up modems that included authentication. While remote dial-up is still provided in some instances, most remote access is provided over the Internet itself and typically uses an encrypted tunnel, or virtual private network (VPN), to protect transmissions from any eavesdroppers. VPNs are so prevalent in remote access technology that the terms VPN and remote access have become synonymous. Remote access architectures are depicted in Figure 4-3.

Two security controls are essential for remote access:

- **Authentication** It is necessary to know who is requesting access to the corporate LAN. Authentication may consist of the same user ID and password that personnel use when onsite, or multifactor authentication may be required.

- **Encryption** Many onsite network applications do not encrypt sensitive traffic because it is all contained within the physically and logically protected corporate LAN. However, because remote access provides the same function as being on the corporate LAN, and because the applications themselves usually do not provide encryption, the remote access service itself usually provides encryption.

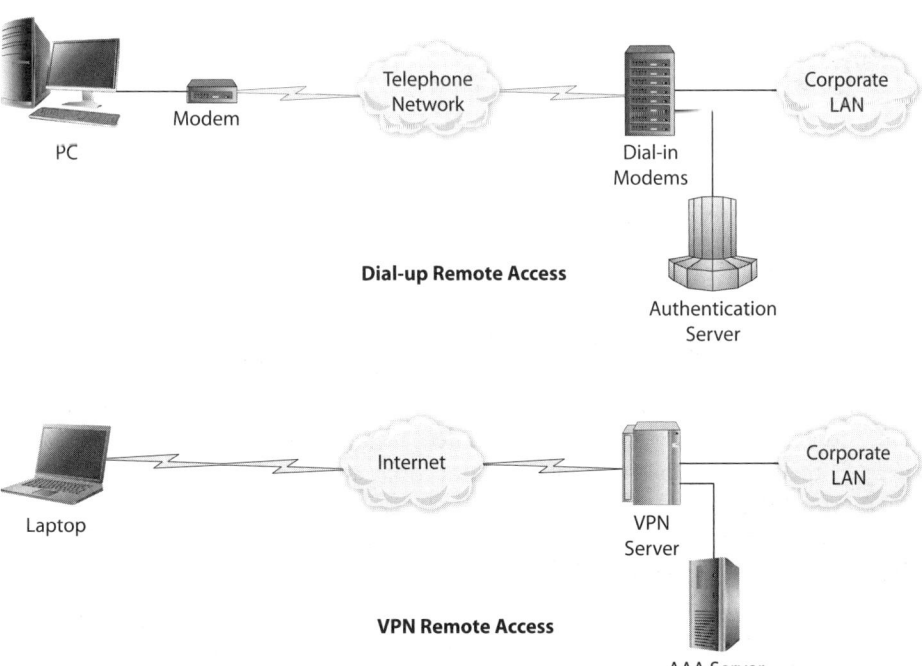

Figure 4-3   Remote access architectures

Authentication and encryption are needed because they are a substitute (or compensating control) for the physical access controls that are usually present to control which personnel may enter the building to use the onsite corporate network. When personnel are onsite, their identity is confirmed through keycard or other physical access controls. When personnel are offsite using remote access, because the organization cannot "see" the person on the far end of the remote access connection, the authentication used is a substitute.

Credential-stealing malware is a significant problem. As a result, organizations are incorporating multifactor authentication for access to the organization's internal resources, regardless of the location of users—whether they are on a corporate network, working from home, in the field, or traveling.

The New Remote Access Paradigm

As organizations migrate their business applications to colocation centers and *XaaS* providers, and after the last internal resource is moved to the cloud, what is the point of remote access? Remote access to *what*?

If we think about this in terms of VPN and the protection afforded through encryption, VPN makes good business sense, by protecting network traffic from potential eavesdroppers (whether the human or malware variety). For this reason, it's preferred to use "VPN" instead of "remote access."

Organizations still need to address several subtopics when considering their VPN architectures in light of cloud migration, such as split tunneling, Internet backhauling, and whether VPN should always automatically activate on workstations away from internal corporate networks.

Client VPN

A client VPN connection can be established using a *VPN client* software utility. A VPN client is generally installed on an endpoint system like any other program, and it can be configured with settings, so that end users do not need to type in the name or IP address of the VPN server every time they want to establish a connection. Further, some VPN clients can be configured by organizations' IT departments with settings like group passwords that end users will not be able to see.

A VPN connection can be invoked at will when an end user is ready to establish a connection to access internal network resources in the organization. Some organizations configure their endpoints so that VPN connections are "always on" whenever endpoints are connected to any noncompany network.

Clientless (SSL) VPN

Clientless VPN is so called because no software—such as a VPN client—needs to be installed on endpoints. Instead, the endpoints' built-in networking capabilities are sufficient for setting up VPN connections. In these cases, a user will navigate to a special

URL from a browser (such as https://remote.company.com, or hopefully something not so easily guessed). The browser will facilitate the establishment of the VPN connection.

Clientless VPN is limited to web protocols such as HTTP and HTTPS. If other protocols such as Remote Desktop Protocol (RDP) or Secure Shell (SSH) are required, additional technologies may need to be deployed, or a VPN client must be used to access these resources in internal networks.

Split Tunneling

Many VPN connections have the ability to establish a *split tunnel*, a characteristic of a VPN connection in which network traffic destined to the organization's internal network will traverse the VPN connection, while network traffic destined to the Internet will proceed directly to the Internet, bypassing the VPN connection. Split tunnels are sometimes preferred by organizations that do not care to carry high-bandwidth Internet traffic through their VPN infrastructures. However, a split tunnel—by design—will bypass an organization's network-based protective controls, potentially permitting an end user (or malware on the endpoint) to exfiltrate sensitive and private information.

NOTE   Organizations that are more privacy or security conscious will not permit split tunneling, and if they are concerned with high-bandwidth network services (such as streaming movies or music), they will direct their workers not to use these services while on VPN (if ever).

System Hardening

System hardening techniques are used to make systems and devices more resistant to attack. Hardening is considered an essential undertaking, particularly for systems that communicate over the Internet. Mainly, hardening refers to a set of configuration settings that are applied to a system. Although these configuration settings can be applied manually, most organizations employ tooling to automate their application to achieve hardening more quickly and more accurately.

Other activities such as vulnerability scanning and patching make up the larger set of activities known as *vulnerability management*. These other activities are discussed in Chapter 6.

Hardening Principles

System hardening principles are general principles describing a number of configuration practices that may be employed to make systems more resistant to attack. The primary principles on the topic of hardening include the following:

- **Single purpose**   Systems should perform a single function or purpose, rather than having multiple roles. For instance, web servers, application servers, and database servers should reside on separate systems. This principle need not be taken to extremes, however: for instance, a DNS server may also serve as an NTP server.

- **Unnecessary subsystems and programs**   Systems should be installed with only the programs and subsystems necessary for the system's purpose. For example, a web server should not have DBMS software installed on it.

- **Listeners**   System utilities should not be placed in network listen mode unless necessary for their function. For instance, a Unix server that occasionally needs to send e-mail may need to have the Sendmail program installed, but Sendmail should not be in listen mode (awaiting inbound e-mail).

- **Permitted connections**   Systems should be configured to permit inbound connections only from authorized systems or networks. For example, inbound administrative SSH connections should be permitted only from administrative workstations or VLANs. Also, production servers should not be able to access the Internet using a browser, FTP, or other protocols, except those that are necessary for their function.

- **Unnecessary data**   Systems should have only the data necessary for their present function on the system. For example, application servers should not have the application's source code stored on the system. Demo or test data should not be present on any production system.

- **Necessary user accounts**   Only those user accounts necessary for the function and maintenance of a system should be present on the system.

- **Default passwords and identifiers**   All default passwords and identifiers on a system should be changed to non-default values. For example, the SNMP community string should be changed from "public" to a value not easily guessed by an outsider.

- **Rename privileged accounts**   Privileged accounts such as root or administrator should themselves be renamed. Direct brute-force login attacks will be more difficult for attackers to carry out successfully if they do not know the names of administrative accounts.

- **Up-to-date software**   The operating system and all software programs should be currently supported and up to date. Outdated and unsupported software should not be used, particularly on mission-critical systems.

Hardening Standards

The development of specific configuration settings to support hardening principles may be time consuming and may, at times, not cover all circumstances. Further, system administrators may lack detailed expertise to apply all needed hardening configuration changes competently. For these reasons, organizations may choose to adopt one of a few good hardening standards that are available:

- **Device and system manufacturers**   Better manufacturers, realizing that many of their business customers will want to harden their systems, will publish principles and/or sets of configuration standards that can be used to harden systems to protect them from attack.

- **Center for Internet Security Benchmarks** The Center for Internet Security (CIS) has published numerous excellent hardening standards for practically every type of system under the sun, from mainframes to smartphones. These benchmarks are available free of charge from https://www.cisecurity.org/cis-benchmarks/. CIS also provides tools and images that support these standards.

- **DoD Cyber Exchange** The DoD Cyber Exchange publishes numerous Security Technical Implementation Guides (STIGs) that include high-quality hardening standards for numerous types of systems. STIGs and other information are available at https://public.cyber.mil/stigs/.

 **EXAM TIP** CDPSE candidates do not need to memorize hardening standards, but you are expected to understand the concepts of system hardening.

Security and Privacy by Design

At every layer and in every instance of infrastructure, it is essential that architectures, designs, and configurations all contribute to and support security and privacy. The *system development life cycle* represents the business processes used to develop and maintain information systems, and must include steps to ensure that new information systems and changes to existing information systems do not impact security and privacy in negative ways. Policies and standards must also contribute to and support security and privacy principles to ensure that personal information is adequately protected and used. Ongoing operations must likewise ensure that the security and privacy of systems are not compromised, and that continuous monitoring is employed to detect security and privacy incidents early so that they may be contained.

 **EXAM TIP** Remember that implementing security and privacy by design involves not only development processes but a change in organizational culture.

Chapter Review

Effective privacy needs effective security. Effective security needs an IT architecture with integrity and protective controls.

The technology stack is the set of technologies that make up a system. It's often expressed in bottom-up notation, such as Linux-Apache-MySQL-Perl, or Windows-IIS-SQL-ASP.Net. LAMP is Linux-Apache-MySQL-PHP/Perl/Python, and WISA is Windows-IIS-SQL-ASP.Net.

Hardware is the physical machinery of computing and communications in IT. Hardware comes in many forms, including mainframes, servers, network devices such as routers and firewalls, laptop computers, and smartphones.

Computers are general-purpose machines that run operating system software, which in turn runs subsystem or application software. Computers consist of a CPU, main storage, secondary storage, one or more buses, and adaptors for communicating over networks or with humans.

Storage systems are configured by organizing one or more volumes, within which file systems are created, and data of some kind is stored within the file systems. Volumes can vary in size, and often their size can be adjusted dynamically.

Networks are the means through which computers communicate with one another. Whether networked computers are in a single room or spread across the globe, and whether or not a telecommunications company is involved, networks facilitate all computer communications, both wired and wireless.

Network devices carry network traffic. The arrangement of network devices leads to the architecture of a network. Security devices on a network protect systems from unwanted traffic.

Computer operating systems are large, general-purpose programs that are used to control computer hardware and facilitate the use of software applications and tools. They also facilitate access to peripheral devices, manage storage, allocate resources, facilitate communications, and protect data.

Computers can be combined in various ways, including in clusters and grids.

Cloud computing refers to dynamically scalable and usually virtualized computing resources that are used internally or provided as a service from a third party.

Virtualization refers to the set of technologies that permits two or more running operating systems (of the same type or different types) to reside on a single physical computer. Virtualization technology enables organizations to use computing resources more efficiently.

Containerization is a form of virtualization whereby multiple applications may inhabit a single running operating system. Each such application is completely isolated from every other and can access only the resources allocated to the container.

A file system is a logical structure that facilitates the storage of data on a digital storage medium such as a hard drive, SSD, optical disc, or flash memory device. The structure of the file system facilitates the creation, modification, expansion and contraction, and deletion of data files. A file system may also be used to enforce access controls to control which users or processes are permitted to access, alter, or create files in a file system.

A database management system, or DBMS, is a software program or collection of programs that facilitates the storage and retrieval of potentially large amounts of structured information. A DBMS contains methods for inserting, updating, and removing data; computer programs and software applications can use these functions to manipulate data in databases.

An application server is a system that runs one or more business software applications that are designed to support one or more business processes. The application(s) run under the control of a server operating system, which in turn either runs directly on server hardware, under the control of a hypervisor as a virtual machine, or in a container.

Infrastructure as a service (IaaS) refers to servers where computing, network, or storage infrastructure is offered as a service to customers. IaaS solutions are generally accessed via the Internet, but they can also be accessed via dedicated network connections.

Platform as a service (PaaS) is a service platform hosted by a service provider and available over the Internet. Service platforms typically have a core service that resembles software as a service (SaaS), but with the ability for other organizations to attach or associate their own software programs and/or databases that are integrated into the core platform.

Software as a service (SaaS) is the model through which software vendors make their applications available to customers. Rather than the software being available to customers to install on their own servers, software vendors run the software on their computers and make it available to customers in a leasing arrangement.

Mobile backend as a service (MBaaS) is a cloud-based service platform that facilitates the use of data-backed mobile apps.

Shadow IT is the phenomenon whereby an organization's departments (or individual end users) procure their own IT services, bypassing corporate IT.

An endpoint is any of several types of end-user devices, including desktop computers, laptop computers, tablet computers, and smartphones.

Laptop and desktop computers are the mainstays of end-user computing in most organizations. For the most part, IT departments issue laptop or desktop computers to employees, and these IT departments utilize a service desk function to support the end users when they have questions about or problems with their computers.

Security and configuration management tools are used to manage the configuration, health, and security of endpoints. Security on endpoints includes antimalware, firewalls, intrusion prevention systems, data loss prevention, application whitelisting, and web content filtering.

A virtual desktop infrastructure (VDI) employs central, server-based desktop operating systems that use remote interaction protocols that give end users the appearance of local computing.

Mobile devices, including tablets and smartphones, provide mobile and convenient computing capabilities for users. Mobile device management (MDM) systems are used by organizations to control mobile devices that are used for business purposes.

Bring-your-own-device (BYOD) is growing trend whereby an organization's workers can use their own personal devices to conduct business. Some organizations permit BYOD, others allow it with a degree of control, and others forbid it altogether.

Connected devices are nonhuman interaction computers and devices used in residential, office, and industrial environments to monitor and control equipment and machinery. Connected devices are sometimes less configurable and more difficult to secure than servers and endpoints; thus, organizations often isolate them on separate networks with strict access controls.

Remote access provides remote connectivity to an internal corporate network through a data link. Remote access is used by many organizations to enable employees who are temporarily or permanently offsite to access internal network resources from their remote location.

Remote access is facilitated by virtual private networks (VPNs) that encapsulate and encrypt network traffic, and that require authentication. VPN clients are the programs used on endpoints to establish VPN connections. Clientless VPN is used when only web access to internal networks is required.

System hardening refers to techniques used to make systems and devices more resistant to attack. Mainly, hardening refers to a set of configuration settings that are applied to a system. Hardening principles include the removal of unnecessary programs, utilities, data, and user accounts; renaming administrative user accounts; permitting only required connections; changing default passwords; and using only up-to-date software. Hardening standards are available from the Center for Internet Security and the DoD Cyber Exchange.

Quick Review

- Increasingly, operational knowledge about server computing hardware is becoming an abstraction with the advent of cloud computing services. It is still important to understand the principles of computer hardware design, however, since end users' devices continue to be used.

- The fundamentals and operation of TCP/IP protocols and services are unchanged, regardless of the media used to transport them. Some media, such as Bluetooth, has inherent vulnerabilities that can be mitigated through controls within and above TCP/IP.

- Encryption is a useful tool for protecting personal information while stored and in transit; however, encryption may also hamper efforts to detect the improper use or compromise of personal information.

- Increasingly, network devices such as firewalls, intrusion prevention systems, web content filters, phishing filters, and cloud access security brokers (CASBs) are consumed as cloud-based servers rather than in the form of physical appliances.

- IaaS, PaaS, and SaaS are foundational cloud services. Newer offerings blur the lines and present more specialized services.

- While IaaS, PaaS, and SaaS offerings relieve organizations of the burden of managing computer hardware and operating systems, organizations need to understand cloud responsibility models and tailor their operations to ensure that all required security services are being performed effectively.

- Shadow IT and citizen IT can be especially difficult to detect and control, unless advanced controls such as CASBs are in place.

- Virtual desktop infrastructure (VDI) can be an effective remedy in a BYOD environment by moving computing and sensitive data to backend servers and out of desktop systems.

- Zero trust is a new approach to security architecture that focuses more on the data being protected and less on the perimeter, which is disappearing anyway.

- Connected devices often have inferior security and require additional protective controls such as network segmentation.

- With the migration of on-premises resources to the cloud, the paradigm of remote access is pivoting into secure communications.

- Hardening systems without automation is difficult, given the sheer number of configuration changes required.

Questions

1. Why is it important for users of corporate laptops to use VPN when communicating on open Wi-Fi hot spots?

   A. VPNs protect stored data on public networks.

   B. A VPN is necessary to reach an internal corporate network.

   C. Traffic on open Wi-Fi networks is not encrypted.

   D. Privacy laws require that corporate data be encrypted in transit.

2. Why is twisted pair considered more secure than Wi-Fi?

   A. Physical security controls must be compromised to reach wired networks.

   B. Twisted pair uses better encryption algorithms.

   C. Physical security controls must be compromised to reach wireless networks.

   D. Twisted pair has higher throughput capability.

3. An organization will be introducing smart TVs and other connected devices into the enterprise network. Which of the following security controls will most effectively protect the enterprise?

   A. Data loss prevention

   B. Annual penetration testing

   C. Adding smart devices to configuration management systems

   D. Network segmentation

4. What risks will an organization with network-based IPS be assuming when its workforce is working remotely?

   A. Remote systems not on VPN will not be protected by the network-based IPS.

   B. Network administrators will not be able to update the IPS as often.

   C. Network-based IPS only protects devices physically in an internal network.

   D. There's no change in risk because network-based IPS systems protect all devices regardless of location.

5. An organization is migrating its servers from physical to virtual. What privacy risks does the organization need to be concerned about concerning this migration?

   A. Guest OS privilege escalation

   B. Eavesdropping of sensitive network traffic

   C. Security hardening of the container layer

   D. Security hardening of the hypervisor layer

6. A privacy officer wants to restrict the direct database queries that analysts can run, so that they can view records only for customers who reside in the United States. Which is the best remedy that will achieve this?

    A. Encrypt the records that the analysts should not be permitted to view.

    B. Provide a weekly extract of only the records they are permitted to view.

    C. Create a database view containing only the records the analysts may view.

    D. Implement a VDI located in the United States.

7. An organization is considering moving its on-premises servers to an IaaS service. Which security controls will the organization need to continue operating?

    A. Operating system and network

    B. Operating system, network, and user access

    C. Physical only

    D. Physical, operating system, network, and user access

8. An organization is going to migrate its on-premises application to a SaaS environment. Which security controls will the organization need to continue operating?

    A. Operating system and network

    B. Physical

    C. User access

    D. Operating system and user access

9. The primary risks of end users being local administrators on their endpoints include all of the following except:

    A. Malware will execute at a privileged level and do more damage.

    B. Malware will not require human intervention to execute.

    C. Malware will be able to move laterally.

    D. Malware will be able to obtain password hashes.

10. The main reason for implementing application whitelisting on endpoints is:

    A. Permits end users to install only approved programs

    B. Prevents end users from installing applications

    C. Prevents end users from installing utilities

    D. Prevents malware from executing

11. A privacy manager is advocating the use of VDI for a call center. What is the primary privacy benefit of using VDI?

    A. Reduces impact of malware

    B. Prevents local programs from being installed

   C. Reduces likelihood of data leakage

   D. Logs all transactions

12. A privacy manager is concerned that there may be excessive instances of PII on unstructured file shares. Which tool would best confirm or refute this suspicion?

   A. NAC

   B. DLP discovery

   C. CASB

   D. EUBA

13. A new security manager is concerned about the increase in connected devices that may be present on the enterprise network. What tool(s) can best determine the extent of this situation?

   A. Network discovery scans

   B. Examine firewall logs

   C. Examine CASB logs

   D. Asset loss prevention plan

14. A new security manager is concerned about the increase in connected devices that are present on the enterprise network. What action would best mitigate this matter?

   A. Implement a SIEM.

   B. Use network segmentation.

   C. Use VLANs.

   D. Use network access controls.

15. What is the primary risk related to split tunneling?

   A. Reduces network traffic visibility

   B. Creates excessive amounts of backhaul traffic

   C. Creates routing loops

   D. Decreases performance

Answers

   1. C. Open Wi-Fi networks do not encrypt over-the-air traffic—it is transmitted in plaintext. Although some protocols such as HTTPS are encrypted, many other protocols are not. Because it is easy to eavesdrop on all users' network traffic on an open Wi-Fi network, it is important to set up a VPN session so that all traffic will be encrypted.

   2. A. Networks over twisted pair cabling is considered more secure than Wi-Fi networks primarily because physical security controls must be circumvented to reach network jacks. On the other hand, Wi-Fi signals generally radiate beyond the physical perimeter of a workplace.

3. D. Connected devices often have inferior security and commonly cannot be hardened or patched. For this reason, it is important to segment these devices onto separate, protected networks so that any hazards associated with connected devices cannot bring harm to the rest of the enterprise.

4. A. Network-based IPS protects only those devices on the internal network. Devices connected via VPN will also be protected during VPN sessions, provided the VPN does not permit split tunneling. A better solution for a remote workforce is agent/cloud-based IPS that is built into each endpoint, so that they are protected regardless of location.

5. D. The primary concern in a physical-to-virtual migration is the security of the hypervisor. If the hypervisor is not hardened, it could be compromised, which could lead to the compromise of guest OSs. Security concerns will be exactly the same for operating systems that are copied into the virtualization environment.

6. C. A database view is the easiest solution for this problem, because it will permit analysts to view only specific records any time they query relevant tables.

7. B. When an organization moves its operating systems to an infrastructure as a service (IaaS) environment, the organization will need to continue performing all operating system, network, and user access security controls as before. The IaaS organization provides only physical security.

8. C. An organization migrating from an on-premises application to a SaaS environment will no longer need to manage physical, operating system, or network security—these will be managed by the SaaS provider. The organization will have to manage user access control, however.

9. B. When end users are local administrators, malware will often be able to execute with elevated privileges, thereby inflicting potentially more damage to the system and the organization through increased access to sensitive information as well as the ability to move laterally more easily. Whether an end user is an administrator or not has little bearing on whether the malware requires human intervention to execute—this is more of a design feature of the malware than a factor related to privilege.

10. D. The primary purpose of application whitelisting is to block all instances of malware from executing on a system. Secondary benefits include the ability to prevent the installation and use of unapproved programs, utilities, and tools.

11. C. In a VDI environment, all data and application programs reside on centralized servers, not on end-user workstations. Because personal information does not reside on end-user workstations, there are fewer opportunities for that data to leak out of the organization.

12. B. DLP discovery is the correct class of tooling to determine the extent to which PII exists on unstructured file shares. DLP discovery tools scan file shares, looking for specific patterns that indicate sensitive or personal information.

13. A. Network discovery scans can be used to detect all of the systems and devices on a network. Manual effort will be required to distinguish connected devices from IT equipment such as endpoints, printers, and so on.

14. D. Using network access controls that isolate connected devices is the best remedy. VLANs and network segmentation do not necessarily imply access controls and are insufficient by themselves. A SIEM will collect log data but does nothing to remedy the presence of connected devices on an enterprise network.

15. A. The main risk associated with VPN split tunneling is the fact that some network traffic will flow from endpoints directly to the Internet, bypassing the organization's network. This amounts to reduced visibility, which may result in security and privacy events going undetected.

Applications and Software

In this chapter, you will learn about

- The concept of privacy and security by design, instead of being an afterthought
- Business processes used to acquire, develop, and maintain business applications
- Techniques used to make applications resilient to attack and misuse
- The ways in which persons' use of technology is tracked in detail

This chapter covers Certified Data Privacy Solutions Engineer job practice 2, "Privacy Architecture," part B, "Applications and Software." The entire Privacy Architecture domain represents 36 percent of the CDPSE examination.

Applications and software relate to privacy in this way: individual users interact with information technology primarily through business applications, which track and record transactions and other activity. At times, this tracking is intrusive, and organizations sometimes fail to protect such tracking data. A privacy professional who wants to be proficient in information security must understand how business applications work. Privacy and security by design are not a part of many organizations' business processes, which has led to serious flaws, leading to invasions of privacy and breaches of personal information.

Privacy and Security by Design

Privacy and security by design is a concept that reinforces the need to have privacy and security considerations incorporated into systems and applications by default. In other words, the product is designed with privacy and security as a priority, along with whatever other functional purposes the product delivers.

Privacy by design is based on seven foundational principles:

- **Proactive not reactive; preventive not remedial**   The privacy by design approach is characterized by proactive rather than reactive measures. It anticipates and prevents privacy events before they happen. Privacy by design does not wait for privacy risks to materialize, nor does it offer remedies for resolving privacy infractions once they have occurred; it aims to prevent them from occurring.

- **Privacy embedded into design**   Privacy by design is embedded into the design and architecture of IT systems as well as business practices. It is not bolted on as an add-on, after the fact. The result is that privacy becomes an essential component of the core functionality being delivered. Privacy is integral to the system without diminishing its functionality.

- **Privacy as the default setting**   Privacy by default seeks to deliver the maximum degree of privacy by ensuring that personal data is automatically protected in any given IT system or business practice. If an individual does nothing to protect her privacy, it should still remain intact. No action should be required on the part of the individual to protect her individual privacy.

- **Full functionality—positive-sum, not zero-sum**   Privacy by design seeks to accommodate all legitimate interests and objectives in a positive-sum "win-win" manner, not through a dated, zero-sum approach, where unnecessary tradeoffs are made. Privacy by design avoids the pretense of false dichotomies, such as privacy versus security, demonstrating that it is possible to have both.

- **End-to-end security—full life-cycle protection**   Privacy by design, having been embedded into the system prior to the first element of information being collected, extends securely throughout the entire life cycle of the data involved—strong security measures are essential to privacy, from start to finish. This ensures that all data is securely retained and then securely destroyed at the end of the process, in a timely fashion. Thus, privacy by design ensures cradle-to-grave, secure life-cycle management of information, end to end.

- **Visibility and transparency—keep it open**   Privacy by design seeks to assure all stakeholders that whatever the business practice or technology involved, it is, in fact, operating according to the stated promises and objectives, subject to independent verification. Its component parts and operations remain visible and transparent to users and providers alike.

- **Respect for user privacy—keep it user-centric**   Privacy by design requires architects and administrators to keep the interests of the individual uppermost by offering such measures as strong privacy defaults, appropriate notice, and user-friendly empowerment options.

It is essential that business application architectures, designs, data flows, and configurations all contribute to and support privacy and security. The *systems development life cycle* represents the business processes used to develop, maintain, and operate business applications, and it must include steps to ensure that new information systems and changes to existing information systems do not impact security and privacy in unexpected ways. Policies and standards must also contribute to and support security and privacy principles to ensure that personal information is adequately protected and properly used. Ongoing operations must likewise ensure that the security and privacy of systems are not compromised, and that continuous monitoring is employed to detect security and privacy incidents early so that they may be contained.

Systems Development Life Cycle

Civil engineers design bridges prior to their construction. Engineers first obtain requirements, which specify things like the number of lanes on the bridge and the amount of weight it is expected to carry. These engineers then consider other factors such as wind, rain, snow, ice, heat, cold, and other weather, as well as the potential weight loads on the bridge, and environmental factors such as landslides and soil stability that threaten the integrity of the bridge pilings and other support structures.

Once the bridge is built, it must be maintained. This includes painting, checking and retightening of fasteners, repairs of the road surface, checking the foundations and abutments, and making careful measurements to ensure that it is staying in place and performing as designed.

Systems development is not so different from designing and building bridges. Organizations gather requirements about a proposed software application (or operating system, tool, or other), which should include security and privacy requirements and any applicable compliance requirements. Design of a software application should not commence until all of these requirements are known. Likewise, development should not commence until the design is complete and reviewed and approved by stakeholders.

This section describes the entire end-to-end process of systems development, including well-known models such as Waterfall, DevOps, and DevSecOps.

The "S" in SDLC Now Stands for "Systems"

IT and security professionals who have been in the business for more than ten years will sooner or later notice the switch in the well-known SDLC acronym. Originally, and for decades, SDLC referred to the *software development life cycle*, for many organizations developed their own business applications and spent considerable effort maintaining and customizing those applications over many years.

Today, because of two changes that have occurred, the *S* in SDLC has been changed to *systems* to represent a broader perspective than just that of software applications: First, fewer organizations develop their own business application software and instead purchase off-the-shelf applications or (more often) subscribe to software as a service (SaaS) services for their primary business applications. Second, the SDLC has been expanded to encompass projects such as the development of infrastructure.

SDLC Phases

The systems development life cycle describes the end-to-end process for developing and maintaining information systems. A common structure for SDLC is a *waterfall*-style framework that consists of several distinct phases:

- Feasibility study
- Requirements definition
- Design

- Development
- Testing
- Implementation
- Post-implementation

Organizations often employ a "gate process" approach to their SDLCs by requiring a formal review at the conclusion of each phase before the next phase is permitted to begin. The review is usually a formal meeting where project managers and other participants describe the status of the project. Management, if satisfied that the current phase of the project has been completed successfully and that all requirements have been met, will then permit the project to proceed to the next phase.

In addition to the waterfall SDLC model, iterative and spiral models are used in SDLC processes. The iterative and spiral models both operate in (visually) circular modes, as opposed to the linear waterfall model. The *spiral* model consists of the development of requirements, design, and one or more prototypes, followed by additional requirements and design phases until the entire design is complete. Similarly, the development in the *iterative* model goes through one or more loops of planning, requirements, design, coding, and testing until development and implementation are considered complete.

The *DevOps* model is also often used for systems development processes. DevOps is an iterative development and operations model that is discussed later in this chapter. *DevSecOps* is a DevOps model that is modified to include essential security and privacy steps.

SDLC in this section is described from the waterfall model's perspective. The activities discussed in this section in the waterfall model are quite similar to those in the iterative and spiral models.

Feasibility Study

The feasibility study is an intellectual effort that seeks to determine whether a specific change or set of changes in business processes and underlying systems is practical to undertake.

Often the purpose of a feasibility study is not to answer the question, "Can a specific type of change be made to the business?" but rather, "Is a specific type of change to the business feasible from a cost and benefit perspective?" In other words, the feasibility study is an analysis of proposed changes to business processes and supporting applications, including the costs associated with making those changes and the benefits that are expected as a result of making those changes. Although there is often a qualitative aspect in the feasibility study, there is almost always a quantitative aspect that states, "These specific changes will cost XXX to build, YYY to maintain, and are anticipated to make a ZZZ impact on revenue."

Organizations don't always make changes to business processes to increase revenue or reduce costs. However, revenue and costs are nearly always the quantitative elements that are considered. For example, if an organization is enacting changes to processes and systems to comply with new regulations, management is still going to be interested in the cost and revenue impact that the changes will bring about.

A feasibility study is not always done as a part of a "*Shall* we comply with this new law?" but rather, "*How* shall we comply with this new law?" In other words, a feasibility study may be used to explore various options for complying with new regulations.

A feasibility study should seek to uncover every reasonable issue and risk that will be associated with the new system. The study should have the appearance and form of impartiality and should not reflect the biases and preferences of those who are taking part in the study or its outcome.

A feasibility study may also include or reference a formal business plan for the proposed new activity. A *business plan* is a formal document that describes the new business activity, its contribution and impact to the organization, the resources required to operate the activity, the benefits from operating the activity, and any risks associated with the activity.

 **NOTE** When the feasibility study has been completed, a formal management review should take place so that senior management fully understands the results and recommendations of the study and can determine whether (or how) the project should proceed or whether any changes to the plan should take place.

Performing the Feasibility Study: Imagination for Improvement A feasibility study is performed when management has decided that some new process is needed or some significant changes are needed in an existing process. Management has decided to initiate the process to develop a new system or update an existing system supporting the process change. Such a decision is made as a response to an event, which could be any of the following:

- **Changes in market conditions** For example, the entrance of a new competitor or the development of a new product or service feature by a competitor may spur management to respond by matching the competitor's capabilities. A competitor can also create a new market through innovation in products or services; this kind of a move sometimes incites an organization to make changes to maintain parity with the competitor. Or perhaps *your* organization is creating a new market through some groundbreaking innovation in the way that it does business or in what it delivers to its customers.

- **Changes in costs or expenses** Dramatic shifts in capital or expense costs may force an organization to make changes. For instance, higher fuel costs may prompt the organization to reduce field service calls, but doing this may require better remote diagnostic and self-healing capabilities. In the 1990s, for example, the shift to software development outsourcing required transformations in development methodologies that prompted organizations to make or buy better defect-management applications. And dropping telecommunications costs and higher bandwidth meant that online service providers began to ratchet up their offerings, most of which required enhancements to existing online service applications, and sometimes brand-new ones.

- **Changes in regulation**   The rise in dependence on technology has resulted in some negative events, which in turn has resulted in new legislation or changes in existing legislation. Relatively recent privacy regulations, including European Union General Data Protection Regulation (EU GDPR) and the California Consumer Privacy Act (CCPA), have up-ended industries that store and process personal information for marketing and other purposes. Other examples of relatively recent and updated regulations include Sarbanes-Oxley, GLBA (Gramm-Leach-Bliley Act), HIPAA (Health Insurance Portability and Accountability Act), FERC/NERC (Federal Energy Regulatory Commission/North American Electric Reliability Corporation) regulations, PCI DSS (Payment Card Industry Data Security Standard), and many others. Many of these regulations require organizations to implement additional safeguards, controls, recordkeeping, and data governance to business processes and information systems. Sometimes this results in an organization opting to discontinue the use of an older information system in favor of making or buying a newer application that can more effectively comply with applicable laws.

- **Changes in risk**   New types of vulnerabilities are discovered with regularity, and new threats are developed in response to vulnerabilities as well as changes in economic conditions and organizational business models. In other words, hackers find new ways to try and attack systems for profit within the growing cyber-criminal enterprises of the world. Applications that were considered safe just a few years ago are now known to be too vulnerable to operate. Reducing risk sometimes means making changes to application logic, and sometimes it requires that an application be discontinued altogether.

NOTE   Recent privacy and security regulations have highlighted the concept of *compliance risk*. Organizations increasingly need to track their compliance with applicable laws and regulations and determine the potential consequences for failing to do so.

- **Changes to business processes**   Privacy laws such as GDPR and CCPA have compelled organizations to change their business models and business processes. Often, this will require organizations to alter their business applications so that they continue to support those changed processes. For example, a change in the way that marketing reaches its customers requires changes in business processes and supporting applications.

- **Changes to legal agreements**   Changes in legal agreements between organizations can compel an organization to make changes to its software applications. There are several possible reasons for this, including changes in risks or regulations imposed upon customer or partner organizations.

- **Changes in customer requirements and expectations**   Changes such as those just discussed will often prompt customers and customer organizations to ask for new features or for changes in existing features in the products and services they buy. Often this requires changes in processes and applications to meet these customers' needs.

It is important to understand that *innovation* is also a valid and frequent reason that an organization chooses to make changes to a business process or software application. Generally, in this case, an organization has developed new features or methods in a business process together with its supporting software applications in an attempt to gain a competitive advantage.

NOTE Internal and external events prompt management to action by initiating changes in business processes, product designs, service models, and, frequently, the software applications that are used to support and manage them. What begins as an informal discussion turns to more formal actions and eventually to the initiation of a project to make changes.

Requirements Definition

Requirements describe the necessary characteristics of a new system or of changes being made to an existing system. They describe how the application should work as well as the technologies that it should support. The types of requirements used in software projects include the following:

- Business functional requirements
- Technical requirements and standards
- Privacy requirements
- Security and regulatory requirements
- Disaster recovery and business continuity requirements

Business Functional Requirements   Nearly every systems development project will include *functional requirements*. These statements describe the required characteristics that the system must have to support business needs. This includes both the way that the system accepts, processes, and produces information, and how users interact with the software in terms of technology, appearance, and user interface function.

Functional requirements should be a part of new systems acquisitions as well as modifications or updates to systems.

Example functional requirements include the following:

- Application supports payroll tax calculations for US federal, states, counties, and cities
- Application supports payment by credit card, electronic check, and virtual currency
- Application encrypts credit card numbers, social insurance numbers, and driver's license numbers in storage and when transmitted

Notice that the preceding examples do not specify *how* the system is to accomplish these things. Business requirements are interested in *what* the system does; the system architect or designer will determine *how* the system will support those requirements.

 **CAUTION** Many organizations are tempted to search for technical solutions prior to defining the business problem and how it should be solved.

In a few circumstances, new business requirements are not needed for a system modification. For example, if a software interface is being upgraded, an existing software program may need to be modified to work with the new interface. A change like this should be transparent to users, and the system should not differ in the way that it supports existing business requirements. So, in a way, it can be argued that business requirements apply even in this case: the system will still be required to adhere to existing business functional requirements.

 **NOTE** It is not unusual for a formal requirements document to span many hundreds of pages. This will be the case especially for larger and more complex systems such as customer relationship management (CRM), enterprise resource planning (ERP), manufacturing resource planning (MRP), or service management systems.

Technical Requirements and Standards To help the organization remain efficient, any new application or system should use the same basic technologies that are already in use (or that are planned on being used in the long term). The details related to maintaining the consistency that is required constitute the majority of technical requirements and standards.

An organization of any appreciable size should have formal technical standards in place. These standards are policy statements that cite the technologies, protocols, vendors, and services that make up the organization's core IT infrastructure. The purpose of standards is to increase technological consistency throughout the entire IT infrastructure, which helps to simplify the environment and reduce costs.

When an organization is considering the acquisition of a new system, the requirements for the new system should include the organization's IT and security standards. This will help the organization select a system that will have the lowest possible impact on capital and operational costs over its lifetime.

Besides IT standards, many additional technical requirements will define the desired new system. These requirements will describe several characteristics of the system, including

- How the system will accept, process, and output data
- Specific data layouts for interfaces to other systems
- Support of specific modules or tools that will supplement or support application functions (for example, the type of tax table that will be used in an invoicing or payroll system)
- Language support

- Specific middleware support
- Client platform support

 **NOTE** The entire body of technical requirements should accomplish two sweeping objectives: ensure that the new system will blend harmoniously with the existing environment, and ensure that the new system will operate as required at the technical level.

Privacy Requirements   In the broadest sense, privacy is about two distinctly different issues. First, privacy has to do with the *protection* of personal information so that it cannot be accessed by unauthorized parties. This aspect of privacy neatly falls into the umbrella of security: security requirements that require access controls or encryption of personal information can be developed. Regulations such as GDPR and CCPA have forced new paradigms upon organizations regarding the handling of personal information in many industry sectors. Organizations that were once free to do practically anything they wanted with personal information have found themselves constrained and also forced to be transparent about their personal information dealings.

Second, privacy is the prevention of *proliferation and misuse* of personal information. This has less to do with security and more to do with how the organization collects, treats, and uses personal information and whether it permits this information to be passed on to other organizations for their own purposes. In this regard, privacy is about business functionality that is specifically related to how the application collects and handles personal information and how to ensure that only the information needed to perform the intended business function is collected.

For example, if a system includes canned reports about customers that are sent to third parties, those reports should be configurable so that they can contain (or omit) certain fields. For instance, customers' dates of birth may be omitted from a report that is sent to a third-party organization to reduce the possibility of the third party using or abusing information to the detriment of individual customers. The rule, in this case, is this: You can't abuse or misuse the information you do not possess or cannot access. Indeed, regulations such as the EU GDPR require that organizations collect sensitive data only as required to perform services and that they retain the data for only as long as it is needed.

Privacy requirements are not always easy to discern from the language of privacy regulations; this has compelled many organizations to seek outside counsel for interpretation and guidance. In the absence of case law and other precedents, many organizations have adopted a "wait and see" attitude before making sweeping changes to business processes and supporting information systems.

Security Requirements   Security requirements must be developed to ensure that the new or updated system will contain appropriate controls and characteristics that will protect personal information and other sensitive information such as intellectual property and internal financial data.

Organizations should have an existing *security requirements document* that can be readily applied to any systems development or acquisition project. These requirements should describe the business and technical controls that address several security topics, including the following:

- **Authentication**   This broad category includes many specific requirements related to the manner in which system users authenticate onto the system. For systems that perform autonomous authentication, this will include all of the password quality requirements (minimum length, expiration, complexity, and so on), account lockout settings, password reset procedures, user account provisioning, and user ID standards. Authentication standards may also include requirements for machine and system accounts in support of automated functions in the application. For systems that use a network-based authentication service such as LDAP (Lightweight Directory Access Protocol), Kerberos, or a single-sign-on (SSO) solution, security requirements should describe how the application must interface with a network authentication service.

- **Authorization**   This category includes requirements related to the manner in which different users are granted access to different functions and data in the application. Authorization requirements may include the way in which roles are established, maintained, and audited. An organization may require that the application support a number of *roles*, which are templates that contain authorization details that can be applied to a user account.

- **Access control**   This category has to do with how the system is configured to permit access to users and/or roles. Unlike authorization, which is about assigning roles to users, access control is concerned with assigning access permissions to objects such as application functions and data stores. Depending upon the way in which a system is designed, permissions assignment may be user-centric, object-centric, role-based, or a combination of these.

- **Encryption**   Really another form of access control, encryption is used to hide data that, for whatever reason, may exist in "plain sight" in some contexts and yet must still be protected from those who do not have the authorization to access it. Encryption standards fall into two broad categories: data requiring encryption in certain settings and contexts and with certain encryption algorithms and key lengths, and key management to be handled in specific ways that permit the system to be operated similarly to other systems in the IT environment.

- **Data validation**   Systems should not blindly trust all input and output data to be properly formed and formatted. Instead, a system should perform validation checks against input and output data, whether a user types data into a system input form or the application receives the data via a batch feed from a trusted source. Data validation includes not only input data but also the results of intermediate calculations and output data. Requirements should also specify what the system should do when it encounters data that fails a validation check.

- **Audit logging**   This is the characteristic whereby the system creates an electronic record of events. These events include changing system or application configuration settings, adding and deleting users, changing user roles and permissions, resetting user login credentials, changing access control settings, and, of course, the actions and transactions that the system is designed to handle. Requirements about audit logging will be concerned with the configuration that is used to control the types of events that are written to an audit log, as well as the controls used to protect the audit log from tampering (which, if permitted, could enable someone to "erase their tracks").

- **Security operational requirements**   Management of passwords, encryption keys, event logs, patching, and other activities is required to maintain a system's confidentiality, integrity, and availability.

- **Misuse and abuse requirements**   This category needs to include the full range of use (and misuse) cases through which a user may—deliberately or not—misuse or abuse the system. This includes malicious input and other methods that may cause the system to malfunction, resulting in an escalation of privilege, exposure of—or tampering with—sensitive data, and exhaustion of system resources. The list of requirements should not merely match the capabilities of any automated or manual testing tools used by the organization.

Regulatory Requirements   Regulatory requirements must be identified and tabulated to ensure that the new or updated system will contain appropriate controls and characteristics that will facilitate the system's compliance with applicable regulations.

Identifying applicable laws and regulations is not an easy undertaking. Organizations need to understand which laws and regulations apply in which circumstances. Considerations include the following:

- Location of the organization's legal entity headquarters, as well as the locations of other parts of the organizations and what operations take place there

- Location(s) where data is stored

- Location(s) where data is processed

- Countries, states, and provinces where data subjects reside

- Industry sector

Taking all of these considerations into account, larger organizations find themselves within the jurisdiction of many countries, states, and provinces for different aspects of their business operations and various segments of their employee or customer base.

Organizations need to identify various classes of business requirements, which include

- **Information technology requirements**   For instance, some laws and regulations are specific about controls, such as encryption.

- **Business process requirements** Some regulations define requirements for business processes such as security incident response.

- **Organization requirements** For example, Article 37 of the GDPR defines requirements for the role of a data protection officer.

An organization's legal counsel typically tracks the laws and regulations that the organization is obligated to comply with. Often, legal counsel will rely upon subscription services that periodically issue bulletins announcing new regulations as well as articles on case law and other developments that help them better understand how to interpret regulations and transform them into business requirements.

Disaster Recovery and Business Continuity Requirements Systems that do—or may in the future—support critical business functions included in an organization's disaster recovery plans need to have certain characteristics. Depending upon recovery targets specified for the business process supported by the system, these requirements may include the ability for the system to run in the public cloud, on a server cluster, on a virtual machine, or in a load-balanced mode; to support data replication; to facilitate rapid recovery from backup tape or database redo logs; or to be installed on a cold recovery server without complicated, expensive, or time-consuming software licensing issues. Requirements could also dictate the ability for the system to be easily recovered from a server or virtual machine image on a storage area network (SAN), to operate correctly in a virtual server environment, and to operate correctly in a cloud environment such as AWS or Azure. A system may also be expected to work with a different brand or version of a database management system or to coexist with other systems, even though it may usually be configured to run on a server by itself.

Organizing and Reviewing Requirements In a systems acquisition or development project in which many individuals are contributing requirements, the project manager or another person should track each requirement back to a specific individual to enable that person to justify or explain those requirements if needed. When all requirements have been collected and categorized, the project manager should check with each contributor to make sure that each requirement is actually a *requirement* and not merely a "nice-to-have" feature. Perhaps each requirement can be weighted or ranked in order of importance. This will help, especially in a request for proposals (RFP) situation, where analysts need to evaluate suppliers' conformance to individual requirements. This helps project personnel determine which vendors are best able to meet the requirements that matter most.

 TIP The teams that develop requirements need to ensure that requirements are measurable and verifiable, because the requirements developed in this phase of the project should flow directly into user acceptance testing plans (for functional requirements) and system testing plans (for technical requirements).

Design

After all functional, technology, security, privacy, regulatory, and other requirements have been finalized, the design of the system can begin. It is assumed that a high-level design was developed in the feasibility study, since an elementary design is necessary to estimate costs to compute the financial viability of the system—but, if not, the high-level design should be developed first.

The design effort should be a top-down process, starting with the major components of the system and then decomposing each module into increasingly detailed pieces.

It is important that data flow diagrams (DFDs), an entity-relationship diagram (ERD), or some other high-level depiction of the system be developed first. Data privacy laws such as the GDPR and CCPA obligate organizations to know and determine where personal information is stored, how it is processed, and how and where it moves between systems and into and out of the organization. Design should start at a high level and graduate to levels of increasing complexity, to the point where database designers and developers have sufficient detail to begin development.

Project team members who represent business owners/operators/customers should review the application design to confirm that the analysts' and designers' concept of the application agrees with that of the business owners. Reviews should be done at each level of design, not just at the top level. Business experts should be able to read and understand both a high-level design and a detailed design and to confirm whether the design is appropriate or not.

Design reviews by privacy and security specialists help the organization understand whether the system's design will comply with policies and applicable regulations. For this reason, it is important that design documents specify not only the technology components and controls in the design, but also the physical locations where personal information will be stored and processed.

Design review by customers can be a step in the process where business customers and designers do not see eye-to-eye and where they may disagree on the design; any disagreements can be attributed either to differences in the understanding of technology or to practical versus abstract thinking. To end the design review prematurely could have costly consequences. The potential consequences of failing to come to an agreement on design are vividly depicted in the classic illustration shown in Figure 5-1.

Key activities in the system design phase include

- The use of a structured software design tool or methodology that records details of data flow and processing flow from high levels to detail levels
- Data flow diagrams that identify the locations of storage and processing, as well as third parties to whom data is sent to or received from
- Generalized and detailed database design at the logical and physical levels
- Storyboards showing user interaction with the system
- Details on reports that can be generated by the system

The system design effort should also include the development of test plans that will be used during the development and test phases of the project. Test plans need to be

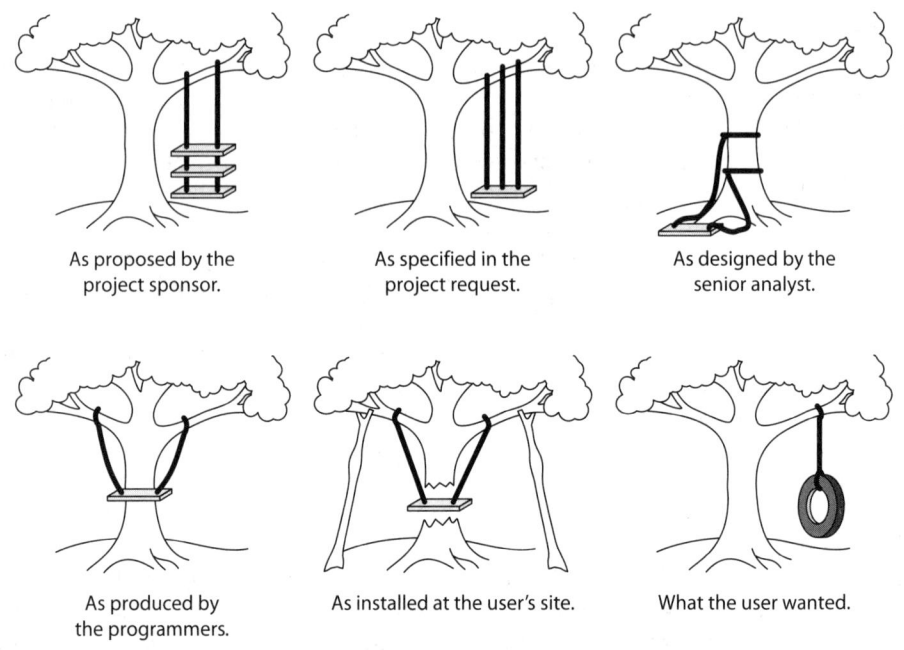

As proposed by the
project sponsor.

As specified in the
project request.

As designed by the
senior analyst.

As produced by
the programmers.

As installed at the user's site.

What the user wanted.

Figure 5-1   The potential consequences of failing to agree on design (Image courtesy of Oxford University Press, Inc.)

developed no later than the design phase, because developers will need to perform unit testing during development as a way of verifying that they have coded software modules properly (and they may need to consult test plan documents for confirmation that they are developing software correctly). If test plans are not developed until the test phase, developers will have to figure out tests on their own, or they might not perform enough testing, which will result in many more defects being discovered during the formal testing phase of the project.

When design reviews have concluded that the design is complete, a "design freeze" should be instituted, whereby no further changes to any level of design will be permitted. With a design freeze in effect, both designers and users are more inclined to think through all of the details of the design and do a better job of confirming whether the design is correct. An organization that does not institute a design freeze will find the design changing throughout the development phase, which will result in different parts of the system conforming to different "versions" of the ever-changing design. This will result in chaos during the development and testing phases and is sure to result in many more reported defects during user acceptance testing and after implementation. Management should strongly assert a design freeze, since changing the design during the development phase will drive up development costs when developers are forced to rework code that was written in conformance to earlier versions of the design.

NOTE Organizations that have internal IT auditors on staff should include them in design reviews so that they can confirm whether the system design will result in a system whose integrity can be confirmed through auditing. Organizations that incur external audits may want to invite external auditors to review the design documents for this same purpose.

Development

Developers and engineers take detailed design documents that were developed in the design phase and begin building the system. These activities are included in the development phase:

- **Coding the application**   Using tools selected for the project, developers will build the application code. Newer development tools may include design elements, code generators, debuggers, or testing tools that will make developers more productive.

- **Designing systems**   Engineers will design the systems upon which applications, database management systems, and other components will reside, so that they will have the necessary resources to function properly.

- **Developing program- and system-level documents**   During development, developers document technical details such as program logic, data flows, and interfaces. This aids other developers later on when modifications to the application are needed.

- **Developing user procedures**   As they develop user interfaces, developers can write the procedure documents and the help text that system users will read. In a more extensive, formal environment, developers may write the essential core of these documents, which will be completed by tech writers. But an even better idea may be this: end-user documentation is written by tech writers who derive procedures from requirements; then, the software developer and engineers will use technical requirements and the completed end-user documentation to guide them on the development of end-user systems.

- **Working with users**   As they develop the parts of the system that interface with users, developers will need to work with them to ensure that the forms, screens, and reports that they build will meet users' needs.

Development in a Software Acquisition Setting   In a software acquisition situation where an organization is purchasing or leasing software instead of developing it in-house, development activities may still be required. In a software acquisition project, software development is often needed to facilitate several needs:

- **Customizations**   Larger off-the-shelf applications make accommodations for customizations that must be developed. These customizations can take many forms, including application code modules, XML documents, and configurations.

- **Integration with other systems**   Applications rarely stand alone. Instead, they accept data from various sources and, in turn, provide data to other systems. Sometimes "bridge programs" or integration gateways need to be written that serve to move and transform data from one environment to another. All of these integrations should have been identified in the design phase.

- **Authentication**   In an effort to improve security or make system adoption easier, organizations often desire that new applications use a system- or network-based authentication service. The primary advantage of this approach is that users do not need to remember yet another user ID and password. An application's authentication can often be tied to LDAP or Microsoft Active Directory, or it can be part of a federated identity environment.

- **Reports**   Complex applications may have a report writer module that is used to create custom reports. Depending upon the underlying technology, a developer may be needed to develop these reports. Even if a report authoring tool is intuitive and easy to use, a developer may still be needed to help users design reports.

 **NOTE**   An organization that is considering acquiring software should develop and enforce policies regarding the extent to which customizations will be permitted. Customizations can be costly when off-the-shelf software upgrades take place, because they may need to be rewritten to work with the upgraded software. The cost savings of using off-the-shelf software can be negated by the additional time required to manage and upgrade customizations.

Source Code Management   In any size development effort, whether the development team is a single developer or 250 developers, an organization should use a source code repository tool. Such a tool has several purposes:

- **Protection**   A source code management tool often includes access controls so that only authorized personnel are permitted to access application source code. This helps to protect the organization's intellectual property and to prevent other persons from learning the secrets of the application's inner workings or performing unauthorized changes to source code, either of which could lead to fraud or misuse of the application later on.

- **Control**   A source code management system utilizes "check-out" and "check-in" functions so that only one developer at a time may work on a specific part of the application. This helps to ensure the integrity of the application's source code.

- **Version control**   A source code management system tracks each version of the code as it is checked in by developers. The system tracks the changes made from version to version and can show the differences in code between versions, and it also permits the reversion to an older version if application problems arise later on.

- **Recordkeeping** A source code management system maintains records related to check-outs, check-ins, and modifications to source code. This makes it possible for management to know what changes are being made to source code and who is making those changes.

Organizations that outsource some of their software development to third parties need to determine the business rules regarding those outsiders' access to source code. Some portions of a software application may be considered intellectual property or may constitute trade secrets. Further, there may be sections that are security related. In such cases, organizations should consider enacting and enforcing business rules that restrict outsourced developer access to these more sensitive portions of code.

 **NOTE** Source code management is not an activity that is limited to the period when the application is first developed; on the contrary, source code management is a vital activity that must continue throughout the lifespan of the application.

Testing

During the requirements, design, and even development phases of a software project, various project team members develop specific facts and behavioral characteristics (reflecting the requirements) about the application. Each of those characteristics must be verified before the application is approved for production use. This concept is depicted in a V-model in Figure 5-2. The V-model is sometimes used to depict the increasing levels of detail and complexity in the SDLC.

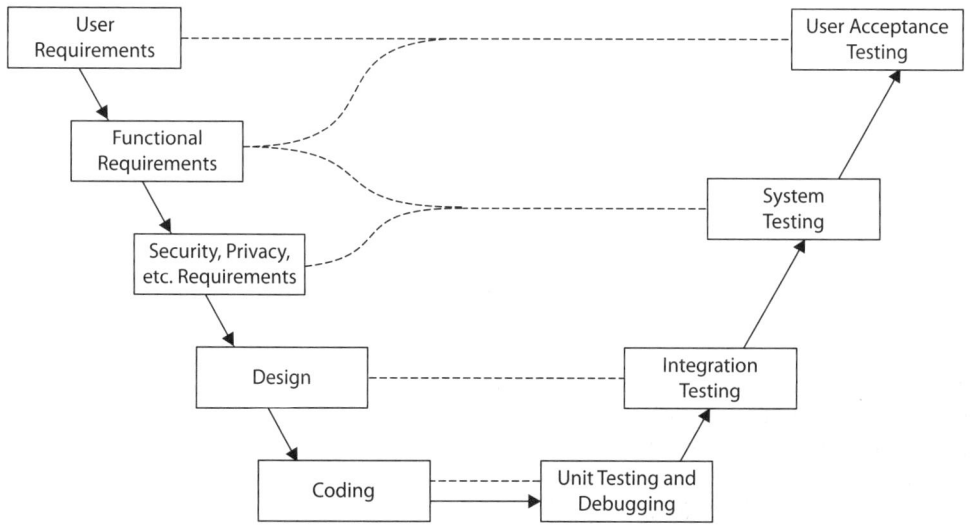

Figure 5-2 Requirements and design characteristics must all be agreed upon and verified through testing.

The stages of testing in a systems development project are unit testing, system testing, functional testing, user acceptance testing, and quality assurance testing. Each stage will be addressed in turn following a brief overview of test plans.

Test Plans   Before testing can take place, it is first necessary to create test plans. Testing, at the overall project level and at the detail level, should be a methodical and repeatable process, not subject to the skills and experience of any individuals who are performing tests.

To a great extent, test plans are going to be derived directly from requirements that were developed prior to development taking place. There may, however, be other sources or types of testing that may not be explicitly stated in requirements, including

- Adequacy of business use cases
- Resistance against misuse and abuse cases
- The degree to which a program's operation and functions are self-evident to the user

Because of the volume and/or complexity of test cases, it may be necessary to create test plans. Test plans may be developed for several reasons, including

- The volume of tests that need to be distributed to several individuals in some logical manner
- Testing performed by one or more outside parties or organizations
- Tests allocated based on the availability of individual testers or test teams
- Tests allocated based on the knowledge or skills of individual testers or test teams
- Tests allocated based on the tools required to perform testing (for example, workload testing or security defect testing)

Unit Testing   Unit testing is usually performed by developers during the coding phase of the software development project, or by engineers as they assemble and integrate various system components. When each developer or engineer is assigned the task of building a section of a system, they are given specifications that include test plans or test cases that will be used to verify that the system components work properly. This is true regardless of whether the part of the system that the developer is working on will be seen and used by end users or will be buried deep within the bowels of the system and never seen by anyone.

In a formal development environment, the unit test plans should be precise and list each test that the developer should undertake. The developer then performs each of the tests and records the results (usually the actual output) of the test. Those test results are then archived so that they can be referred to later if needed.

The archiving of unit testing records sometimes proves valuable when later phases of testing are taking place and some problem is found. Developers trying to isolate the cause of later testing problems can refer back to test plans and results at the unit testing phase to determine whether the test plans and other unit-testing activities were performed correctly, or whether they contained appropriate test cases. This evidence can save the project team a lot of time by eliminating the need to repeat unit testing.

Unit testing should be a part of the development of each module in the application. When a developer is assigned a programming task in a software development project, unit testing should be performed immediately after coding and debugging have taken place. In some organizations, developers work in pairs—the senior developer writes code, and the junior developer performs testing. This gives junior programmers an opportunity to learn more about advanced programming by observing the senior developer and by testing his or her code.

NOTE   It can be easily argued that unit testing for a software module should not be performed by the developer who wrote the module. The developer may be under time pressure to complete development and testing and may overlook test cases or gloss over errors as irrelevant. Also, a developer can be too familiar with his or her code to be capable of objectively testing it. The methodology of "written by one and tested by another" has the advantage of objective testing, but it can be more difficult to carry out in smaller organizations where only a single developer may be writing all of the code.

System Testing   As various parts of a system are developed and unit-tested, they will be installed into a test environment. When a sufficient number of modules or components has been completed, it will eventually become possible to begin end-to-end (or at least partial end-to-end) testing. In this way, it will be possible to test several components as a whole to verify whether they work together properly.

System testing includes *interface testing* to confirm that the system is communicating properly with other systems. This will include real-time interfaces as well as batch processing.

System testing also includes *migration testing*. When one system is replacing another, data from the old application is often imported into the new system to eliminate the need for both old and new systems to function at the same time. Migration testing ensures that data is being properly formatted and imported into the new system. This testing is often performed several times in advance of the real, live migration at cutover time.

As with unit testing, system testing should have pre-prepared test plans that were developed at the system design phase. And as with unit testing, system testing should probably not be performed by the developers and engineers who developed the modules under test or by the integrators who set them up in the test environment. Further, system testing results should be formally documented and archived in case they are needed later.

Functional Testing   *Functional testing* is primarily concerned with the verification of functional requirements that were developed earlier in the project.

Each functional requirement must be expressed in a way that makes it inherently verifiable. When each functional requirement is developed, one or more tests should also be developed, which are conducted during the functional testing phase of the project.

Functional tests should be formally recorded, including test input and test results. All of this should be archived in case it's needed if the application is suspected of malfunctioning. Often functional test results can verify whether the malfunction was present during the functional testing before the application went live.

User Acceptance Testing   Before business users will formally approve and begin using a new (or updated) system, *user acceptance testing* (UAT) often occurs. UAT should consist of a formal, written body of specific tests that permits system users to determine whether the system will operate properly.

The detailed output of user acceptance testing should be archived, as it may be needed in the future.

UAT is often a stage in the acceptance of a leased or purchased system, as well as in a system that is developed by a third-party organization. User acceptance testing determines whether the customer organization will accept (and pay for, as the case may be) the system and begin formal use of it.

NOTE   Acceptance criteria for UAT should be developed by end users and not by developers or designers; otherwise, internal or external customers are liable to end up with a system that does not function as desired or expected.

Quality Assurance Testing   *Quality assurance testing* (QAT) is a formal verification of system specifications and technologies. Users are usually not involved in QAT; instead, this testing is typically performed by IT or IS departments.

Like UAT, QAT should be a "gatekeeper" test in any situation where the organization is purchasing off-the-shelf systems, or the system is being developed by an external organization. The results of QAT should also determine whether the organization will formally accept and pay for the system.

Implementation

In the implementation phase of the project, the completed system is placed into the production environment and started. Implementation must be started before UAT and QAT begin. UAT and QAT should be performed on the environment that is anticipated to become the in-use production environment once approvals to use the system are obtained.

From the very day that construction of the implementation environment begins, that environment should be as controlled as a production environment. This means that all changes to the environment should go through a change management process. Also, administrative access to the production environment should be restricted to personnel who will be supporting the environment after it goes live. The implementation timeline, in relation to other phases of the software development project, is depicted in Figure 5-3.

NOTE   Because the production environment is the environment where UAT and QAT testing usually take place, this environment must be pristine and free from the possibility of being accessed by developers and other personnel. This reduces the likelihood of unauthorized changes to the system that could compromise its security or privacy.

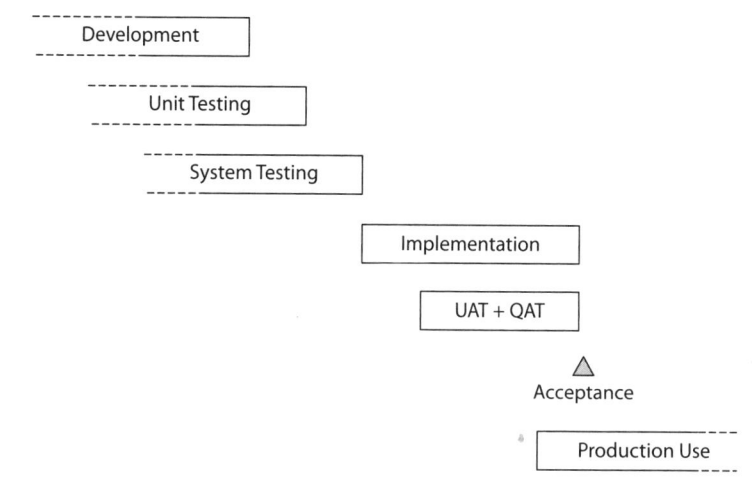

Figure 5-3   Implementation involves preparing the production environment prior to UAT and QAT.

Implementation Planning   Implementation is a complicated undertaking that requires advance planning. Some activities may be associated with a long lead time, requiring some implementation activities to begin during development or earlier.

- *Prepare physical space for on-premises production systems.* For organizations implementing a system on physical servers, an existing data center may be used for a system's servers and other equipment. But if there isn't room, or if an existing data center's available space is insufficient, the organization may need to consider expanding an existing data center or using a collocation center. More often, however, an organization will be implementing its system in a hosted cloud environment, where the service provider offers virtual machines onto which the organization will build and configure operating systems.

- *Build production systems.* The actual servers (whether physical or virtual) that the application will use must be built and configured. If the organization does not have the necessary servers available, the hardware systems must be leased or purchased; depending upon the type of hardware, considerable lead time may be required. If the public cloud will be used, the organization needs to select a public cloud vendor (this should be done at design time or earlier!) and implement server operating systems there. Once the hardware or virtualization platform is available, personnel will need to install and configure operating systems and possibly other subsystems such as database management systems or application management systems. Supporting infrastructure such as configuration management systems, monitoring systems, routers, switches, firewalls, and so forth (whether physical or virtual) will also be implemented at this time.

- *Install application software.* Once the systems are ready for the application software, it can be installed and configured.

- *Migrate the data.* For environments in which an existing system will be retired, data from the former environment usually needs to be transferred to the new environment. Often this procedure requires the development of one or more custom programs to extract, convert, and insert the data into the new environment. This procedure is usually performed more than once: it must be rehearsed at least one time to make sure that it works properly. Also, migrated data is often needed for functional testing, UAT, and training before the actual cutover.

 **NOTE** As each phase of implementation is completed, the newly completed component should be locked down immediately and treated as though it is already in production. Usually, this is the only way to ensure the integrity of the entire environment.

Training　The success of the entire development or acquisition project hinges on the knowledge and skills of several different people in the organization. The following are among those who may need training:

- **End users**　Personnel who will be using the system need to be trained so that they will know how to operate it properly.

- **Customers**　If outside customers will be using the new system, they will need an appropriate amount of information so that they will understand how to use it. In other cases, customers will not be using the system directly, but a new system can still influence how they interact with the organization. If customer service or sales personnel are using a new system for taking orders or for looking up customer data, they may be asking different questions or presenting different information to the customer.

- **Support staff**　Personnel who provide customer service to users and customers need to be trained in the workings of the system, as well as on administrative "back office" tools that they may use to assist users.

- **Trainers**　Organizations that employ a training organization will need to "train the trainers" so that, in turn, they will be able to train users and customers correctly.

The purpose of a system may require that others also receive training. This could include internal or external auditors or regulators who have oversight over the organization.

Data Migration　In the context of the SDLC, the purpose of a data migration is to transfer data from an older, soon-to-be-retired system to a new system. Depending upon the nature of the old and new systems, the purpose of the data migration may be to make historical records that originated in the older system available in the newer system. From a privacy perspective, it is critical that the organization keep track of data movement and ensure that any temporary copies of data are indeed temporary and deleted once no longer needed.

In some cases, an organization will continue to keep the older system running to facilitate access to historical data. In some circumstances, it may require fewer resources to keep the old system running than to migrate the historical data to the new system.

Data migration often requires the development of programs that extract data from the old system, perform required transformations, and then format the data and import it into the new application. This is frequently a complex task, as there may be differences so significant between the data models of the old and new systems that the *meaning* of stored data differs between them. In some cases, it will be necessary to create some parts of the database in the new system by extracting data from the old system and then performing calculations to create the data necessary in the new. Careful analysis is required in all cases to make sure that the *meaning* of data in each system is known so that the migration will be done properly.

Following are some techniques and considerations that ensure a successful migration:

- **Record counts** Programs or utilities should be used to count the number of records in counterpart tables in the old and new environments. This will confirm the completeness of the migration programs that move data from the old environment to the new one.

- **Batch totals** Data records with numeric values can be added together in the old and new databases. This will help to confirm the integrity of key data elements in the old and new environments.

- **Checksums** Programs that compute checksums can be run against old and new databases to ensure the accuracy of migrated data. Developers do need to be aware of the methods used to store data, which could lead to differences in checksums. For instance, an address field in one system may pad the field with spaces, but in the other, it may be padded with nulls. Also, the way that dates are stored can vary between systems. While using checksums can be valuable, developers and analysts must be familiar with any differences in data representation between the old and new environments.

NOTE Like other software projects, the migration programs themselves must be carefully designed and tested, and the results of tests must be analyzed to make sure that they are working properly. Often it is necessary to perform a test migration—well in advance of the scheduled cutover date—to give enough time to make sure that the migration programs have been properly written.

Executing the Cutover When the production system has been constructed, applications loaded, data migrated, and all testing performed and verified, the project team has reached the cutover milestone. Often, management review and approval are required to verify that all necessary steps have been completed correctly.

Depending upon the nature of the system as well as external influences such as regulation or business requirements, an organization may transfer processing to the new environment in one of several ways:

- **Parallel cutover** The organization may operate both the old and new systems in parallel for a time, making careful comparisons between the two to ensure that the new system is working properly.

- **Geographic cutover**   In an application used throughout large geographic regions, such as a retail point-of-sale system, the organization may migrate individual locations to the new system instead of moving all locations at once.

- **Module-by-module cutover**   The organization may migrate different parts of the system at different times. In a financial management system, for instance, the organization could move accounts receivable to the new environment, later move accounts payable, and still later move general ledger. During and between each of these phases, the organization must keep track of exactly which business information resides in which system.

- **All-at-once cutover**   An organization may elect to migrate the entire environment at one time.

The project team must analyze all available methods for a cutover and choose the method that will balance risk, efficiency, and cost-effectiveness.

Analysts may discover problems in data in the old environment that necessitate a cleanup be performed prior to the migration or as a part of the migration. Examples of the types of problems that can be found include duplicate records, incomplete records, or records that contain values that violate one or more business rules. Analysts who discover data inconsistencies such as these need to alert the project team to the matter and then help the project team decide how to remedy the situation.

Rollback Planning   Sometimes an organization will migrate a system from an old environment to a new one, and shortly afterward will discover a serious problem in the new environment that requires a return to the old environment. *Rollback planning* is a safety net that provides a last-resort path away from a situation where the organization cannot continue using the new environment.

A rollback is a serious undertaking and would be considered only when a problem in the new environment is so serious that it cannot be easily remedied. Rollback planning is recommended in environments where the availability and integrity of a system are critical to the organization, even if a rollback is never needed.

   **NOTE**   To ensure success of the project and that the lifeline will function if needed, rollbacks should also be designed and tested in advance.

Post-implementation

The software project is not completed when the system cutover has taken place. Several other activities still must take place before the project is closed. This section describes these final tasks.

Post-implementation Review   After the implementation of a new system, one or more formal reviews must take place. The purpose of these reviews is to collect all known open issues and to identify and discuss the performance of the project. Because the organization is likely to undertake similar projects in the future, it is valuable to identify what parts of the project went well and what could have been done better. The implementation review should consider the following:

- **System adequacy**   The project team should work with the users of the new system and collect issues and comments, which are then discussed in the implementation review. Any issues requiring further attention should be identified.

- **Privacy review**   The system's privacy features and controls need to be discussed and any problems identified. Any temporary copies of data should be purged once stability of the new environment is confirmed.

- **Security review**   The system's access controls and other security controls should be discussed and any issues or problems identified.

- **Audit review**   The system's ability to be audited, as well as any early audit results, should be discussed.

- **Issues**   All known problems regarding the new environment should be identified, including user feedback, operations feedback, and the accuracy and completeness of documentation and records. The project team needs to discuss each issue and assign it to one or more individuals who will address and remedy it.

- **Return on investment**   If the purpose of implementing the application was to establish or improve return on investment (ROI) or efficiency, then initial measurements need to be taken. The project team needs to recognize that several business cycles may be required before an accurate ROI can be determined.

More than one post-implementation review may be needed. To hold a single post-implementation review shortly after going live and then calling it good is probably inadequate for most organizations. Instead, a series of reviews may be needed, perhaps stretching over years.

NOTE   IS auditors should be involved in every phase of the SDLC, including post-implementation reviews, to ensure that the system is functioning according to whatever control or regulatory requirements are attended to by auditors. Auditor feedback must be included in the body of issues and comments that is reviewed in the initial and subsequent reviews.

Software Maintenance   Immediately after implementation, the system enters the maintenance phase. From this point forward, all changes to the environment must be performed under formal processes, including incident management, problem management, defect management, vulnerability management, change management, and configuration management. All of these processes should have been developed and modified as necessary to accommodate the new application when the cutover was completed.

Software Development Risks

Software development is not a risk-free endeavor. Even when management provides adequate resources for a software development project and supports a viable methodology, there are still many more paths to failure than to success.

Several specific risks are associated with software development projects:

- **System inadequacy**   The system may fail to support all business requirements. During the requirements and specifications phases of a software development project, some business requirements may have been overlooked, disregarded, misunderstood, or unappreciated. For whatever reason, if a system falls short of meeting all business requirements, the system may be underutilized or even abandoned.

- **Security and privacy defects**   The system may contain security or privacy defects that permit various forms of misuse and abuse, including denial of service, escalation of privilege, data disclosure, and data corruption.

- **Project risk**   If the system development (or acquisition) project is not well run, the project may exceed spending budgets, time budgets, or both. This may result in significant delays and even abandonment of the project altogether if management considers the project a failure.

- **Business inefficiency**   The system may fail to meet business efficiency expectations. In other words, the system itself may be difficult to use, it may be exceedingly slow, or business procedures may require additional manual work to meet business needs. This can result in critical business tasks taking too long or requiring additional resources to complete.

- **Regulatory changes**   In the time between the development of requirements (reflecting regulations at that time) and the implementation of a system, new regulations may have been enacted that may require significant design changes to the system. In the modern world of extraterritorial regulations, this can include laws passed outside of the city, state, province, or country where the organization is located.

- **Market changes**   Between the time that a software development project is approved and when it is completed, sudden or unexpected changes in market conditions can spell disaster for the project. For instance, drastic supply or price shocks in a macro environment can have an adverse effect on costs that may make a new business activity no longer viable. Changes in the market can also result in reduced margins on products and services, which can turn the ROI of a project upside down.

 **NOTE**   Management is responsible for the business decisions that it makes; in ideal situations, management makes these decisions with sufficient information at hand. Usually, however, there are some unknowns.

Alternative Software Development Approaches and Techniques

For decades, the waterfall approach to software development was used by most organizations. Breakthroughs and changes in technology and practices in the 1970s and 1980s have led to new approaches in software development that can be every bit as effective as the waterfall model and, in many cases, more efficient and faster.

DevOps

DevOps is the growing movement that utilizes agile development methodology coupled with tighter integration of development teams, software QA, and IT operations. DevOps isn't complete without tools facilitating more effective (often automated) testing.

In DevOps, the lines between software development, QA, and IT operations are somewhat blurred. It is essential for organizations to ensure that access control models and capabilities continue to support regulatory and compliance requirements such as

- **Data segregation**   Developers should never have access to production data.
- **Separation of duties**   Critical processes such as change control still require administrative and technical controls so that no one person (such as a developer) can make unauthorized changes in production environments.

The relationship between development, software QA, and IT operations is depicted in Figure 5-4.

DevSecOps

DevSecOps is an offshoot of (others would say an improvement to) DevOps. DevSecOps represents the best of DevOps and includes security design and testing capabilities that are a part of the rapid development and automated testing process. Often, static and/or dynamic code-scanning capabilities are integrated into the software build environment so that security defects are identified as soon as possible. Further automated testing can

Figure 5-4
DevOps is the integration of development, software QA (testing), and IT operations.

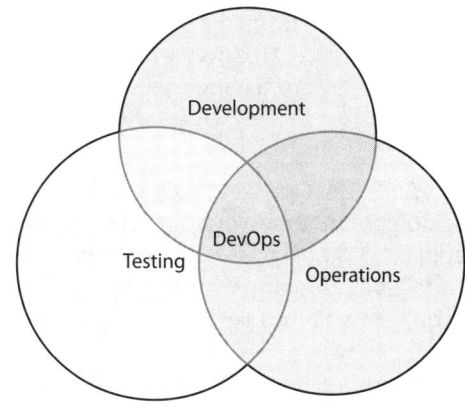

be performed on production environments to reveal exploitable defects that can be remediated by developers in subsequent sprints.

Continuous Integration/Continuous Deployment

Continuous integration/continuous deployment (usually expressed as CI/CD) combines the concept of continuous integration (where developers' code is integrated into a single shared mainline several times each day) and continuous deployment (where software updates are delivered iteratively through automated deployments). The result is similar to DevOps, where software applications are undergoing a near-constant stream of small updates, versus the traditional waterfall methodology, where there may be just a few updates per year.

Agile Development

The agile software development model is referred to as an alternative methodology that is appropriate for some organizations. The agile methodology utilizes the Scrum project methodology. In an agile development project, a larger development team is broken up into smaller teams of five to nine developers and a leader, and the project deliverables are divided into smaller pieces that can each be attained in just a few weeks.

EXAM TIP   Agile is one of several techniques that were developed as alternatives to the traditional waterfall methodology that was considered too slow and that deprived organizations of agility.

Web-based Application Development

The creation of the HTML content-display standard and the HTTP communications protocol has revolutionized application development. The web browser is ubiquitous and has become the universal client platform and is not unlike the intelligent display terminals or client-server architectures from earlier eras.

The Web, as it is popularized now, came along just in time: two-tier and three-tier client-server computing, the great new application development paradigm that was developed in the 1990s, was not living up to its promise, particularly in the areas of network performance and upkeep of client software. Web software has dramatically simplified software development from the perspective of the user interface (UI); though the developer has a little less control over what and how data will be displayed on a user workstation, the trade-off in not having to maintain client-side software is viewed as an acceptable compromise.

NOTE   Agile, DevOps, DevSecOps, and CI/CD can all be considered rapid development methodologies where small, incremental changes are made to applications on a near-continuous basis, as opposed to the traditional waterfall methodology, where significant updates are made to applications only one or two times per year.

From a development methodology perspective, web application development can be performed within virtually all of the development frameworks, including waterfall, DevOps, agile, RAD (rapid application development), DOSD (data-oriented software development), CI/CD, and OO (object-oriented), some which were discussed in this chapter. Primarily, it's the target technology that differentiates web-based application development from its alternatives.

Important standards have been developed that facilitate communications between web-based applications, including JSON-RPC, SOAP (Simple Object Access Protocol), and Web Services Description Language (WSDL). JSON-RPC is an XML (eXtensible Markup Language)–based protocol coded in JavaScript Object Notation (JSON) and used by a client system to request a method from a remote system.

SOAP is an XML-based application programming interface (API) specification that facilitates real-time communications between applications using the HTTP and HTTPS protocols. Functionally, SOAP operates similarly to RPC, wherein one application transmits a query to another application, and the other application responds with a query result. SOAP messages are based on the XML standard.

WSDL serves as a specification repository for the SOAP services available in a particular environment. This enables an application to discover what services are available on an application server.

System Development Tools

Application developers can create source code using tools ranging from simple text editors to advanced tools such as computer-aided software engineering (CASE) and fourth-generation languages (4GLs). While there's little reason to discuss text editors such as vi, Notepad, or Emacs, advanced development tools are worthy of attention and are discussed in this section.

Integrated Development Environment

Integrated development environments (IDEs) are a class of desktop software development tools that incorporates source code editing, source code version control, compilation, and debugging in a single tool. An IDE enables a developer to write, test, and debug code without having to switch between programs.

IDEs typically have multiple windows, or panes, that enable the software developer to view and edit code, run code and observe execution, and view the source code library. Other functions may be available as well.

Some IDEs offer connectivity to external tools such as source code scanning tools that look for security and quality defects.

Computer-Aided Software Engineering

CASE represents a broad variety of tools used to automate various aspects of application software development. CASE tools cover three basic realms of development:

- **Upper CASE**   This includes activities ranging from requirements gathering to the development of data models, data flow diagrams, and interfaces.
- **Middle CASE**   This involves the development of detailed designs, including screen layouts, report definitions, data design, and data flows.
- **Lower CASE**   This involves the creation of program source code and data schemas.

These terms are loosely used to classify various CASE tools. Some CASE tools are strictly Upper CASE, while others include Middle CASE and/or Lower CASE, but many cover the entire range of functionality and can be used to capture specifications, create data structure and flow diagrams, define program functions, and generate source code.

CASE tools do not usually create source code that is ready for implementation and testing. Instead, they are used to create the majority (in the best cases) of code for a given program; then, the developer(s) would add details and specific items that the CASE tool did not cover. CASE tools are not used to replace the work of a developer, but they help make the coding part of a development project take less time, improve consistency, and enhance program quality.

CASE tools often contain *code generators* that create the actual program source code.

 **NOTE** CASE tools do not eliminate the need for any of the essential phases of the SDLC. With or without CASE tools, it is still necessary for a project team to create requirements, specifications, and design. CASE does help to automate some of these activities, however.

Acquiring Cloud-based Infrastructure and Applications

Organizations often choose to acquire business applications that are hosted in a cloud or SaaS environment, as opposed to hosting the applications on their own systems. This section discusses issues that organizations should understand when considering this option.

The common options available for cloud-based application environments are

- **Software as a Service**   In SaaS, an application service provider is hosting its application software on its own infrastructure, often located in a data center and used by several customers. Users access the application in much the same way that they would if the application were hosted within the organization's own IT environment.

- **Infrastructure as a Service**   In IaaS, a cloud service provider is providing an environment in which its customers build and operate virtual machines. While the client organization is relieved of the burden of purchasing network, system, and storage hardware, it still needs to create a network architecture, security architecture, systems architecture, and application architecture, and it must install and manage operating systems, virtual network devices such as switches and routers, and virtual security tools such as firewalls, intrusion prevention systems, and data loss prevention systems.

- **Platform as a Service**   In PaaS, a cloud service provider is providing an application-based or data-based platform on which customers can develop and/or integrate their applications. PaaS services are typically organized around a business theme—for example, Salesforce Lightning for sales enablement or SAP Concur for expense and travel management. Other examples include AWS Elastic Beanstalk and Azure App Service for deploying and scaling web applications, and Heroku for the development and deployment of applications.

Regardless of the cloud model that is chosen, the organization needs to understand many details that are related to the manner in which the cloud provider provides its services to the organization, such as the following:

- **Access control**   The cloud service provider must have an effective access control plan to ensure that only authorized personnel have access to infrastructure components and virtual machines. Often, the organization using cloud services will manage access control in upper layers (such as in operating systems, database management systems, and applications that it may install and maintain on cloud servers), while the cloud provider will manage access control in lower levels (such as in virtual machine hypervisors and via physical access).

- **Environment segregation**   The cloud service provider must effectively separate systems and data between customers so that no cloud customer is able to access the systems and/or data of other customers.

- **Physical security**   The cloud services provider must provide adequate physical security so that only authorized personnel will have physical access to all cloud environment infrastructure and facilities.

- **Regulation**   The cloud service provider must provide controls that will meet applicable regulatory needs for its customers. However, customers must also identify applicable regulations it must implement in a cloud environment. For example, an organization that manages end user profiles in a SaaS environment is accountable for its uses of that user profile information and must disclose those uses in its privacy policies.

- **Privacy**   The cloud service provider (and, indeed, the customer organization using cloud services) must implement safeguards to ensure appropriate protection and handling of personally identifiable information (PII) stored in cloud environments.

- **Legal jurisdiction**   The cloud service provider and its customers must have a firm understanding regarding the physical location of stored data, relative to the location of the data subjects. This will enable legal counsel to understand the applicability of security and privacy laws governing the use of stored data. This is particularly important in the context of data privacy and data sovereignty laws, some of which are extraterritorial.

- **Availability**   The cloud services provider must deliver the availability of services to customers at a level to meet customers' expectations. This applies not only to the steady availability of services, but also to on-demand availability.

- **Audit**   Many standards, regulations, and legal agreements require some level of auditing of systems, applications, and supporting controls. The cloud environment must be verifiable in this regard.

The "Cloud Responsibility Models" sidebar in Chapter 4 shows a typical cloud responsibility model that illustrates which party is responsible for implementing and operating which aspects of security in a cloud environment.

NOTE The Cloud Security Alliance (https://cloudsecurityalliance.org) is a high-quality resource for controls and guidance for cloud services providers as well as organizations using cloud-based services.

Applications and Software Hardening

The concept of resilience is not always considered in the design, development, and management of software applications. Resilience, however, is an essential characteristic of any software program that is to run in the real world, with the potential for users who may use the program correctly or incorrectly or attackers who may attempt to trick the application into performing in ways not intended by its designers.

CAUTION Hardening is not an activity that should be applied to a system once it has been implemented, but as a part of its initial design.

Hardening at the application layer is not altogether different from hardening at the network or system layer. The concept is the same: hardening refers to techniques intended to make systems more resistant to misuse and attack or, to put it another way, to prevent unintended consequences resulting from unexpected stimulus.

NOTE Security professionals need to pay close attention to hardening in the entire technology stack. Refer to the "System Hardening" section in Chapter 4.

Application Hardening Principles

Application layer hardening addresses several perspectives about software applications, including the following:

- **Input validation** Software programs must carefully examine all input prior to acting upon it. Input attacks are among the most popular because of their ease of implementation, and also because so many software programs do an inadequate job of examining input data. The types of attacks that may occur because of inadequate input checking include

  - **Buffer overflow** Input data deliberately exceeds the program's input storage capacity and attempts to overwrite the program's instructions with code of the attacker's choice, usually to take over control of the program.

  - **Injection attacks** Input data is crafted in such a way that it results in arbitrary commands sent to the backend database management system (a SQL injection attack) or to an end user's browser (a JavaScript injection attack).

- **Denial-of-service attack**   Input data is crafted in such a way that it results in the malfunction of the target program or system, causing it to cease operations.

- **Data corruption**   A program that does not perform boundary and type checking on input data may malfunction or attempt to process invalid data.

- **Authentication and session management**   Software applications need to have robust authentication functions to resist authentication bypass attacks. For logged in users, software applications must have layers of defense against attempts to break into other users' sessions.

- **Temporary files**   As a part of routine operation, programs may create temporary files that may contain sensitive data. An attacker may be able to access a temporary file easily and steal sensitive data or alter the contents of a temporary file, altering the processing of data.

- **Logging and monitoring**   Software applications need to generate security events that are sent to a central log repository or SIEM. Events such as logins (successful or not), creation of new privileged accounts, changes in configuration settings, creation of new customers (in SaaS environments), and others should result in event messages sent to a log management and alerting system. Someone should be monitoring these logs and have detailed instructions to follow when alerts are generated.

- **Verbose error messages**   Software application developers need to be careful about the creation of error messages so that they are not too verbose. For instance, a credit card payment function should not be sending the full details of a failed transaction to the display or to an event log; instead, sensitive data such as names and credit card numbers should be truncated or not included at all. Also, in user interfaces, error messages should not reveal too much about the inner workings of software programs and the databases behind them. Error messages that are too verbose could give an attacker vital information that would help him conduct an attack on the application.

- **Improper encryption**   Application developers should not use outdated encryption algorithms, poor key management (such as easily found keys and hard-coded encryption keys in apps), and unsalted hashed data, and they should avoid improper uses of cryptography. Also, web sites should use HTTP instead of HTTPS, to avoid permitting sensitive information to be transmitted without encryption.

- **Components with known vulnerabilities**   Software applications often use components that originate in other organizations. Examples include open source code libraries, unsupported operating systems (as well as subsystems and network devices), insecure and unsafe browser extensions, and outside SaaS services with unpatched vulnerabilities.

- **Cross-site scripting and cross-site request forgery**   These two common web application vulnerabilities can give attackers control of sessions and inject malware into users' computers.

> **NOTE** The Open Web Application Security Project (OWASP), at https://owasp.org, is an excellent resource for web developers and cybersecurity professionals' concerns about web application security.

Testing Applications

To ensure that applications are free of exploitable vulnerabilities that could result in the compromise of personal information or the malfunction of applications, various tools and techniques for testing applications need to be regularly used. Application hardening testing is used to confirm that an application is reasonably free of exploitable vulnerabilities that would permit an attacker to compromise the application. Functional testing—that is, confirming that design requirements were properly implemented—is separate from hardening testing.

Following are types of application hardening testing that are available:

- **Code reviews** This is a manual review of application source code that has been updated recently. Generally, when a developer is given a task to make a change to an application's source code module(s), another developer will examine the code to confirm that the code changes have not introduced any new vulnerabilities. Code reviews are not performed solely for the purpose of vulnerability analysis; other reasons include compliance with organization coding policies.

- **Static code scans** Static application security testing (SAST) is an automated scan of source code that intends to determine whether there are any exploitable defects in the application's source code. Code scans can be performed in at least three ways:

  - In the IDE, as a developer is working on source code

  - On demand, as directed by someone who is running a source-code scanning tool

  - As a part of a daily build cycle in the build system

- **Dynamic code scans** Dynamic application security testing (DAST) is an automated scan of a running application. A dynamic code scanning program "runs" the application as though it were a human user, clicking on pages, filling in forms, and seeing the results of stimulus such as filling in forms and manipulating cookies. Dynamic code scans are typically run on demand on all or a part of an application, as well as being integrated into an application's build environment where all or parts of an application can be scanned automatically, usually overnight.

- **Manual tests** Various tools can be used to test individual web forms, fields, and cookies, similar to the work performed by dynamic code scanners, but more on a "one page at a time" or even "one field at a time" basis. Many such tools exist for manually testing various types of exploits and various components of applications. Pen tests use these tools to go beyond the capabilities of automated scanning tools.

- **Penetration testing**   This involves the use of an external party to conduct a broad range of tests against a target application. External parties are often used because most organizations lack this expertise internally; also, external parties are generally more objective and unafraid to conduct tests and produce test results without the being concerned with internal politics.

- **Bug bounty**   In this arrangement, an organization will solicit outside security researchers to perform security tests on a target system. Researchers will be compensated for identifying exploitable vulnerabilities, with higher rates paid for more serious vulnerabilities. There are several "bug bounty as a service" organizations that organize the effort into a security testing "marketplace," bringing together organizations that need testing and security researchers who want to earn additional income.

APIs and Services

In addition to human-interactive systems are legions of software applications and tools that perform automated tasks without any human interaction. Whether these tools and applications handle regularly scheduled batch jobs or run on demand, they work quietly on systems, doing whatever processing, data transfer, or gateway services are needed.

An API, or application programming interface, is an automated computer interface developed to perform some dedicated function. An API is generally always running on a computer and listening for incoming messages or transactions that it can process in some predetermined way. An API is a part of a software program that is designed to accept input data through a communications interface; then it processes that data somehow, perhaps storing it, performing calculations, converting it to a different format, or performing some other task.

From a privacy and security perspective, these backend systems are every bit as important as human-interactive systems, and they store and process potentially vast amounts of data, including personal information. Here are some examples of these types of systems:

- **Bulk file transfer**   Data feeds from one organization to another, copied with the old standby, File Transfer Protocol (FTP), or hopefully one of the secure versions of FTP—FTPS or SFTP

- **Month/quarter/year-end financial processing**   Automated programs that perform financial period closes and other tasks

- **Call-outs to other software platforms**   Mobile apps that integrate with social media programs, mapping applications, messaging applications, and more

- **Backend calculation services**   Commercial applications that perform tasks such as calculating sales tax and shipping costs, and location services

- **Banking services**   Commercial applications that process funds transfers and payments

- **Payment services**   Commercial applications that accept credit cards and other forms of payment on behalf of organizations that sell products and services

All of these and more utilize interfaces of various kinds to communicate with other systems on demand, in bulk, or both. The majority of these types of services utilize the global Internet for communicating with other organizations' and users' systems, although some of the highly sensitive services between companies are still transmitted over private, dedicated communications channels.

Many different technologies are used to operate these services, including FTP and related batch file transfer protocols, web services, and somewhat older interfaces such as CORBA and SOAP.

Contrary to what many believe, plenty of security- and privacy-related issues that are associated with the use of these services need to be addressed. Among them are the following:

- **Application, interface, and system security**   These interfaces are backed by applications and run on operating systems over networks, and they are subject by many of the same kinds of attacks that human-interactive systems put up with. Secure development processes, monitoring, and testing are required. A skilled hacker can inflict just as much damage hacking an API as she can hacking a human-interactive application. Both are windows into backend data that may prove valuable on the black market. Many of the same types of tools are used to test software security in APIs and in human-interactive systems.

- **Privacy**   These interfaces are often a part of other processing ecosystems. Organizations using outside services need to understand how the organizations running their services protect and use personal information.

- **Data sovereignty**   Organizations using other parties' interfaces for processing of personal information need to understand the physical locations where this processing occurs so that organizations can comply with privacy laws that impose cross-border data transfers.

Online Tracking and Behavioral Profiling

The World Wide Web and mobile applications are the engines of commerce in many industries. Today, measuring business is all about measuring what happens in information systems. In their zeal for insight, some of these measurements intrude into persons' privacy. This section describes a variety of techniques used to track individual users' activities, as well as ways in which tracking can be limited.

NOTE   The 2020–21 COVID-19 pandemic, with its lockdowns and work-from-home (WFH) shift, has accelerated the transformation of many organizations into digital businesses, resulting in the proliferation of usage tracking data.

Advertising Tracking

An old joke in the advertising business goes like this: "Did you hear about the marketer who could not sleep at night? He was worried that he was wasting half of his advertising budget, but he didn't know which half." Cue rim shot.

In the traditional advertising world, when companies purchased ad space on billboards, buses, airports, and radio and television commercials, there was no direct way of knowing whether their ads were influential, never mind which individual persons were responding to those ads. The Internet and the World Wide Web have changed all of that. With ads served to individuals on their laptops, tablets, and smartphones, it is now fairly simple to distinguish individuals from one another, deliver data-driven targeted advertising, and know the outcomes with more certainty based on shared tracking data.

This leap in technology has resulted in citizens saying, "Enough!", resulting in laws to curb this tracking and its uses and potential abuses. Indeed, the existence of this book, and your interest in it, is a result of tracking, plus the accumulation and abuse of personal information that has gone too far in many cases.

Tracking Techniques and Technologies

Numerous techniques and technologies are used to track the activities and locations of Internet-connected devices and their owners. Information systems log various types of events that give system owners better insight into how, how much, and by whom their systems are used. Some of this logging is highly detailed and often includes, directly or indirectly, the identities of the persons using these devices; this is may be considered unnecessary and can represent an invasion of privacy.

IP Addresses

Every endpoint—smartphone, tablet, laptop, or desktop computer, or connected device such as home surveillance camera, voice assistant, printer, and more—all have an IP address. Discussed briefly in Chapter 3, an IP address is a unique numeric value assigned to a network and the devices within it.

Most web sites, as well as many mobile applications, log basic activities such as authentication and meaningful transactions. Because IP addresses on the public Internet are unique and provide approximate geographical location information (sometimes no better than an entire country), IP addresses are often a part of these log entries.

Techniques such as network address translation mean that a public IP address will represent an organization's network or a residential network, but not the individual devices within that network. This means that separate individuals on their own devices visiting the same web site will have the same public IP address associated with them. To the uninformed, these could appear as a single user, unless log entries include some other uniquely identifying information.

 **NOTE** In some jurisdictions, an IP address is considered an element of PII.

Device Identifiers

Individual devices such as laptop computers, tablet computers, and smartphones have internal device identifiers that uniquely identify them. Device serial numbers are stamped on these devices and are also available electronically. Also, mobile phones have an IMEI (International Mobile Equipment Identity) number that mobile network service providers use to uniquely identify devices throughout the world. Some of the activity tracking performed by mobile network operators and Internet service providers include identifiers like these. Often, these identifiers can be associated with their owners, giving network operators unique insight into the detailed usage of their devices.

Web Tracking

Web tracking refers to general practices associated with measuring and observing users who visit web sites. Web site operators track individual user sessions on their web sites for three primary reasons:

- **Session integrity**   The nature and design of Internet protocols and multi-user applications require that each user's session be uniquely identified. This is necessary to distinguish each user from every other. For instance, e-commerce sites need to identify individual user sessions properly and uniquely, so that each user is able to browse through and purchase products and services. This session integrity also gives each user a relative feeling of privacy, knowing that no other user is able to know what products they are viewing and purchasing.

- **Usage statistics**   Web sites and applications want to accumulate analytics regarding their use: how many users are visiting (and at what times of the day, days of the week, and so on), what pages are they viewing in what sequence, and how long are they remaining on individual pages. This information helps organizations design web sites and applications that are easier and simpler to use.

- **Advertising tracking**   Advertising and its revenue fuels a significant portion of the Internet. Thus, advertisers and web site operators track not only the numbers of visitors, but they track and uniquely identify visitors to distinguish one from another. The technologies in play here give rise to the "creepy factor"—for example, when a person does an Internet search for a specific thing, and then for days afterward, on every page he visits, he sees advertisements from various companies for those very things.

Cookies   Cookies are small pieces of data that web sites create and store in a user's browser. Generally, a cookie is used by the web site to uniquely distinguish users from one another, and also to remember unique users' preferences such as preferred language and display or usage settings. Several types of cookies are used for various purposes, including

- **Session cookies**   These are used to uniquely identify a user's session. A session cookie is assigned when a user logs in to a web site and is removed when the user logs off or closes the browser.

- **Persistent cookie**   These cookies remain on a user's browser and are sometimes used to remember users' preferences such as preferred language, country, and preferred landing page. Persistent cookies are also known as advertising cookies because they are used to distinguish users from one another.

- **First-party cookies**   These cookies are placed by the domain the user is visiting and identified with that domain. For instance, if a user is visiting www.company .com, a first-party cookie will be associated with the domain company.com.

- **Third-party cookies**   These cookies are placed by the domain the user is visiting, where the cookie is associated with a different domain. For example, if a user is visiting www.company.com, that web server could attempt to place a cookie from www.cooltrackinging.com for advertising purposes.

- **Super cookies**   These are cookies with an origin of a top-level domain such as .com or .co.uk. They are used for tracking users across many domains.

- **Flash cookies**   The once-popular Adobe Flash program has a feature, Flash Local stored object, that functions much like a cookie.

- **Zombie cookies**   These cookies are created by various means and designed to be difficult or impossible to detect or delete; they regenerate themselves when possible.

- **HTML5 Web storage**   The now popular HTML5 standard includes specifications for local storage of information. One such use mimics the function of cookies.

Web Beacons   A *web beacon* is a technique used by web servers to track the viewing of web pages and e-mail messages. Web beacons generally take the form of a 1×1 pixel image that is essentially invisible to users. Because web servers log details of the downloading of every object, including images, web beacons can function much like cookies and can enable collection of information such as whether the recipient has opened an e-mail and whether it was opened by others (presumably after being forwarded). This is particularly true if a web site utilizes uniquely named web beacon image files that are each sent only to a specific user.

Location   On devices with GPS or similar device location capabilities, web servers may request specific device location via the user's web browser. In modern operating systems, web sites are not permitted to obtain a device's location unless the user specifically consents to it. These consents are generally persistent, and it may be difficult for a user to view the web sites for which she has approved for sending location information.

Device Information and Name   When a user visits a web site, the site's web servers are able to obtain a limited amount of information about the user's device, including

- **Device name**   For a personally owned device, usually the name that the user assigned to the device when she purchased it and set it up

- **Browser**   The name and version of the browser used to visit the site

- **Operating system**   The name and version of the operating system

- **Viewport width**   The width of the device's display in pixels

Location

Virtually all mobile devices, as well as many laptops, have built-in GPS receivers and, thus, make precise locations available to applications. Many mobile apps utilize location information directly for navigation but also often for business reasons. For instance, many web sites and e-commerce apps from organizations with "brick-and-mortar" stores tell users where the closest stores are located. Other uses for location information support the delivery of location-relevant advertising to users.

With location tracking enabled, some of these apps and web sites accumulate a detailed location history for users. This becomes a point of contention with citizens who believe that this constitutes overreach—perhaps this is information could be used against them in some way. Many also believe that the manufacturers of mobile devices likewise accumulate detailed location history that includes an excessive amount of personal information that could lead to misuse or abuse.

Eavesdropping

A growing concern among privacy and security professionals is the increase in consumer devices and mobile apps that eavesdrop on their users in various ways. Many mobile apps access sensitive data on users' mobile devices, often with no good reason, and many "smart" consumer devices collect more information from us than may seem reasonable. For instance, there is no reason for a camera or photo-editing app to access a user's contact list; when installing such an app, users often "click through" the permission dialogue without thinking about what they're being asked to permit.

Similarly, some apps are able to sneak around a mobile device's controls to obtain such information anyway. The operators of mobile app stores (primarily Apple and Google) do a pretty good job of preventing illicit eavesdropping, but some app developers are clever and find ways around the controls.

Some consumer devices are designed to eavesdrop on their customers, with only vague clues in the standard terms and conditions that indicate what is really going on. For instance, one brand of smart TVs advises that customers should not have conversations on sensitive topics in the presence of the television!

Applications can access and abuse users' privacy and security in the following ways:

- **Camera**   Many apps for both mobile devices and laptops request access to the device's camera. For "honest" applications, it's evident when the camera is being used, but apps could access the camera at other times as well. On many laptop computers, a small indicator light illuminates when the camera is in use, but not all mobile devices have this feature. (Personally, I am suspicious about whether new "smart" televisions have built-in cameras in their bezels.)

- **Photos**   Many apps request access to stored photos on mobile devices. Although this is a legitimate need for photo-editing apps and social media apps for purposes of posting photos or updating profile pictures, users should pay attention to permission requests to access stored photos.

- **Microphone**   Apps that need to record a user's voice or other sounds will need to request access to a device's microphone. This includes videoconferencing and audio calling applications. Others, such as health apps (for observing sleep, for example) may also request mic access. Remember that any voice-activated product has a built-in microphone that is essentially listening all the time (likely even when the product is switched off).

- **Location**   Some applications need to know the location of the device in order to be useful to users. While mapping, navigation, and travel-related applications obviously require location services, others may not.

- **Contacts**   Mobile device users will sometimes be asked if certain applications are allowed to access stored or cloud-based contact lists. Users should be especially careful with this permission, as unscrupulous vendors' applications may harvest others' contact information for marketing purposes.

- **Voice assistant**   Many mobile devices and laptop computers are equipped with "Siri," "Alexa," and "Hey Google" voice assistants. Users need to be aware of whether these voice assistants, when activated, are listening and potentially uploading all speaking that takes place within range of the device.

- **Local and cloud storage**   Mobile device users should be wary of applications that request permission to access local and cloud-based storage. Miscreant apps may attempt to exfiltrate those contents for who knows what purposes.

- **Social media accounts**   Many mobile apps provide "value add" services as an adjunct to popular social media services such as Facebook, Twitter, LinkedIn, and Instagram. Those apps request permission to log in to users' social media accounts to provide their services. Sometimes this access is misused or abused, with more personal information being sent to these other services than most users would consider reasonable.

- **Paste buffer (clipboard)**   The paste buffer (known as the clipboard) on Apple mobile devices can, at times, contain highly sensitive information such as passwords, URLs, e-mail addresses, and phone numbers. Apps in some mobile devices have free access to the paste buffer on mobile devices, resulting in leakage of sensitive information.

Is Big Tech the New Big Brother?

Companies like Apple, Google, Microsoft, Samsung, Vizio, and scores of others have designed numerous high-tech products that are revolutionary in the ways in which they make our work lives and personal lives easier and richer. But in recent years, we're learning of some of the practices that may represent overreach in their capabilities. Here are some examples:

- It was revealed that Google Nest thermostats are equipped with secret microphones.

- Samsung suggests that the owners of their smart televisions should take sensitive conversations away from the television lest voice-recognition software hear what's said.

(continued)

- Several brands of children's smart toys, including toys from Mattel and Genesis Toys, eavesdrop on children and their families' conversations—by design.

- Google keeps a detailed dossier about the specific movements of users who use Android devices or who use Google Maps and other navigation apps.

- Employees of Amazon's Ring doorbell and ADT's security systems have been caught eavesdropping on customers' homes.

- And, finally, voice assistants hear everything.

Consumers are growing wary of big tech and the potential for overreach. Privacy professionals have long been concerned. The pendulum of acceptable tracking versus intrusions into privacy continues to swing, even as new technologies reveal even more about the lives of the people using them.

Facial Recognition

Combined with advancements in mobile device and CCTV optical capabilities, facial recognition software is going mainstream. Many products from Apple, Microsoft, and others utilize facial recognition for logging in to mobile devices. Commercial facial recognition products are also enabling corporations and law enforcement to recognize people, such as wanted criminals.

Privacy rights advocates are rigorously opposing facial recognition capabilities in public places such as airports, shopping malls, and city streets out of concerns that it could be abused by authorities aspiring to create a surveillance state. Some cities, states, provinces, and countries are passing laws forbidding the use of facial recognition capabilities in public places, and some larger technology organizations are refusing to sell these capabilities to governments. Public-setting facial recognition cannot be "uninvented," however, and it is likely to continue to be used secretly by corporations and governments despite regulations forbidding its use.

Biometrics

Aside from facial recognition, other forms of biometrics have been in use for years and even decades. Numerous companies manufacture fingerprint and palm scan readers for use on mobile devices, as well as for building and secure zone entrance control. Iris scanning is also fairly common, as high-resolution cameras can obtain a quality image from a few feet away.

Static signature recognition, which is the task of verifying whether a signed document is genuine, has been used for centuries and is still used in banking to confirm signatures on checks. More intrusive biometric techniques such as voice recognition, retinal scan, and dynamic handwriting scanning are no longer in common use.

Figure 5-5
Contact tracing is built into Apple mobile devices. (Source: author)

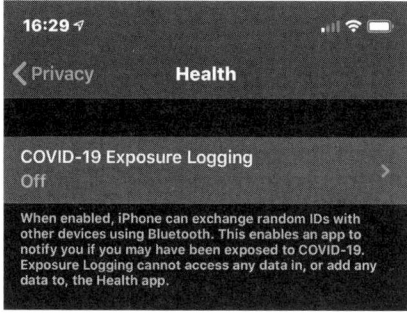

Contact Tracing

Contact tracing has been used for disease control for decades. Historically, contact tracing has been a manual process consisting of interviews with confirmed case patients to learn about their recent contacts with other individuals. Being highly manual, it has not been the most efficient tool to assist in reducing the spread of disease. However, the proliferation of smartphones that can provide proximity information may prove useful for contact tracing. As a result of the COVID-19 pandemic, Apple and Google introduced support for "COVID-19 apps" that could use their smartphone's Bluetooth radio signals to notify a user who comes into close proximity with someone who is also using a "COVID-19 app" and has recorded in the app as testing positive for an infectious disease.

While health authorities view contract tracing as a valuable tool for infectious disease control, privacy advocates consider it an overly intrusive process that is subject to abuse by police states. Indeed, contact tracing can be used to discover associations between people who meet only face-to-face. Critics of contract tracing point to numerous "false positives" that would result. For instance, two hotel guests sleeping in adjacent hotel rooms might be identified as being in close proximity for several hours.

In mid-2020, Apple and Google included contract tracing as a standard feature of the iOS and Android operating systems; this feature is not activated by default and must be explicitly turned on (see Figure 5-5).

Tracking in the Workplace

Organizations that conduct part or all of their business through the use of computers need to enact a variety of controls to reduce the likelihood and impact of attacks. Indeed, this is the whole point of cybersecurity, as well as a substantial portion of information privacy. Some of the controls associated with cybersecurity involve the management and logging of activities on computing devices (laptops, desktops, tablets, and smartphones) used by its workers. After all, anomalous behavior on any of these devices may be signs of an attack. To detect and prevent such attacks, it is necessary to track and centrally log many types of activities on computers, including the following:

- Web sites visited
- Files created, viewed, updated, transferred to other media, and deleted
- E-mail messages sent and received

- Contents of network communications
- Location of said devices (for device theft detection and remote data destruction)

Organizations with more mature cybersecurity programs will track and log most or all of these activities and use analytics of these records to detect anomalies that may be indications of security or privacy breaches.

Organizations undertaking such tracking and recording often include notices on these devices and in company policy, stating that such measures are taken in the name of data protection, and that any personal use of these devices (or networks) is subject to these practices, resulting in "no expectation of privacy."

 **NOTE** In many countries, security tracking must be approved by Works Councils and similar employment bodies, even if the organization claims legitimate interest for the protection of its information.

Internet Access History

Web content filtering is used to prevent users from visiting web sites known to be malicious, which when visited will attempt to install malware or spyware on visitors' computers, and it's used to prevent users from visiting web sites whose subject matter is not business related (such as weapons, gambling, and pornography sites). Many such web content filtering systems log the web sites and web pages viewed by organization personnel, often associating web activity with specific workers by name. This log data can prove invaluable in a security or privacy breach investigation. Cloud access security brokers (CASBs) are implemented to prevent the use of unauthorized cloud services. Logging capabilities in these systems is frowned upon or even illegal in some countries.

SSL Decryption

To detect and prevent the leakage of the company's sensitive information, some organizations undertake a practice known as *SSL decryption*. Encrypted network traffic is decrypted so that the contents of network traffic can be examined for evidence of a security or privacy breach. Because so much Internet traffic is encrypted, organizations lacking SSL decryption are blind to many types of threats.

Legal problems with SSL decryption arise when workers occasionally use organization-issued computers to conduct personal business, such as accessing personal e-mail, making personal purchases, accessing healthcare services, conducting personal banking, and so on. Though a small number of organizations prohibit and actively block all such personal uses, most permit a minimal amount of personal use and will make an effort not to decrypt traffic from sites believed to be low business risk that could transmit personal information; however, they warn workers that all activities, whether business or personal, are monitored for security and privacy purposes.

Some organizations "whitelist" the use of personal banking and other, similar, activities so that an employee's personal use of organization-issued computers is not examined in some cases. Indeed, personal banking is an unlikely path for exfiltrating sensitive data and represents a low risk for security and privacy breaches.

E-mail Archiving

Internal e-mail represents an ongoing conversation in most organizations. For this reason, many organizations continuously archive all e-mail communication on separate e-mail archive servers. If the organization receives a legal request for specific e-mail messages, search capabilities on e-mail archive servers streamline the data collection effort. Any personal uses of organization e-mail accounts are naturally going to be included in such archiving. Again, employees are generally cautioned through visible notices that monitoring is taking place.

Tracking Prevention

Users of mobile devices and smart products have limited abilities to prevent tracking and eavesdropping. Various tracking prevention remedies are discussed in this section.

Cookie Opt-Out

Visitors to web sites are often informed of the use of cookies for tracking their preferences. Users are free to decline the use of cookies. Sometimes this will mean that a user's preferences won't be remembered between visits; if the cookie opt-out includes the use of session cookies, however, a visitor may be unable to conduct transactions with the organization from their browser. Although many browsers provide a function to remove all cookies, some browsers permit users to remove individual cookies.

Cookie Blocking

Most browsers always permit users to reject third-party cookies. This will sometimes break the normal function of some web sites, depending upon their architecture. Some browsers permit users to permit and/or block cookies by specific domain, which gives them more granular control over cookie-based web tracking.

Cookie Removal

Web browsers on mobile devices and laptop computers permit users to remove all cookies from their browser. This will result in all logged in sessions being effectively logged out, and any web site preferences such as preferred language or postal code will be removed.

Do Not Track

The Do Not Track web browser setting can be used to disable web server tracking for a user. Note that Do Not Track is a request that lacks specific controls for enforcement; web site operators must voluntarily implement features that result in the user's visits not being tracked. Do Not Track has not been widely adopted by the industry, in part because of the lack of legal mandates for its use. Do Not Track is the web version of the US Do Not Call legislation that was enacted in 2003 as a result of the scourge of annoying telemarketing calls.

Privacy Mode Browsing

Many browsers have a privacy mode, sometimes called incognito mode, in which the tracking of web site visits is not included in browsing history. This may be useful on shared computers if a user does not want other users to know about their browsing history.

PART II

Many people are unaware of the fact that privacy mode browsing does not diminish or affect the full logging that web content filters, cloud access security brokers, and web sites themselves perform. Indeed, web sites make no distinction regarding privacy mode browsing in their activity logs.

Tor Browsers

Users who don't want their locations to be tracked online can use a Tor browser, which employs network routing through the Tor network. The Tor network is designed to conceal the IP address and, thus, the physical location of the device using the Tor browser. Tor browsers also do not retain cookies or browsing history. Use of the Tor network is limited to the Tor browser. Users who want to anonymize their IP addresses for other programs turn to the use of private VPN services.

 **NOTE** it is generally believed that the details of many of the "exit nodes" of the Tor network have been identified by government law enforcement or intelligence agencies, resulting in Tor use not being as anonymous as it once was.

Private VPN Services

Persons who are concerned with the protection of their network traffic or who want to anonymize their IP address can use a private VPN service. These services, available on mobile devices as well as laptop and desktop computers, enables users to "hide" behind a relatively anonymous IP address, which will help to conceal their location.

VPN services do not anonymize a user's web browser. For instance, if an e-mail user logs in to his webmail service and then activates a VPN, his webmail session will probably continue uninterrupted, since the user's identity is asserted through session cookies. That said, web sites with better security regimens may alert users or even block access if they are seen to be logging in from faraway countries. In a similar vein, my online banking app on my smartphone blocks VPN function because the GPS location and the VPN IP address location contradict each other.

Faraday Bags

Users of mobile phones and other small devices can purchase Faraday bags, which are small pouches that include a metallic material that blocks RF signals. Placing a mobile device into a Faraday bag essentially causes it to "disappear" from cellular, Wi-Fi, and Bluetooth networks. Figure 5-6 shows a mobile device Faraday bag.

Mobile device Faraday bags can also be used for building access cards to prevent them from being cloned by attackers. Smaller versions of Faraday bags are made for key fobs used to lock, unlock, and remotely start automobiles. Because of the relatively poor security associated with key fobs, people concerned with automobile theft utilize these bags. Figure 5-6 shows a key fob Faraday bag. A disadvantage of a Faraday bag is that the mobile device is unavailable for any use while inside the bag.

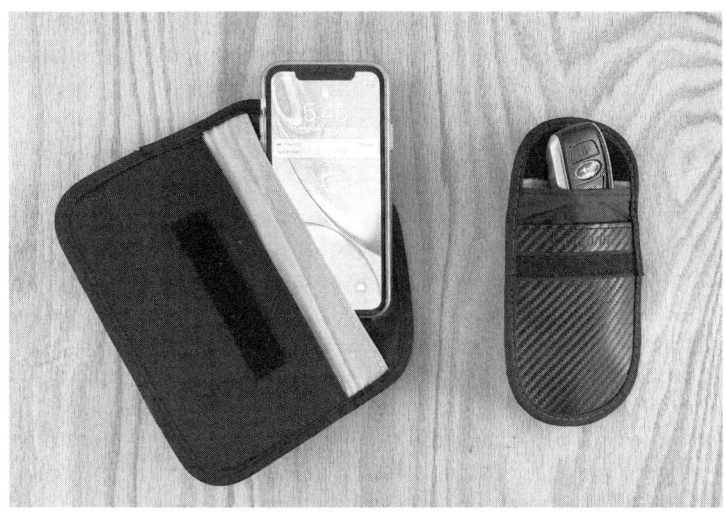

Figure 5-6 Faraday bags protect mobile devices and key fobs from eavesdropping and tracking. (Source: author)

Chapter Review

The practice of privacy and security by design will result in systems and applications that are more likely to comply with internal policy as well as evolving privacy and security laws and regulations.

The systems development life cycle (SDLC) describes the end-to-end process for developing and maintaining information systems. The phases of the traditional waterfall SDLC are feasibility study, requirements definition, design, development, testing, implementation, and post-implementation.

The feasibility study is an intellectual effort that seeks to determine whether a specific change or set of changes in business processes and underlying systems is practical to undertake. A feasibility study is an analysis of proposed changes to business processes and supporting applications, including the costs associated with making those changes and the benefits that are expected as a result of those changes.

Requirements describe the necessary characteristics of a new system or of changes being made to an existing system. They describe how the application should work as well as the technologies that it should support. In addition to developing functional requirements, organizations need to develop security and privacy requirements to ensure that information systems will continue to comply with applicable regulations.

While a conceptual design may have been developed during the feasibility study, the design phase requires more detail, enough that engineers and developers can build the entire system.

Data flow diagrams (DFDs) and entity-relationship diagrams (ERDs) should be included in system design documentation so that security and privacy professionals can ensure the protection and proper use of personal information.

Development takes place when requirements and designs have been completed and approved. Development takes on many forms, from software development to software integration and infrastructure implementation.

Test plans are directly derived from requirements. Formal testing is performed to ensure that all requirements are met in the new or updated system. Tests are performed in layers, including unit testing, integration testing, system testing, functional testing, and user acceptance testing.

Implementation is a complicated undertaking that requires advance planning. Some activities may be associated with long lead times, requiring some implementation activities to begin during development or earlier.

Training is needed for personnel who will be using, operating, monitoring, and supporting new systems and applications. Training content needs to be recorded so that personnel who join the organization after implementation can receive the same training as those who were present for initial training.

In the context of the SDLC, the purpose of a data migration is to transfer data from an older, soon-to-be-retired system to a new system. Depending upon the nature of the old and new systems, the purpose of the data migration may be to make historical records that originated in the older system available in the newer system.

When the production system has been constructed, applications loaded, data migrated, and all testing performed and verified, the project team has reached the cutover milestone. Often, management review and approval are required to verify that all necessary steps have been completed correctly.

A rollback is a serious undertaking and would be considered only when there is a problem in the new environment that is so serious that it cannot be easily remedied. Rollback planning is recommended in environments where the availability and integrity of a system is critical to the organization, even if a rollback is never needed.

Some time after implementation, a post-implementation review should be conducted for a new system so that the project and its expected benefits can be discussed.

Software development is not a risk-free endeavor. Even when management provides adequate resources for a software development project and supports a viable methodology, there are still many more paths to failure than to success.

The waterfall SDLC model contains all of the elements found in other development methodologies, although the structure, sequence, and roles of other methodologies vary.

DevOps is the growing movement that utilizes the agile development methodology coupled with tighter integration of development teams, software QA, and IT operations. DevOps isn't complete without tools facilitating more effective (often automated) testing.

Continuous integration/continuous deployment (CI/CD) combines the concept of continuous integration (where developers' code is integrated into a single shared mainline several times each day), and continuous deployment (where software updates are delivered iteratively through automated deployments).

Integrated development environments (IDEs) are a class of desktop software development tools that incorporates source code editing, source code version control, compilation, and debugging in a single tool. An IDE enables a developer to write, test, and debug code without having to switch between programs.

Organizations often choose to acquire a business application that is hosted in a cloud or SaaS environment, as opposed to hosting the application on their own systems. Their choices include Software as a Service (SaaS), Platform as a Service (PaaS), and Infrastructure as a Service (IaaS).

Cloud responsibility models are depictions of the distribution of security- and privacy-related responsibilities in a cloud environment between the cloud service provider and the customer.

Hardening at the application layer is not altogether different from hardening at the network or system layer. The concept is the same: hardening refers to techniques intended to make systems more resistant to misuse and attack—or, to put it another way, to prevent unintended consequences resulting from unexpected stimulus.

To ensure that applications are free of exploitable vulnerabilities that could result in the compromise of personal information or the malfunction of applications, various tools and techniques for testing applications need to be regularly used. The types of testing include code reviews, static and dynamic code scanning, manual tests, penetration testing, and bug bounties.

An API, or application programming interface, is an automated computer interface developed to perform some dedicated function. An API is generally always running on a computer and listening for incoming messages or transactions that it can then process in some predetermined way.

Tracking schemes are established so that web site, API, and application providers can observe and measure their systems and to enrich the web site or application experience for its users. Some of the logging and other techniques may represent overreach and an invasion of privacy.

Tracking identifiers include IP addresses, device identifiers, HTML cookies, web beacons, location, and the ability for some apps to covertly obtain other information on devices, such as the names of a user's contacts and the contents of data on the user's device.

Facial recognition and biometrics may have some useful purposes, although there is growing concern that these capabilities will be misused to the detriment of private citizens. Several prominent organizations and governments have sought to ban its use.

Contact tracing represents an attempt to track the spread of infectious diseases. However, contact tracing data can also be misused for political and nefarious purposes.

Organizations use internal tracking tools and techniques to reduce the probability and impact of attackers and malware. Tools include web content filters, cloud access security brokers, e-mail archiving, and SSL decryption.

End users can limit some tracking through several techniques, such as opting out of cookies, blocking and removing cookies, and setting the Do Not Track flag in their browsers.

Quick Review

- The *systems* development life cycle (SDLC) is a new way of expressing the traditional *software* development life cycle to emphasize the shift from pure software development to two separate activities: the design and management of IT infrastructure, and the acquisition of off-the-shelf and SaaS-based business applications.

- Data flow diagrams (DFDs) and entity-relationship diagrams (ERDs) should be included in system design documentation so that security and privacy professionals can ensure the protection and proper use of personal information.

- DevSecOps represents the best of DevOps and includes security design and testing capabilities that are a part of the rapid development and automated testing process.

- Many newer IDEs are extensible; modules for security code scanning and other features can be integrated to IDEs so that these other tools can run within the IDEs.

- Tracking techniques in the workplace are generally in place to protect the organization from attackers and malware. However, in many locations, Works Councils and other bodies must approve tracking.

- Privacy mode browsing does not diminish or affect the full logging that web content filters, cloud access security brokers, and web sites themselves perform. Indeed, web sites make no distinction regarding privacy mode browsing in their activity logs.

Questions

1. An organization wants to fast-track the development of a consumer social media product and skip the requirements definition. What is the likely privacy-related consequence of this?

   A. Users will have to be re-registered.

   B. Audit logs will need to be scrubbed of PII.

   C. Rework will be necessary to comply with privacy laws.

   D. Personal information will need to be encrypted in storage.

2. A US state has enacted a sweeping new extraterritorial privacy regulation that focuses on cross-state border data transfer. What is the first step that an online social media vendor should take?

   A. Retain expert privacy counsel to opine on applicability and interpretation.

   B. Wait until the law takes effect to begin making changes to systems.

   C. Wait until there is sufficient case law to see whether the law is enforceable.

   D. Direct developers to make changes to the system to comply with the new law.

3. An organization will be introducing voice-command smart TVs into the enterprise network. What is the primary risk associated with the introduction of such devices?

    A. Data leakage

    B. Unencrypted network traffic revealing PII

    C. Many smart devices cannot be patched

    D. Eavesdropping on private conversations

4. How frequently should an organization revise its security and privacy standards?

    A. Annually and whenever significant new laws have been enacted

    B. Quarterly and whenever significant new laws have been enacted

    C. Annually

    D. Quarterly

5. A privacy officer wants to better understand where personal information appears in a system. Which design element should the privacy officer examine?

    A. Physical network diagram

    B. Logical network diagram

    C. Entity-relationship diagram

    D. Data flow diagram

6. A privacy officer wants to better understand where personal information appears in a system—in particular, which individual personal information elements (such as date of birth, address) exist. What design element should the privacy officer examine?

    A. Physical network diagram

    B. Logical network diagram

    C. Entity-relationship diagram

    D. Data flow diagram

7. At which stage of the life cycle of a software application is source code management no longer necessary?

    A. After the application is designed

    B. After the application is retired

    C. After initial implementation

    D. After formal requirements definition

8. At what point in the software development life cycle can detailed test plans be created?

    A. After coding has been completed

    B. After design has been finalized

    C. After requirements have been finalized

    D. After tests have been completed

9. Who performs unit testing, and what is its purpose?

    A. End users perform unit testing to confirm module functionality.

    B. Developers perform unit testing to confirm module functionality.

    C. Management performs unit testing to confirm developer competency.

    D. End users perform unit testing to confirm screen-object placement.

10. Which of the following statements is true about data migration programs?

    A. Data migration programs are used for cross-border data transfers.

    B. Data migration programs are provided by software vendors.

    C. Data migration programs become a permanent part of the new system.

    D. Data migration programs transfer information from an old system to a new system.

11. An organization is replacing an internally developed, on-premises ERP application with a SaaS application. What must the organization do to make legacy data available on the SaaS platform?

    A. Migrate data from the SaaS platform to the legacy platform.

    B. Import the data into the new application.

    C. Write a migration program.

    D. Develop a data flow diagram.

12. An organization has determined that its waterfall SDLC does not provide sufficient agility for the organization to respond to rapidly changing market forces. What steps should the organization take?

    A. Migrate to a SaaS application.

    B. Move to a DevSecOps development model.

    C. Migrate to a PaaS platform.

    D. Migrate to an object-oriented system.

13. To improve software quality, an organization wants to incorporate code scanning into the process so that developers will get immediate feedback during development. What tooling should be used to fulfill this purpose?

    A. Code scanning built into the IDE

    B. Code scanning built into the build system

    C. Code scanning built into the check-in process

    D. Code scanning performed quarterly by an outside firm

14. A software development manager is developing a policy and a set of principles that will result in better software hardening. Which organization should the software development manager use as the best source for software hardening information?

    A. DISA

    B. SANS

    C. EFF

    D. OWASP

15. Which of the following do web applications use to manage and distinguish users from one another?

    A. Session cookies

    B. Persistent cookies

    C. Flash local storage

    D. Web beacons

Answers

    1. C. When the requirements phase of a project is skipped because of a "fast track" mandate, there is a high likelihood that considerable rework will be required to bring the product or service into compliance with privacy requirements (and applicable laws), security requirements, technical standards, and functional requirements. From a privacy perspective, it will be necessary to assess every aspect of the system, as well as supporting processes, to identify compliance gaps. The rework would be expensive.

    2. A. When a new privacy regulation is passed, particularly one that is extraterritorial, the best first step for an organization to take is to obtain qualified expert legal counsel to help the organization better understand the applicability of the new regulation and to interpret the regulation so that the organization can understand what changes to processes and systems are necessary to comply.

    3. D. Many voice-activated consumer devices listen to virtually every spoken word in their vicinity, and often this audio stream is transmitted to the product vendor's central servers for processing and interpretation. A breach in the vendor's network would mean that intruders could have access to live, or even stored, conversations.

    4. A. Security and privacy standards should be reviewed and revised at least annually. Further, such standards should be revised when significant new applicable regulations have been enacted to ensure alignment with and compliance with those new regulations.

5. **D.** A data flow diagram, or DFD, illustrates the flow of information in an organization between and among its systems, as well as flows to and from external parties.

6. **C.** An entity-relationship diagram (ERD) is a logical data model that depicts the details of information stored in a system.

7. **B.** Source code management is a critical activity throughout the life of an information system, from the start of development until the application is retired.

8. **C.** Detailed test plans can be created as soon as requirements have been developed and finalized. Test plans should be derived directly from requirements, because they function as a tool to determine whether a system properly meets those requirements. Therefore, requirements must be written in a way that permits them to be verified.

9. **B.** Unit testing is the first level of testing. It is performed by developers who are writing individual modules or sections of code. Unit testing confirms that the modules or sections are properly coded and function correctly.

10. **D.** Data migration programs are typically custom programs that will be used once to migrate data from an older system to a newer system.

11. **C.** An organization that wants to get historical data from a legacy system to a new SaaS platform will have to write a migration program that reads data out of the legacy system and transforms it as necessary so that it can be imported into the new SaaS system.

12. **B.** While the waterfall SDLC model is time proven and can result in a high-quality system, its downside is that it does not provide much agility to an organization. Migrating to a DevSecOps model can result in a development organization that has far more agility and can help the business better respond to changing market forces.

13. **A.** To give developers immediate feedback on code quality, code scanning tools should be built into the IDEs (integrated development environments), which are the tools that developers use on their computers to write and test code.

14. **D.** OWASP, the Open Web Application Security Project, is the de facto source for detailed information on the principles of secure web applications. The OWASP Top Ten vulnerabilities are published annually and represent the ways in which web applications often exhibit weakness.

15. **A.** Session cookies are identifiers set by an application and exchanged during each user session to distinguish users from one another.

Technical Privacy Controls

In this chapter, you will learn about

- Controls, control frameworks, and their support of privacy and security
- Communications network protocols used to transmit personal information
- Encryption and key management techniques used to protect personal information
- Logging and monitoring of privacy and security events
- Identity and access management

This chapter covers Certified Data Privacy Solutions Engineer job practice 2, "Privacy Architecture," part C, "Technical Privacy Controls." The entire Privacy Architecture domain represents 36 percent of the CDPSE examination.

This chapter contains a variety of technical subjects that are all related to the security of networks and systems containing personal information. These complex topics are presented in summary form. Several references are included that refer the reader to more complete discussions of these topics.

Controls

Before we dive into the topic of technical privacy controls, we first need to discuss the concept and application of controls in general. This section contains a brief discussion of controls; a comprehensive discussion of controls and control frameworks may be found in *CISM Certified Information Security Manager All-In-One Exam Guide*.

The policies, procedures, mechanisms, systems, and other measures designed to reduce risk are known as *controls*. An organization develops controls to ensure that its business objectives will be met, risks will be reduced, and errors will be prevented or corrected. Controls are used in two primary ways in an organization: they are implemented to ensure desired outcomes and to avoid unwanted outcomes. In the context of privacy and information security, controls should be defined and implemented to ensure the protection and proper handling of personal information.

Control Objectives

Control objectives are statements of desired states or outcomes from business operations to mitigate risks. When building a security program, and preferably before selecting a control framework, you need to establish high-level control objectives. Example control objective subject matter includes the following:

- Protection of IT assets
- Accuracy of transactions
- Confidentiality and privacy of sensitive information
- Availability of IT systems
- Controlled changes to IT systems
- Compliance with corporate policies
- Compliance with applicable regulations and other legal obligations

Control objectives are the foundation for one or more controls. For each control objective, one or more *control activities* (more often regarded to as controls) will exist to ensure the realization of the control objective. For example, the "availability of IT systems" control objective could be implemented via several control activities, including these:

- IT systems will be continuously monitored, and any interruptions in availability will result in alerts sent to appropriate personnel.
- IT systems will have resource-measuring capabilities.
- IT management will review capacity reports monthly and adjust resources accordingly.
- IT systems will have antimalware controls that are monitored by appropriate staff.

Together, these four (or more) controls contribute to the overall control objective of IT system availability. Similarly, other control objectives will include one or more controls that will ensure their realization.

After establishing control objectives and defining the control activities that will support the objective, your next step is to design controls. This can be a considerable undertaking when done in a vacuum. A better approach is to utilize one of several high-quality, industry-accepted control frameworks that are discussed later in this section as a starting point.

If an organization elects to adopt a standard control framework, the next step is to perform a risk assessment to determine whether controls in the control framework adequately meet each control objective. Where there are gaps in control coverage, additional controls must be developed and put in place.

 **NOTE**　An IT organization supporting many applications and services will generally have some controls that are specific to each application. However, IT will also have a set of controls that apply across all applications and services. These are usually called its *IT general controls* (ITGCs).

Privacy Control Objectives

Privacy control objectives resemble ordinary control objectives but are set in the context of privacy and information security. Following are some examples of privacy control objectives:

- Protection of personal information from unauthorized personnel
- Protection of personal information from unauthorized modification
- Integrity of personal information
- Controlled use of personal information
- Operational compliance with the privacy policy

An organization will probably create several additional information systems control objectives on other basic topics such as malware, availability, and resource management, many of which directly or indirectly contribute to the protection and proper use of personal information.

Control Frameworks

A *control framework* is a collection of controls that are organized into logical categories. Well-known control frameworks such as ISO/IEC 27001, NIST SP 800-53, and the CIS Controls are intended to address a broad set of information risks common to most organizations. The standards bodies that publish these frameworks are now publishing privacy-centric frameworks that build upon the information security frameworks, such as ISO/IEC 27701 and the NIST Privacy Management Framework (PMF).

Standard control frameworks have been developed to streamline the process of control development and adoption within organizations. If there were no standard control frameworks, organizations would have to assemble their controls using other, inferior, sources such as the following:

- Gut feeling
- Prior experience in another organization
- A security practitioner in another organization
- An Internet search
- A deficient or incomplete risk assessment

A security manager could perform a comprehensive risk assessment and develop a framework of controls based upon identified risks, and indeed this would not be considered unacceptable. However, with the variety of high-quality control frameworks that are freely available (with the exception of ISO/IEC standards that must be purchased), an organization could start with a standard control framework to simplify and accelerate efforts.

PART II

Selecting a Control Framework

Several high-quality control frameworks are available for organizations that want to start with a standard control framework as opposed to starting from scratch or other means. Table 6-1 lists commonly used control frameworks.

Control Framework	Description	Industry Use
ISO/IEC 27001/27002	Broadly adopted international controls	All
ISO/IEC 27701	Published in August 2019 as an extension to ISO/IEC 27001 and 27002	All
NIST SP 800-53	Broadly adopted US-based controls	Government, private industry
CIS Controls	Broadly adopted US-based controls	All
Payment Card Industry Data Security Standard (PCI-DSS)	Controls for protection of credit card data	Retail, hospitality, entertainment, banking, credit card processing
NIST Cyber Security Framework (CSF)	Emerging US-based controls	All
NIST Privacy Framework	Published in January 2020 as a companion of the NIST CSF	All
Health Insurance Portability and Accountability Act (HIPAA)	Controls for the protection of electronic protected health information (ePHI)	Medical services including delivery, billing, and insurance
COBIT 5	Broadly adopted international controls	All
Committee of Sponsoring Organizations of the Treadway Commission (COSO)	Controls for preserving the integrity of financial information and financial statement reporting	All US public companies and private companies requiring similar controls
North American Electric Reliability Corporation (NERC) Reliability Standards	Controls for the protection of electric generation and distribution infrastructure	Electric utilities
Cloud Security Alliance (CSA) Controls	Controls for use by cloud-based service providers	All
SOC 1 (System and Organization Controls Report)	Bespoke controls for use by financial service providers	All
SOC 2 (System and Organization Controls Report)	Standard controls for use by cloud-based service providers	All
SIG (Standardized Information Gathering)	Controls used in the assessment of third parties	All

Table 6-1 Commonly Used Control Frameworks

There is ongoing debate regarding which control framework is best for an organization. In the author's opinion, the fact that there is debate on the topic reveals that many do not understand the purpose of a control framework or the risk management life cycle. The common belief is that once an organization selects a control framework, the organization is "stuck" with a set of controls, and no changes will be made to the controls used in the organization. Instead, as is discussed throughout this book, the selection of a control framework represents a starting point, not the perpetual commitment. Once a control framework is selected, the risk management life cycle is used to understand risk in the organization, resulting in changes to the controls used by the organization. In fact, it may be argued that an organization practicing effective risk management will eventually arrive at more or less the same set of controls, regardless of the starting point.

A different and valid approach to control framework selection has more to do with the structure of controls than the controls themselves. Each control framework consists of logical groupings based on categories of controls. For instance, most control frameworks have sections on identity and access management, vulnerability management, incident management, and access management. Some control frameworks' groupings are more sensible in certain organizations based on their operations or industry sector.

There is also nothing wrong with a security manager selecting a control framework based on her familiarity and experience with a specific framework. This is valid to a point, however: selecting the PCI-DSS control framework in a healthcare delivery organization might not be the best choice.

 **EXAM TIP** CDPSE candidates are not required to memorize COBIT or other frameworks, but familiarity with them will help the CDPSE candidate better understand how they contribute to effective privacy and security governance and control.

Mapping Control Frameworks

Frequently, organizations find themselves in a position where more than one control framework needs to be selected and adopted. The primary factors driving this are as follows:

- Multiple applicable regulatory frameworks
- Multiple operational contexts

Organizations with multiple control frameworks often crave a simpler organization for their controls. Often, organizations will "map" their control frameworks together, resulting in a single control framework with controls from each framework present. Mapping control frameworks together is time-consuming and tedious, although in some instances, the work has already been done. For example, Appendix H in NIST SP 800-53 contains a forward and reverse mapping between NIST SP 800-53 and ISO/IEC 27001. Other controls mapping references can be found online. Some must be built manually.

Working with Control Frameworks

Once an organization selects a control framework and multiple frameworks are mapped together (if the organization has decided to undertake that), security managers will need to organize a framework of activities around the selected/mapped control framework.

Risk Assessment   Before a control can be designed, the privacy or security manager needs to have some idea of the nature of risks that a control is intended to address. In a running risk management program, a new risk may have been identified during a risk assessment that led to the creation of an additional control. In this case, information from the risk assessment is needed so that the control will be properly designed to handle these risks.

If an organization is implementing a control prior to a risk assessment, the organization may not design and implement the control properly. Here are some examples:

- A control may not be rigorous enough to counter a threat.
- A control may be too rigorous and costly (in the case of a moderate or low risk).
- A control may not counter all relevant threats.

In the absence of a risk assessment, the chances for one of these undesirable outcomes are quite high. If an organization is implementing a control, a risk assessment needs to be performed. If an organization-wide risk assessment is not feasible, then a risk assessment that is focused on the control area should be performed so that the organization will know what risks the control will be intended to address.

Control Design   An early step in control use is its design. In a standard control framework, the control language itself appears, as well as some degree of guidance. The privacy or security manager, together with personnel who have responsibility for relevant technologies and business processes, need to determine what activities should occur. In other words, they need to figure out how to operationalize the control.

Proper control design will potentially require one or more of the following:

- New or changed policies
- New or changed business process documents
- New or changed information systems
- New or changed business records

Control Implementation   After a control has been designed, it needs to be put into service. Depending upon the nature of the control, this could involve operational impact in the form of changes to business processes and/or information systems. Changes with greater impact will require greater care so that business processes are not adversely affected. For instance, an organization may implement a control that requires production servers and other devices to be hardened from attack to comply with recognized standards such as CIS Benchmarks. After the hardening standards are developed (no easy task, by the way), they need to be tested and implemented. If a production environment is affected, it could take quite a bit of time to ensure that none of the hardening standard items adversely affects the performance, integrity, or availability of affected systems.

 **NOTE** Further detailed discussion on control development and assessment is found in Chapter 6 in *CISM Certified Information Security Manager All-In-One Exam Guide*. Detailed discussion on audits of controls is found in *CISA Certified Information Systems Auditor All-In-One Exam Guide*.

Control Monitoring   After an organization has implemented a control, it needs to monitor the control. For this to happen, the control needs to have been designed so that monitoring can occur. In the absence of monitoring, the organization will lack methodical means for observing the control to determine whether it is being operated correctly and whether it is effective.

Some controls are not easily monitored. For instance, a control addressing abuse of intellectual property rights includes the enactment of new acceptable use policies (AUPs) that forbid employees from violating intellectual property laws such as copyrights. There are many forms of abuse that cannot be easily monitored.

Control Assessment   Any organization that implements controls to address risks should periodically examine those controls to determine whether they are working as intended and as designed. There are several available approaches to control assessment:

- **Security review**   One or more information security staff members examine the control along with any relevant business records.

- **Control self-assessment (CSA)**   Control owners answer questions and include any relevant evidence.

- **Internal audit**   The organization's internal auditors (or information security staff) perform a formal examination of the control.

- **External audit**   An external auditor formally examines the control.

An organization will select one or more of these methods, guided by any applicable laws, regulations, legal obligations, and results of risk assessments.

Communication and Transport Protocols

Networks are the means through which computers communicate with one another. Whether the computers are across the room or halfway around the world, networks facilitate all computer communications, whether wired or wireless and whether telecommunications services are used or not. This section contains a brief treatise on network technology.

Network Media

Network media is the material through which network traffic travels, whether wired or wireless. That is, the network connection consists of copper wires or fiber-optic cables for a wired connection or a radio frequency (RF) protocol for wireless.

Another perspective of network media is whether it's provided by the user or by a telecommunications provider. For instance, a home or office user would set up a wireless Wi-Fi network, and a telecommunications provider would set up a 5G or LTE network.

Network media types include the following:

- **Twisted pair**   This ubiquitous network cabling is used in homes and businesses for connecting systems through networks.

- **Fiber-optic**   Used by larger businesses in their data centers as well as telecommunications providers to connect businesses and data centers to the Internet, this high-capacity glass fiber media carries signals in the form of light pulses and is impervious to magnetic and electric field interference.

- **Broadband cable**   The mainstay of television signals for decades, cable is increasingly used as the medium for residential and business Internet connection.

- **4G/5G/LTE/GSM/CDMA**   These over-the-air wireless protocols are provided by telecommunications carriers the world over. Our smartphones and, occasionally, tablets use these carrier protocols to communicate online. These protocols are provided as part of a fee-based voice and/or data service.

- **Wi-Fi**   This ever-popular wireless network standard is used in homes, businesses, and even outdoors in urban areas. These days, devices in homes are usually connected to a residential Wi-Fi network that is, in turn, connected to a telco-provided Internet connection delivered over DSL, cable, or fiber.

- **Bluetooth**   This short-range wireless protocol is a handy replacement for formerly wired connections for devices such as printers, keyboards, mice, and headphones.

- **Near-field communication (NFC)**   This ultra–short range (up to 4cm) protocol is used for contactless payment and other uses.

This list represents contemporary media types. Many others—too numerous to mention in this book—are still in use or have been entirely deprecated.

Network Protocols

Generally, the term *network protocols* refers to numerous individual standards, as well as families and suites of standards, whereby information is carried over network media discussed in the preceding section. For the sake of simplicity, this section will discuss lower level protocols, followed by higher level protocols.

Notable low level protocols include these:

- **Ethernet**   This is the dominant protocol used for wired home and office networks, carried over twisted-pair cable. Data in the Ethernet protocol is transported over *frames*, which are individual messages being sent from one station to another.

- **T-Carrier** and **E-Carrier**   These telecommunications standards are used to transport voice and data over telecommunications cables. These families include scores of individual protocols such as T-1 (with a data rate of 1.544Mbps) that can be multiplexed into 24 individual voice or data channels, or as a single logical data pipe.

- **Digital subscriber line (DSL)**   This is broadband Internet delivered over copper telephone landlines.

- **Synchronous Optical Networking (SONET)**   This family of telecommunications standards is used to transport voice and data over long-distance optical fiber. Individual protocols such as OC-1 and OC-3 carry voice and/or data at 51.84Mbps and 155.52Mbps, respectively.

High-level protocols in use today are all a part of the TCP/IP suite of protocol standards. Invented in the 1970s as a resilient, long-distance, packet-carrying network, TCP/IP is the foundation for virtually all data networking everywhere, including the global Internet.

The TCP/IP protocol suite employs a technique known as *encapsulation*, whereby messages in higher level protocols are encapsulated within messages in lower level protocols, which in turn are encapsulated in messages in the physical network medium. For example, when a user loads a web page onto his browser, the HTTPS protocol is encapsulated within the TCP protocol, which is encapsulated within the IP protocol, which is encapsulated within frames in the Ethernet protocol, which is carried on electrical signals on twisted-pair network cable. Figure 6-1 illustrates this encapsulation.

Notable protocols used in the TCP/IP suite include

- **IPv4 and IPv6**   These are the lowest level protocols in the TCP/IP suite, notable on account of their schemes for assigning numeric addresses to devices. When TCP/IP was first invented, the base IP protocol was designed for a maximum of about 4.3 billion devices worldwide. In the 1990s, it was realized that this was no longer sufficient and led to the IPv6 standard, which can accommodate 2^{128}, or 3.4×10^{38}, addresses. Numerous other improvements are included in IPv6.

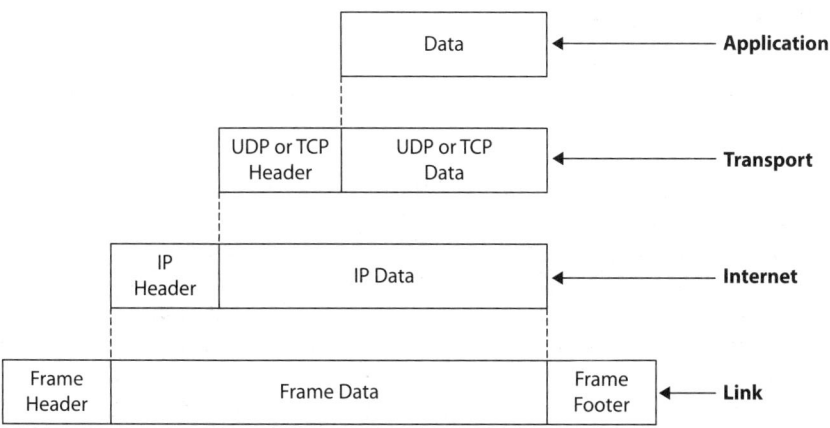

Figure 6-1   TCP/IP network encapsulation

- **Routing protocols** These protocols are used by routers, network devices that organize and keep network traffic moving toward its respective destinations. These protocols include Routing Information Protocol (RIP), Border Gateway Protocol (BGP), Intermediate System-to-Intermediate System (IS-IS), Open Shortest Path First (OSPF), Interior Gateway Routing Protocol (IGRP), and Enhanced Interior Gateway Routing Protocol (EIGRP).

- **HTTP** and **HTTPS** These two protocols are used by web browsers to request and receive information between web servers and web browsers.

- **DNS (Domain Name System)** This is a key protocol used to translate domain names into numeric addresses needed to determine the endpoints of all network communications in TCP/IP.

- **FTP (File Transfer Protocol)** This and newer versions (FTPS and SFTP) are used for bulk file transfers between systems.

- **SSH and TELNET** Secure Shell and TELNET are command line interface protocols often used by administrators to manage systems and devices.

- **SMTP (Simple Mail Transport Protocol)** This protocol is used to transport e-mail between e-mail servers.

- **NTP (Network Time Protocol)** This protocol is used to ensure that the time clocks on systems and devices are accurate to the millisecond. This is important because the clocks in computers tend to "drift."

- **VoIP (Voice over IP)** This family of protocols is used for transporting live "telephone" and video communications over TCP/IP networks.

 **NOTE** This section has barely scratched the surface of network technology. For a complete explanation of network protocols, readers are referred to any of the classic texts on networks such as *TCP/IP Illustrated*.

Network Architecture

Metaphorically, networks are like roads, with varying capacities and with intersections and controls such as traffic signals. Organizations design their internal data networks so that workers with their laptops, desktops, smartphones, or tablet computers can communicate with key business applications.

The architecture of a network has less to do with where the cabling is, and more to do with a network's logical design. Some organizations build large, "flat" networks, where every computer can freely communicate with every other computer. In contrast, other organizations employ the technique of *network segmentation*, in which the network is divided into security zones, and controls between them (such as firewalls) restrict network traffic only to that which is considered necessary. *Microsegmentation* is another means for protecting systems and critical systems, in which individual systems employ firewalls that block all but essential communications with other systems.

The concept of *zero trust* may be viable to many organizations that have a sizeable mobile or remote workforce, and with organizations with a large number of customers or constituents. Zero trust is a model whereby portions of the network (or systems outside of the network) are considered to be untrusted. Zero trust is a way of thinking about the network so that appropriate security controls can be implemented to protect highly sensitive or secretive systems and information.

Encryption, Hashing, and De-identification

Encryption is the technique used to hide information in plain sight. It works by scrambling the characters in a message using a method known only to the sender and receiver, making the message useless to any party that intercepts the message. Encryption plays a crucial role in the protection of personal information. In some situations, it is not practical or feasible to prevent third parties from having logical access to data—for instance, data transmissions over public networks.

This technique can also be used to *authenticate* information that is sent from one party to another. This means that a receiving party can verify that a specific party did, in fact, originate a message and that it is authentic. This enables a receiver to know that a message is genuine and that it has not been forged or altered in transit by any third party.

With encryption, best practices call for system designers to use well-known, robust encryption algorithms. Thus, when a third party intercepts encrypted data, the third party can know which algorithm is being used but still not be able to read the data. What the third party does not know is the *key* that is used to encrypt and decrypt the data. How this works will be explained later in this section.

 **NOTE**   Encryption can be thought of as another layer of access protection. Like user ID and password controls that restrict data access for everyone but those with login credentials, encryption restricts access to (plaintext) data to everyone but those with encryption keys.

Encryption

Encryption is a reversible process, whereby plaintext can be converted to illegible ciphertext and back. Only those who possess an encryption key can read the ciphertext.

Terms and Concepts Used in Cryptography

Several terms and concepts used in cryptography are not used outside of the field. Privacy and security professionals must be familiar with these to be effective in understanding, managing, and auditing information systems that use cryptography:

- **Plaintext**   An original message, file, or stream of data that can be read by anyone who has access to it
- **Ciphertext**   A message, file, or stream of data that has been transformed by an encryption algorithm and rendered unreadable
- **Encryption**   The process of transforming plaintext into ciphertext, as depicted in Figure 6-2

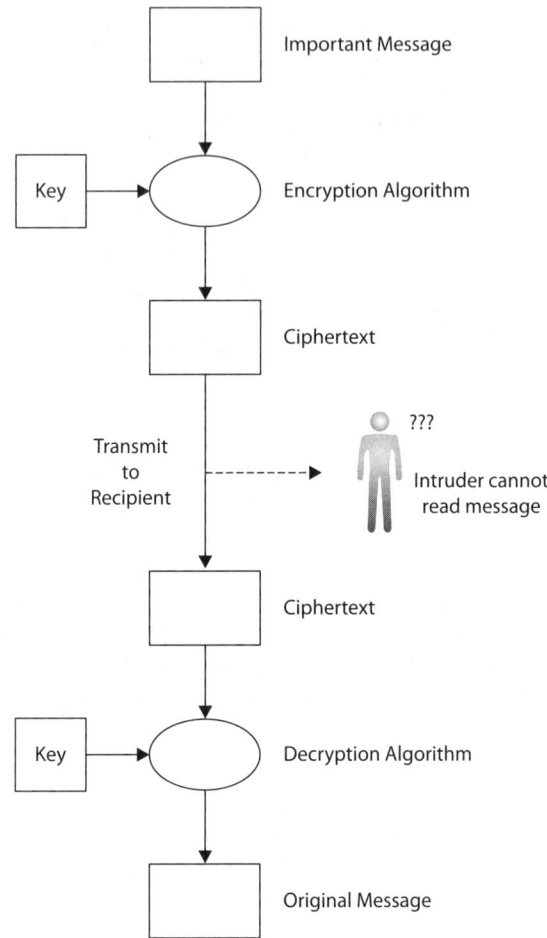

Figure 6-2
Encryption and
decryption utilize
an encryption
algorithm and an
encryption key.

- **Hash function**   A cryptographic operation on a block of data that returns a fixed length string of characters used to verify the integrity of a message

- **Message digest**   The output of a cryptographic hash function

- **Digital signature**   The result of encrypting the hash of a message with the originator's private encryption key, used to prove the authenticity and integrity of a message, as depicted in Figure 6-3

- **Algorithm**   A specific mathematical formula that is used to perform encryption, decryption, message digests, and digital signatures

- **Decryption**   The process of transforming ciphertext into plaintext so that a recipient can read it

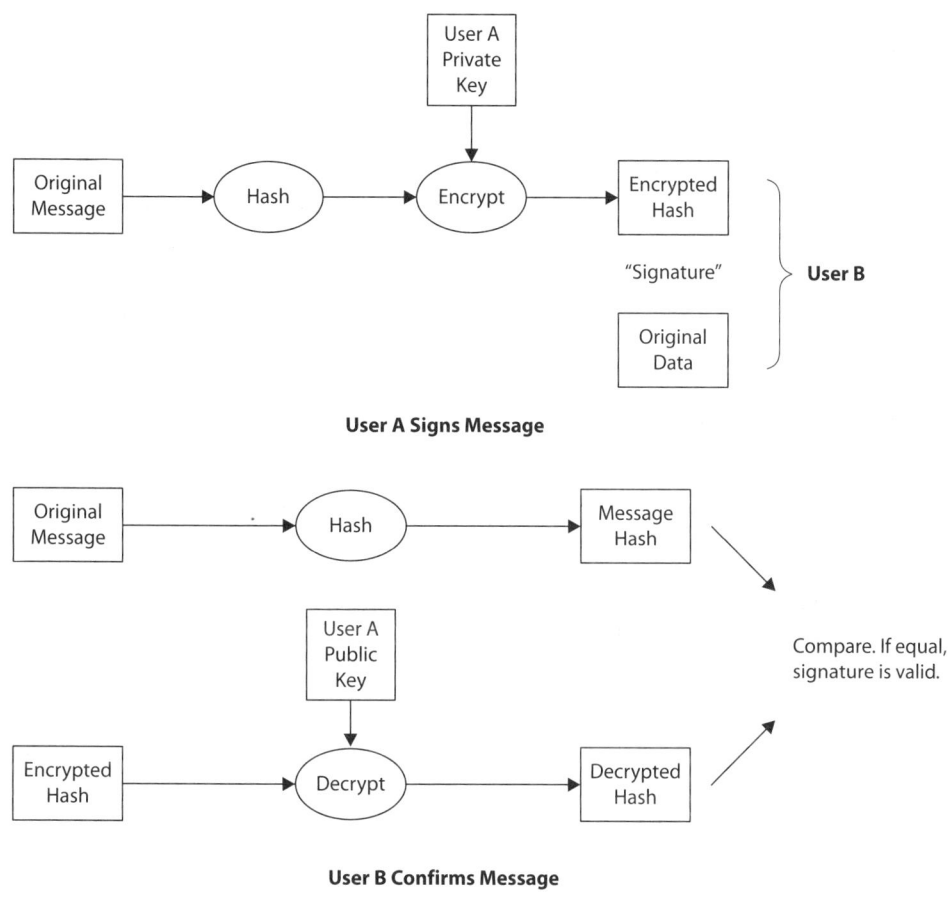

Figure 6-3   Digital signature used to verify the integrity of a message

- **Cryptanalysis**   An attack on a cryptosystem in which the attacker is attempting to determine the encryption key that is used to encrypt messages
- **Encryption key**   A block of characters used in combination with an encryption algorithm to encrypt or decrypt a stream or blocks of data; also used to create and verify a digital signature
- **Key encrypting key**   An encryption key used to encrypt another encryption key
- **Key length**   The size (measured in bits) of an encryption key; longer encryption keys can take considerably more effort to attack a cryptosystem successfully
- **Block cipher**   An encryption algorithm that operates on blocks of data
- **Stream cipher**   A type of encryption algorithm that operates on a continuous stream of data such as a video or audio feed

- **Initialization vector (IV)**   A random number that is needed by some encryption algorithms to begin the encryption process

- **Symmetric encryption**   A method for encryption and decryption in which it is necessary for both parties to possess a common encryption key

- **Asymmetric encryption or public key cryptography**   A method for encryption, decryption, and digital signatures that uses pairs of encryption keys, consisting of a *public key* and a *private key*

- **Key exchange**   A technique used by two parties to establish a symmetric encryption key when no secure channel is available

- **Nonrepudiation**   The property of digital signatures and encryption that can make it difficult or impossible for a party to deny having sent a digitally signed message, unless they admit to having lost control of their private encryption key

NOTE   Poor implementations of cryptography can result in insufficient protection, leading to the compromise of personal information. Privacy and security professionals should be sufficiently familiar with cryptography so that they would recognize whether a cryptosystem was properly designed and is appropriately managed.

Private Key Cryptosystems

A *private key cryptosystem* is based on a symmetric cryptographic algorithm. The primary characteristic of a private key cryptosystem is the necessity for both parties to possess a common encryption key that is used to encrypt and decrypt messages.

The two main challenges with private key cryptography are

- **Key exchange**   An "out-of-band" method for exchanging encryption keys is required before any encrypted messages can be transmitted. This key exchange must occur over a secure channel; if the encryption keys were transmitted over the primary communications channel, anyone who intercepted the encryption key would be able to read any intercepted messages, provided they could determine the encryption algorithm used. For instance, if two parties want to exchange encrypted e-mail, they would need to exchange their encryption key first via telephone or fax, provided they are confident that their telephone and fax transmissions are not being intercepted.

- **Scalability**   Private key cryptosystems require that each sender-receiver pair exchange an encryption key. For a group of 4 parties, 6 encryption keys would need to be exchanged; for a group of 10 parties, 45 keys would need to be exchanged. For a large community of 1000 parties, many thousands of keys would need to be exchanged.

Some well-known private key algorithms in use include AES (Rijndael), Blowfish, Triple DES, RC4, Serpent, Skipjack, and Twofish.

Secure Key Exchange

Secure key exchange methods are used by two parties to establish a symmetric encryption key securely without actually transmitting the key over a channel. Secure key exchange is needed when two parties that are previously unknown to each other need to establish encrypted communications where no out-of-band channel is available.

With the right method, two parties can perform a secure key exchange even if a third party intercepts their entire conversation. This is because algorithms used for secure key exchange utilize information known only by the two parties but not transmitted between them.

The most popular algorithm is the Diffie-Hellman Key Exchange Protocol. Another algorithm in limited use is quantum key distribution (QKD).

Public Key Cryptosystems

Public key cryptosystems are based on *asymmetric*, or *public key*, cryptographic algorithms. These algorithms use two-part encryption keys that are handled differently from encryption keys in symmetric key cryptosystems.

Key Pair   In public key cryptography, *public* and *private* encryption keys are used. Each user of a public key cryptosystem has these two keys in his or her possession. Together, the public and private keys are known as a *key pair*. The two keys require different handling and are used together but for different purposes, as explained in this section.

When a user generates his or her key pair, the key pair will physically exist as two separate files. The user is free to publish or distribute the public key openly; it could even be posted on a public web site. This is in contrast to the private key, which must be well protected and never published or sent to any other party—like keys in a private key cryptosystem. Most public key cryptosystems will utilize a password mechanism to protect the private key further; without its password, the private key is inaccessible and cannot be used. A public key infrastructure (PKI) system can be used to publish public keys and make them accessible to users.

Message Security   Public key cryptography is an ideal application for securing messages, particularly e-mail, because users do not need to establish and communicate symmetric encryption keys through a secure channel. With public key cryptography, users who have never contacted each other can immediately send secure messages between them. Public key cryptography is depicted in Figure 6-4.

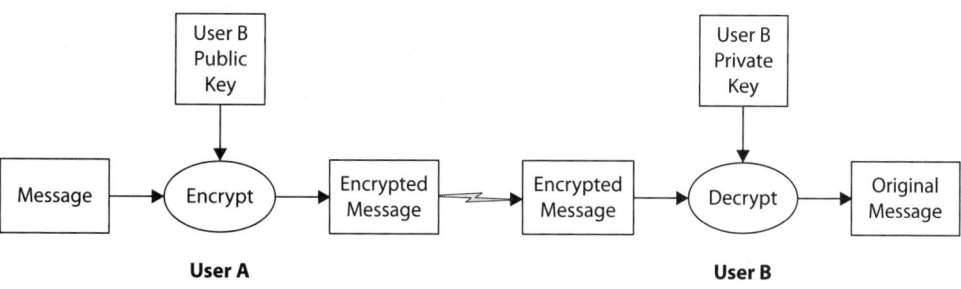

Figure 6-4   Public key cryptography used to transmit a secret message

Every user is free to publish his or her public encryption key so that it is easily retrievable. PKI servers on the Internet can make public keys available to anyone in the world. Public key cryptography is designed so that open disclosure of a user's public key does not compromise the secrecy of the corresponding private key: a user's private key cannot be derived from the public key.

When User A wants to send an encrypted message to User B, the procedure is as follows:

1. User B publishes his public key to the Internet at a convenient location.

2. User A retrieves User B's public key.

3. User A creates a message and encrypts it with User B's public key and sends the encrypted message to User B.

4. User B decrypts the message with his private key and is able to read the message.

Note that only User B's encryption key is used in this example. This method protects the message from eavesdroppers and is not used to verify the authenticity of the message.

Public key cryptography can also be used to verify the authenticity and integrity of a message. This is used to verify that a specific party did, in fact, create the message. The procedure is as follows:

1. User A publishes his public key to the Internet at a convenient location.

2. User B retrieves User A's public key and saves it for later use.

3. User A creates a message and digitally signs it with his private key, and then sends the signed message to User B.

4. User B verifies the digital signature using User A's public key. If the message verifies correctly, User B knows that the message originated from User A and has not been altered in transit.

In this example, only the authenticity and integrity of a message are assured. The message is not encrypted, which means that any party who intercepts the message can read it.

Public key cryptography can be used both to encrypt and to digitally sign a message, which will guarantee its confidentiality as well as its authenticity. The procedure is as follows:

1. User A and User B publish their public encryption keys to convenient places.

2. User A retrieves User B's public key, and User B retrieves User A's public key.

3. User A creates a message, then signs it with his private key and encrypts it with User B's public key, and then sends the message to User B.

4. User B decrypts the message with his private key and verifies the digital signature with User A's public key.

Public key cryptography also supports encryption of a message with more than one user's public key. This enables a user to send a single encrypted message to several recipients that is encrypted with each of their public keys. This method does not compromise the secrecy of any user's private key since a user's private key cannot be derived from the public key.

Elliptic Curve Cryptography

Elliptic curve cryptography (ECC) is attracting interest for use in public key cryptography applications. ECC requires less computational power and bandwidth than other cryptographic algorithms and is thought to be more secure as well. Because of its low power requirements, it is used extensively in mobile devices.

Verifying Public Keys

It is possible for a fraudster to claim the identity of another person and even publish a public key that claims the identity of that person. Four methods are available for verifying a user's public key as genuine:

- **Certificate authority (CA)**   A public key that has been obtained from a trusted, reputable certificate authority can be considered genuine.

- **E-mail address**   Public keys used for e-mail will include the user's e-mail address. If the e-mail address is a part of a corporate or government domain (for example, *adobe.com* or *seattle.gov*), then some level of credence can be attributed to the successful exchange of messages with that e-mail address. However, since e-mail addresses can be spoofed, this should be considered a weak method at best.

- **Directory infrastructure**   A directory services infrastructure such as Microsoft Active Directory, LDAP, or a commercial product can be used to verify a user's public key.

- **Key fingerprint**   Many public key cryptosystems employ a method for verifying a key's identity, known as the key's fingerprint. If a user wants to verify a public key, she retrieves the public key and calculates the key's fingerprint. The user then contacts the claimed owner of the public key, who runs a function against her private key that returns a string of numbers. The user also runs a function against the owner's public key, also returning a string of numbers. If both numbers match, the public key is genuine.

 **NOTE**   When issuing a public key, it is essential that the requestor of the new public key be authenticated, such as by viewing a government-issued ID or by contacting the owner at a publicly listed telephone number.

Hashing and Message Digests

Hashing is the process of applying a cryptographic algorithm on a block of information that results in a compact, fixed-length "digest." The purpose of hashing is to provide a unique "fingerprint" for the message or file—even if the file is very large. A message digest can be used to verify the integrity of a large file, thus assuring that the file has not been altered.

Some of the properties of message digests that make them ideally suited for verifying integrity include

- Any change made to a file—even a single bit or character—will result in a significant change in the hash.
- It is computationally infeasible to make a change to a file without changing its hash.
- It is computationally infeasible to create a message or file that will result in a given hash.
- It is infeasible for any two messages to have the same hash.

One common use of message digests is on software download sites, where the computed hash for a downloadable program is available so that users can verify that the software program has not been altered (provided that the posted hash has not also been compromised).

Digital Signatures

A *digital signature* is a cryptographic operation in which a sender "seals" a message or file using his or her identity. The purpose of a digital signature is to authenticate a message and to guarantee its integrity. Digital signatures do not protect the confidentiality of a message, however, as encryption is not one of the operations performed.

Digital signatures work by encrypting hashes of messages; recipients verify the integrity and authenticity of messages by decrypting hashes and comparing them to original messages. In detail, a digital signature works like this:

1. The sender publishes his public key to the Internet at a location that is easily accessible to recipients.
2. The recipient retrieves the sender's public key and saves it for later use.
3. The sender creates a message (or file) and computes a message digest (hash) of the message, and then encrypts the hash with his private key.
4. The sender sends the original file plus the encrypted hash to the recipient.
5. The recipient receives the original file and the encrypted hash. The recipient computes a message digest (hash) of the original file and sets the result aside. She then decrypts the hash with the sender's public key. The recipient compares the hash of the original file and the decrypted hash.
6. If the two hashes are identical, the recipient knows that the message in her possession is identical to the message that the sender sent, the sender is the originator, and the message has not been altered.

The use of digital signatures is depicted earlier in this chapter in Figure 6-3.

Digital Envelopes

Two aspects of symmetric (private key) and asymmetric (public key) cryptography that have not been discussed yet are the computing requirements and performance implications of these types of cryptosystems. In general, public key cryptography requires far more computing power than private key cryptography. The practical implication is that public key encryption of large sets of data can be highly compute-intensive, which makes its use infeasible in some instances.

One solution to this is a *digital envelope* that utilizes the convenience of public key cryptography with the lower overhead of private key cryptography. This practice is known as *hybrid cryptography*. The procedure for using digital envelopes works like this:

1. The sender and recipient agree that the sender will transmit a large message to the recipient.

2. The sender selects or creates a symmetric encryption key, known as the *session key*, and encrypts the session key with the recipient's public key.

3. The sender encrypts the message with the session key.

4. The sender sends the encrypted message (encrypted with the session key) and the encrypted session key (encrypted with the recipient's public key) to the recipient.

5. The recipient decrypts the session key with his private key.

6. The recipient decrypts the message with the session key.

Encryption Applications

Several applications utilize encryption algorithms. Many of these are well known and in everyday use.

Secure Sockets Layer/Transport Layer Security (SSL/TLS)   These encryption protocols are used to encrypt web pages requested with the HTTPS (Hypertext Transfer Protocol Secure). SSL and its successor, TLS, have become de facto standards for the encryption of web pages as they transit the Internet between web servers and web browsers. TLS is also commonly used to encrypt e-mail between e-mail servers, so that messages sent from one organization to another will be encrypted as they transit the Internet.

Secure/Multipurpose Internet Mail Extensions (S/MIME)   This e-mail security protocol provides sender and recipient authentication and encryption of message content and attachments. S/MIME is most often used for encryption of e-mail messages.

Secure Shell (SSH)   This multipurpose protocol is used to create a secure channel between two systems. The most popular use of SSH is the replacement of the TELNET and R-series protocols (rsh, rlogin, and so on), but it also supports tunneling of protocols such as X Window System and FTP.

Internet Protocol Security (IPsec)   This protocol is used to create a secure, authenticated channel between two systems. IPsec operates at the Internet layer in the TCP/IP suite; hence, all IP traffic between two systems protected by IPsec is automatically encrypted.

IPsec operates in one of two modes: Encapsulating Security Payload (ESP) and Authentication Header (AH). If ESP is used, all encapsulated traffic is encrypted. If AH is used, only IPsec's authentication feature is used.

Blockchain   This distributed ledger is used to record cryptographically linked transactions in a peer-to-peer network. Once recorded, transactions in a blockchain cannot be altered or removed. Blockchain is decentralized by design. Implementations of blockchain include the Bitcoin cryptocurrency as well as emerging uses in financial services and supply chain management.

EXAM TIP   CDPSE candidates must understand how encryption and hashing contribute to effective privacy and must be able to determine whether encryption or hashing would be a better control in particular use cases.

Key Management

Key management refers to the various processes and procedures used by an organization to generate, protect, use, and dispose of encryption keys over their lifetime. Several common key management practices are described in this section.

NOTE   The integrity of a cryptosystem and the confidentiality of the information encrypted by the system are only as strong as the management of its encryption keys.

Key Generation

The start of an encryption key life cycle is its generation. At first glance, it would appear that this process should require little scrutiny, but further study shows that this is a critical process that requires safeguards.

The system on which key generation takes place must be highly protected. If keys are generated on a system that has been compromised or that is of questionable integrity, it would be difficult to determine whether a bystander could have electronically observed key generation. In many situations, it would be reasonable to require that systems used for key generation be highly protected, isolated, and used by as few persons as possible. Regular integrity checks would need to take place to ensure that the system continues to be free of any anomalies.

Furthermore, the key generation process needs to include some randomness (or, as some put it, *entropy*) so that the key generation process cannot be easily duplicated elsewhere. If key generation were not a random event, it could be possible to duplicate the conditions related to a specific key and then regenerate a key with the very same value. This would instantaneously compromise the integrity and uniqueness of the original key.

Key Protection

Private keys used in public key cryptosystems and keys used in symmetric cryptosystems must be continuously and vigorously protected. At all times, they must be accessible *only* to the parties that are authorized to use them. If protection measures for private encryption keys are compromised (or suspected to be), it will be possible for a key compromise to take place, enabling the attacker to view messages encrypted with these keys. In commercial environments, keys are often protected in a hardware security module (HSM).

A *key compromise* occurs when a private encryption key has been disclosed to any unauthorized third party. When a key compromise occurs, it will be necessary to re-encrypt all materials encrypted by the compromised key with a new encryption key.

 CAUTION   In many applications, an encryption key is protected by a password. The length, complexity, distribution, and expiration of passwords protecting encryption keys must be well designed so that the strength of the cryptosystem (based on its key length and algorithm) is not compromised by a weak password scheme protecting its keys.

Key Encrypting Keys

Applications that utilize encryption must obtain their encryption keys in some way. In many cases, an intruder may be able to examine the application in an attempt to discover an encryption key, so that the intruder may decrypt communications used by the application. A common remedy for this is the use of encryption to protect the encryption key. This additional encryption requires a key of its own, known as a *key encrypting key*. Of course, this key also must reside someplace; often, features of the underlying operating system may be used to protect an encryption key as well as a key encrypting key.

Key Custody

Key custody refers to the policies, processes, and procedures regarding the management of keys. This is closely related to key protection but is focused on *who* manages keys, *where* they are kept, and *how* they are used.

Key Rotation

Key rotation is the process of issuing a new encryption key and re-encrypting data protected with the new key. Key rotation may occur when any of the following occurs:

- **Key compromise**   When an encryption key has been compromised, a new key must be generated and used.
- **Key expiration**   In some situations, encryption keys are rotated on a schedule.
- **Rotation of staff**   In some organizations, if any of the persons associated with the creation or management of encryption keys transfers to another position or leaves the organization, keys must be rotated.

Key Disposal

Key disposal refers to the process of decommissioning encryption keys. This may occur upon receipt of an order to destroy a data set that is encrypted with a specific encryption key—destroying an encryption key can be as effective (and a whole lot easier) than destroying the encrypted data itself. Key disposal can present some challenges, however. If an encryption key is backed up to tape, for instance, disposal of the key will require that backup tapes also be destroyed. Hence, it is crucial to dispose of an encryption key *only* after it is determined that it is no longer needed.

NOTE A novel method for data disposal is the destruction of encryption keys.

Public Key Infrastructure

One of the issues related to public key cryptography is the safe storage of public encryption keys. Although individuals are free to publish public keys online, doing so in a secure and controlled manner requires some central organization and control. A PKI is designed to fulfill this and other functions.

A PKI is a centralized function that is used to store and publish public keys and other information. Some of the services provided by a PKI are

- Digital certificates
- Certificate authority (CA)
- Registration authority (RA)
- Certificate revocation list (CRL)
- Certification practice statement (CPS)

EXAM TIP CDPSE test takers are not expected to memorize key management procedures, but you should understand these concepts.

De-identification

De-identification refers to any of several techniques that all serve a single purpose: to remove from business records any traceable reference to a specific natural person. De-identification is a key concept in data privacy, because this is an effective means for reducing risks associated with the storage of large amounts of personally identifiable information (PII).

For several valid reasons, organizations may not completely remove older business records containing PII. Often significant value can be derived from older records, including the charting of long-term trends. However, in many cases, it is no longer necessary for organizations to continue to relate older business records with specific individuals. This is where organizations can use de-identification techniques to remove specific references to individuals while retaining other aspects of these records.

Techniques often used in de-identification include

- **Anonymization**   ISO 25237 (Health Informatics – Pseudonymization) defines anonymization as any "process by which personal data is irreversibly altered in such a way that a data subject can no longer be identified directly or indirectly, either by the data controller alone or in collaboration with any other party." PII fields can be removed or hashed so that the data cannot be associated with specific data subjects.

- **Pseudonymization**   In the European Union, General Data Protection Regulation (GDPR) Article 4 defines pseudonymization as "the processing of personal data in such a manner that the personal data can no longer be attributed to a specific data subject without the use of additional information, provided that such additional information is kept separately and is subject to technical and organisational measures to ensure that the personal data are not attributed to an identified or identifiable natural person." Here, specific identifying fields such as name, address, phone number, e-mail address, and financial account numbers are removed and replaced with pseudonyms.

NOTE   Privacy professionals need to understand an organization's de-identification techniques to be sure that records cannot be re-identified with specific persons.

Monitoring and Logging

The logging of security-related events, the centralized collection of these logs, and the proactive monitoring of these logs with correlation engines are considered essential practices in cybersecurity. These activities help an organization detect an array of activities, from misbehavior by an organization's workers to active attacks by cybercriminal organizations.

Monitoring activities related to data access can help an organization identify improper uses of personal information. This is a newer branch of cybersecurity practiced by few organizations, although it is gaining in popularity on account of new privacy laws such as GDPR and California Consumer Privacy Act (CCPA).

Event Monitoring

Event monitoring is the practice of examining the events occurring on information systems, including applications, operating systems, database management systems, end-user devices, file servers, and every type and kind of network device, and being aware of what is going on throughout the entire operating environment. The types of events of interest to privacy and security managers include the following:

- Successful and unsuccessful logins
- Unexpected system or device reboots

- Changes made to security configurations
- Changes made to operating system files
- Queries to databases
- Changes made to access permissions of sensitive files on a file server
- Anomalous movement of sensitive files

Historically, it was considered sufficient to review system event logs on a daily basis. Mainly this was a review of yesterday's events (and the weekend's events on a Monday) to ensure that no privacy or security incidents warranted further investigation. Those days are mostly gone, however.

Today, most organizations perform *real-time* event monitoring. This means organizations need to have systems in place that will immediately inform them if events are occurring anyplace in the environment that warrant attention. Although the technology available today that enables real-time event monitoring is impressive, the vast amounts of information that are collected and analyzed can create meaningless alerts if the systems are not properly tuned. Staff must understand the logs that are being collected and take the time to define the use cases that warrant alerts and investigation. Organizations that do not invest the time and resources required to tune the system will find teams overwhelmed with alerts, which will often be ignored.

Log Reviews
A *log review* is an examination of an event log in an information system to determine whether any privacy, security, or operational incident has occurred in the system. A log review is an examination of yesterday's activities in a system. Most organizations, however, conduct *continuous log reviews* by sending log data into a security information and event management (SIEM) system, discussed a bit later.

Centralized Log Management
Centralized log management involves sending event logs on various systems over the network to a central collection and storage point, called a *log server*. There are two primary uses for a log server: for archival storage of events that may be used at a later date in an investigation, and for storage of events to be reviewed on a daily basis or in real time. Generally, real-time analysis is performed by a SIEM system.

Security Information and Event Management
A SIEM system collects and analyzes log data from many or all systems in an organization and produces alerts to inform personnel of specific events. A SIEM has rules to correlate events from one or more devices to provide additional detail about an incident. For instance, an attacker performing a brute-force password attack on a web server may be generating alerts on the web server itself and also on the firewall and intrusion detection system. A SIEM would portray the incident using events from these and possibly other devices to give personnel a richer depiction of the incident.

For a SIEM to be effective, the timestamps in log entries from various devices must be accurate. The SIEM must be able to discern the actual sequence of events, which is based on each log entry's timestamp. Because computer time clocks are notoriously inaccurate on their own, configuring computers to synchronize their clocks with a standard *time source*, or time server, is an essential practice.

 **NOTE** Despite its name (*security* information and event management), a SIEM system is often used not only to inform personnel of security events, but also to inform them of operational and privacy-related events.

Orchestration

In the context of security information and event management systems, *orchestration* refers to a scripted, automated response that is automatically or manually triggered when specific events occur. Orchestration systems can be stand-alone systems or may exist as part of the SIEM.

For example, suppose an organization has developed "run books," or short procedures for personnel who manage the SIEM for actions to perform when specific types of events occur. The organization, desiring to automate some of these responses, implements an orchestration tool that includes scripts that can run automatically when specific events occur. The orchestration system can be configured to run some scripts immediately, while other scripts can be set up and run when an analyst "approves" them.

The advantage of orchestration is twofold: first, repetitive and rote tasks are automated, relieving personnel of boredom and improving accuracy, and, second, response to some types of events can be performed much more quickly, thereby blunting the impact of certain types of incidents.

Data Loss Prevention (DLP)

Organizations intent on proactively protecting sensitive information, including personal information about customers, constituents, and employees, may implement one or more types of DLP systems. For many organizations, policy alone is an insufficient means for protecting personal information. Instead, any of several types of controls can be introduced to protect specific data containing personal information, sensitive information, and intellectual property. Several tools and techniques are available for DLP:

- **Document scanning** Tools can be used to scan stores of unstructured data to determine the extent of the presence of sensitive and personal information.

- **Document tagging** During document scanning, tools tag files if they contain data matching specific patterns such as social insurance numbers, credit card numbers, financial account numbers, and others.

- **Document marking** Once tagged, documents can be marked or watermarked, which introduces human-readable content into files to remind people that these files contain sensitive information of some type.

- **E-mail restrictions**   DLP tools can be integrated into an organization's e-mail system to monitor and block the practice of e-mailing files containing sensitive information. These tools can be configured to read files' tags or scan the contents of the files themselves to determine whether they violate e-mail policy. When violations occur, users can be alerted; optionally, they can be given a choice on whether to proceed with their intended activity.

- **Storage restrictions**   DLP tools can be integrated into end-user devices to monitor their handling of sensitive data files. These tools can merely observe data movement or intervene when specific policies are violated. Users can be warned of or forbidden from storing sensitive files or using external storage devices and/ or cloud-based storage and messaging services.

All of these forms of DLP tools can be configured to send their events to the organization's log servers or SIEM so that privacy and security personnel can be alerted when data handling policy violations occur.

 **NOTE**   DLP tools should be used carefully, because false positives can occur, which could disrupt legitimate business activities. Also, false negatives may occur, where forbidden activities are not detected.

Threat Intelligence

Modern SIEMs can ingest threat intelligence feeds from various external sources. This enables the SIEM to better correlate events in an organization's systems with various threats experienced by other organizations.

Organizations can subscribe to one or more machine-readable threat intelligence sources that help the organization better understand which security events in their environment may represent intrusions. Some of these sources are open source, while others are fee-based commercial services.

For example, suppose another organization is attacked by an adversary from a specific IP address in a foreign country. This information is included in a threat intelligence feed that arrives in your organization's SIEM. This helps your SIEM be more aware of activity of the same type or from the same IP address. This can alert the organization to incidents occurring elsewhere that could occur in the organization's network.

Threat Hunting

For many organizations, it's no longer sufficient to wait for attacks to manifest themselves in their SIEM or other monitoring system. Instead, organizations go on the offensive to look for clues of possible intrusions in their environment. *Threat hunting* is the practice of conducting searches—typically in SIEM logs and configuration management databases—to see whether traces of intrusions are present in their systems.

For example, an organization may have received an advisory from a national law enforcement organization with specific intelligence on a new strain of malware. The advisory contains the filenames of some of the malware's artifacts. Threat hunters in the organization can scan log files or configuration management databases (CMDBs) in a search for the presence of those files on their systems to determine whether a similar attack in their own network may be occurring.

Security Advisories

Numerous organizations, including law enforcement, publish human-readable advisories on various cybersecurity events. Security teams in companies often subscribe to one or more of these advisories to be better informed on events occurring around the world. Sometimes these advisories compel security teams to request that their IT departments take action, which could include any of the following:

- Blocking specific IP addresses on external firewalls
- Installing specific security patches
- Making configuration changes to systems or devices
- Threat hunting to look for signs of intrusion
- Blocking e-mail from specific domains, IP addresses, or accounts
- Issuing advisories to the workforce to be on the lookout for signs of suspicious activities

Organizations should actively subscribe to security advisories from the manufacturers of the hardware and software products they use in their environments. Organizations should also subscribe to two or more non-vendor advisory sources.

Identity and Access Management

Identity and access management comprises a collection of activities in an organization that are concerned with the following:

- Management of an accurate inventory of workers in the organization, whether full-time employees, part-time employees, temporary workers, contractors, consultants, or employees of other organizations performing services requiring access to networks, systems, or data
- Management of all of these workers' access rights into networks, systems, data, applications, and places where business operations take place

Identity and access management is getting more difficult. As organizations shift from on-premises to cloud-based computing, the traditional fallback controls of building access and network firewalls are no longer relevant. Only identity and access management processes are available to distinguish persons authorized to access systems and data from those who are not.

Part of the duality of privacy is security. Increasingly, identity and access management is becoming central to security and, therefore, to privacy as well.

Access Controls

Access controls are used to determine whether and how *subjects* (usually persons, but also running programs and computers) are able to access *objects* (usually systems and/or data). Logical access controls work in a number of different ways:

- **Subject access**   A logical access control uses some means to determine the *identity* of the subject requesting access. Once the subject's identity is known and verified beyond a reasonable doubt, the access control performs a function to determine whether the subject should be allowed to access the object. If the access is permitted, the subject can proceed; if the access is denied, the subject cannot proceed. An example of this type of access control is an application that first authenticates a user by requiring a user ID and password before permitting access to the application.

- **Service access**   A logical access control is used to control the types of messages that are allowed to pass through a control point. The logical access control is designed to permit or deny messages of specific types (and may possibly permit or deny based upon origin and destination) to pass. Examples of this type of access control include a firewall, screening router, intrusion protection system (IPS), web content filter, or cloud access security broker that makes pass/block decisions based upon the type of traffic, its content, its origin, and its destination.

These two types of access are like a concert hall with a parking garage. The parking garage (the service access) permits cars, trucks, and motorcycles to enter but denies oversized vehicles from entering. Next door at the concert box office (the subject access), persons are admitted if they possess a photo identification with a name that matches a list of prepaid attendees. Further, certain persons are granted "backstage access" if they possess the required credentials and are not carrying dangerous objects such as weapons.

Access Control Concepts

In discussions about access control, security and privacy professionals often use terms that are not used in other disciplines, including these:

- **Subject, object**   These pronouns refer to access control situations. A *subject* is usually a person, but it could also be a running program, a device, or a computer. In typical security parlance, a subject is someone (or some*thing*) that wants to access something. An *object* (which could be a computer, application, database, file, record, or other resource) is the thing that the subject wants to access.

- **Fail open, fail closed**   This refers to the behaviors of automatic access control systems when they experience a failure of some kind. For instance, if power is removed from a keycard-based building access control system, will all doors be locked or unlocked? The term *fail closed* means that all accesses will be denied if the access control system fails; the term *fail open* means that all accesses will be permitted upon its failure. Generally, security and privacy professionals like access control systems to fail closed because it is safer to admit no one than to admit everyone.

- **Need to know** Individual users should only have access to information required for them to carry out their duties.

- **Least privilege** According to this concept, an individual user should have the lowest privilege possible that will still enable him or her to perform required tasks.

- **Segregation of duties** This concept specifies that single individuals should not have combinations of privileges that would permit them to conduct high-value operations on their own. The classic example is a business accounting department where the functions of creating a payee, requesting a payment, approving a payment, and making a payment should rest with two or more separate individuals. This will prevent any one person from being able to embezzle funds from an organization without notice. In the context of information technology, functions such as requesting user accounts and provisioning user accounts should reside with two different persons so that no single individual could create user accounts on his or her own.

- **Split custody** This is the concept of splitting knowledge of a specific object or task between two or more persons. One example is splitting the password for a critical encryption key between two parties: one person has the first half and the other has the second half. Similarly, the combination to a bank vault can be split so that two persons have the first half of the combination while two others have the second half. In some industries, this practice is known as *dual control.*

Access Control Threats

Because access controls are often the only means of protection between protected assets and users, access controls are often vigorously attacked. Indeed, the majority of attacks against computers and networks containing valuable assets are against access controls in attempts to trick, defeat, or bypass them. Threats represent the intent and ability to do harm to an asset. In the context of privacy, these threats represent the desire for an adversary to access personal information in order to steal it, expose it, corrupt it, or destroy it.

Threats against access controls include malware, eavesdropping, logic bombs, back doors, and scanning.

 **NOTE** The potency and frequency of threats on a system are directly proportional to the perceived value of assets that the system contains or protects.

Social Engineering Is the Initial Attack Vector

Research and numerous surveys reveal that more than 90 percent of successful cyberattacks begin with social engineering—when personnel in an organization are tricked into performing actions that enable an adversary to attack the organization successfully. The most common form of social engineering is phishing, but several other techniques are used as well. Attack techniques used by adversaries are almost always aided by an initial social engineering attack that gives the adversary the beachhead needed to break in to the environment.

Access Control Vulnerabilities

Vulnerabilities are the weaknesses that may be present in a system and that enable a threat to be more easily carried out or to have greater impact. Vulnerabilities alone do not bring about actual harm. Instead, threats and vulnerabilities work together. Most often, a threat exploits a vulnerability, because it is easier to attack a system at its weakest point. Common vulnerabilities include

- **Unpatched systems**   Security patches are designed to remove specific vulnerabilities. A system that is not patched still has vulnerabilities, some of which are easily exploited. Attackers can easily enter and take over systems that lack important security patches.

- **Default system settings**   Default settings often include unnecessary services that increase the chances that an attacker can find a way to break into a system. The practice of *system hardening* is used to remove all unnecessary services and to make security configuration changes on a system to make it as secure as possible.

- **Default passwords**   Some systems are shipped with default administrative passwords that make it easy for a new customer to configure the system. One problem with this arrangement is that many organizations fail to change these default passwords. Hackers have access to extensive lists of default passwords for practically every kind of computer and device that can be connected to a network.

- **Incorrect permissions settings**   If the permissions for access to files, directories, databases, application servers, or software programs are incorrectly set, this could permit access—and even modification or damage—by persons who should not have access.

- **Vulnerabilities in utilities and applications**   System utilities, tools, and applications that are not a part of the base operating system may have exploitable weaknesses that could permit an attacker to compromise a system successfully.

- **Application logic**   Software applications—especially those that are accessible via the Internet—that contain inadequate session management, resource management, and input testing controls can potentially permit an intruder to take over a system and steal or damage information.

Remote Access

Remote access is defined as the means of providing remote connectivity to a corporate LAN through a data link. Remote access is provided by many organizations so that employees who are temporarily or permanently working offsite can access LAN-based resources from their remote locations.

Remote access was initially provided using dial-up modems that included authentication. Although remote dial-up is still provided in some instances, most remote access is provided over the Internet and typically uses an encrypted tunnel, or *virtual private network* (VPN), to protect transmissions from any eavesdroppers. VPNs are so prevalent in remote access technology that the terms *VPN* and *remote access* have become synonymous. Remote access architectures are depicted in Figure 6-5.

Two security controls are essential for remote access:

- **Authentication** It is necessary to know *who* is requesting access to the corporate LAN, and with *what* device. Authentication may consist of the same user ID and password that personnel use when onsite, or multifactor authentication may be required. Authentication may also include a digital certificate or other means for authenticating the device, thereby preventing remote access from assets not owned by the organization.

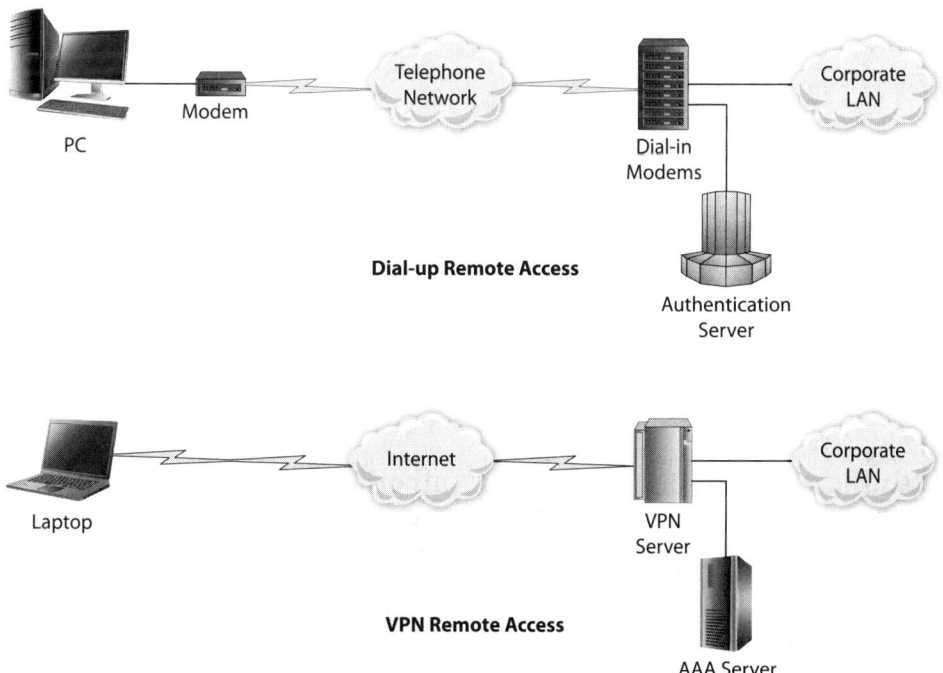

Figure 6-5 Remote access architectures

- **Encryption**   Many onsite network applications do not encrypt sensitive traffic because it is all contained within the physically and logically protected corporate LAN. However, because remote access provides the same function as the corporate LAN, and because the applications themselves sometimes do not provide encryption, the remote access service itself usually provides encryption. Encryption may use SSL, IPsec, L2TP (Layer 2 Tunneling Protocol), or PPTP (Point-to-Point Tunneling Protocol).

These controls are needed because they are a substitute (or *compensating control*) for the physical access controls that are usually present to control which personnel may enter the building to use the onsite corporate LAN. When personnel are onsite, their identity is confirmed through keycard or another physical access control. When personnel are offsite using remote access, because the organization cannot "see" the person on the far end of the remote access connection, the authentication used is the next best thing.

The migration of corporate resources from internal networks to cloud-based networks is changing the notion of remote access. Organizations are incorporating multifactor authentication for access to the organization's cloud-based resources, regardless of the location of users—whether they are on a corporate LAN, working from home, in the field, or traveling.

The New Remote Access Paradigm

As organizations migrate their business applications to colocation centers and XaaS providers, and after the last internal resource is moved to the cloud, what is the point of remote access? Remote access to *what?*

If we think about this in terms of VPN and the protection afforded through encryption, VPN still makes good business sense for the purpose protecting network traffic from potential eavesdroppers (whether the human or malware variety). For this reason, it's preferred to say "VPN" instead of saying "remote access."

Organizations still need to address several subtopics when considering their VPN architectures in light of cloud migration, such as split tunneling, Internet backhauling, and whether VPN should always automatically activate on workstations away from internal corporate networks.

Identification, Authentication, and Authorization

Access to computing resources is protected by mechanisms that ensure that only authorized subjects are permitted to access protected information. Generally, these mechanisms first identify who (or what) wants to access the resource, and then they will determine whether the subject is permitted to access the resource and either grant or deny the access.

Several terms, including *identification*, *authentication*, and *authorization*, are used to describe various activities and are explained here.

Identification   *Identification* is the act of asserting an identity without providing any proof of it. This is analogous to one person walking up to another and saying, "Hello, my name is _____." Because it requires no proof, identification is not usually used alone to protect high-value assets or functions.

Identification is often used by web sites to remember someone's profile or preferences. For example, a bank's web application may use a cookie to store the name of the city in which the customer lives. When the customer returns to the web site, the application will display some photo or news that is related to the customer's location. But when the customer is ready to perform online banking, this simple identification is insufficient to prove the customer's actual identity.

Identification is just the *first* step in the process of gaining entry to a system or application. The next steps are authentication and authorization, which are discussed next.

Authentication   *Authentication* is similar to identification, where a subject asserts an identity. In identification, no proof of identity is requested or provided, but with authentication, some form of proof of the subject's identity is required. That proof is usually provided in the form of a secret password or some means of higher sophistication and security, such as a token, biometric, smart card, or digital certificate. Each of these is discussed later in this section.

When the user presents his or her user ID plus a second factor, whether a password, token, biometric, or something else, the system will determine whether the login request will be granted or denied. Regardless of the outcome, the system will record the login event in an event log.

 **NOTE**   Multifactor authentication is quickly becoming the norm for all human subjects' authenticating to networks and systems containing sensitive information.

Authorization   After a subject has been authenticated, the next step is *authorization*. This is the process by which the system determines whether the subject should be permitted to access the requested resource in the requested manner. To determine whether the subject is permitted to access the resource, the system will perform some type of a lookup or other reference to a business rule. For instance, an access control table associated with the requested resource may have a list of users who are permitted to access it. The system will read through this table to determine whether the user subject's identity appears in the table. If so (and if the type of requested access matches the type permitted in the table), the system will permit the subject to access the resource. If the user's identity does not appear in the table, he or she will be denied access. Whether the login is successful or not, a record of the access attempt (and its disposition) is recorded in an event log.

Typically, permissions are centrally stored by the operating system and administered by system administrators, although some organizations permit the owners of resources to administer user access.

EXAM TIP The terms *identification*, *authentication*, and *authorization* are often misused by business professionals who may not realize the differences between them. For the exam, and in your professional capacity, you need to understand the differences.

User IDs and Passwords   User IDs and passwords are the most common means for users to authenticate to a resource—whether it is a network, server, or application.

User IDs   In most environments, a user's user ID will not be a secret; in fact, user IDs may be a derivation of the user's name or an identification number. Some of the common forms of a user ID include combinations of the user's first and last name or an employee ID number.

 **CAUTION**   Confidential numbers such as social insurance (Social Security in the United States) or driver's license numbers should not be used as user IDs, because these identifying numbers are generally meant to be kept confidential.

Passwords   Whereas a user ID is not necessarily kept confidential, a password *always* is kept confidential. A password, also known as a *passphrase*, is a secret combination of letters, numbers, and other symbols known only to the actual user. End users are typically advised the following about passwords:

- Select a strong password or passphrase that is easy to remember but difficult for others to guess.
- Passwords must never be shared or used by others.
- Passwords must never be transmitted over any network.
- Passwords should be stored in a secure password vault.
- Each system should have a unique password.
- Passwords used for personal accounts should not be used for any work-related account.

User Account Provisioning   When a user is issued a new computer or system user account, he or she needs to know the password to access the resource. Generating and transmitting an initial password to a user can be tricky, because passwords should never be sent in an e-mail message. A sound practice for initial user account provisioning would involve the use of a limited time, one-time password that would be securely provided to the user; upon first use, the system would require that the user change the password to a value that no one else would know.

NOTE Ideally, users will be required to change their password as soon as they have their new user account, but some systems don't even permit this. Privacy and security professionals should understand an environment's capabilities as well as the risks and value of the assets being protected. Any recommendations should reflect system capabilities and asset value.

Risks with User IDs and Passwords Password-based authentication is among the oldest in use in information systems. Although password authentication is still quite prevalent, a number of risks are associated with its use because of the different ways in which passwords can be discovered and reused by others. Some of these risks involve the following:

- Eavesdropping
- Key logging
- Phishing
- Finding a password written down
- Finding a stored password
- Exploiting a browser's password store

These follow the same theme: user IDs and passwords are static and, if discovered, can be used by others. For this reason, other, more secure means for authentication have been developed, including biometrics, tokens, smart cards, and certificates, all of which are collectively known as *multifactor authentication*.

Multifactor Authentication Multifactor authentication (MFA) is so called because it relies not only on "something you know" (namely, a user ID and password), but also upon "something you have" (such as a key card or smart card) and/or "something you are" (such as a fingerprint). MFA requires a user's user ID and password, but the user must also possess something or use a biometric to form a part of the authentication. Several technologies are used for MFA, including tokens, soft tokens, SMS tokens, smart cards, digital certificates, and biometrics.

Users of MFA systems need to be trained on their proper use. For example, they need to be told not to store their tokens or smart cards with their computers, and to keep their smartphones or mobile devices locked except when in use.

NOTE SMS-based MFA is increasingly being considered unsafe because of risks associated with SIM fraud. Other methods are preferred as a result.

Biometrics A number of different biometrics authentication technologies have a common theme: all use some way of measuring a unique physical characteristic of the person who is authenticating. Some of the technologies in use are

- Fingerprint
- Handprint
- Voice recognition
- Iris scan
- Facial scan

Reduced Sign-On In a reduced sign-on environment, several applications use a centralized directory service such as LDAP (Lightweight Directory Access Protocol), RADIUS (Remote Authentication Dial-in User Service), Diameter, or Microsoft Active Directory for authentication. The term comes from the result of changing each application's authentication from stand-alone to centralized and the resulting reduction in the number of user ID–password pairs that each user is required to remember.

 EXAM TIP The terms "reduced sign-on" and "single sign-on" are often used interchangeably. Many times, a reduced sign-on environment is labeled as single sign-on. For the exam, remember that they are not the same.

Single Sign-On In a single sign-on (SSO) interconnected environment, applications are logically connected to a centralized authentication server that is aware of the logged-in/logged-out status of each user. At the start of the workday, when a user logs in to an application, he or she will be prompted for login credentials. When the user logs in to another application, the application will consult the central authentication server to determine whether the user is logged in and, if so, the second application will not require the user's credentials. The term refers to the fact that a user needs to sign on only one time, even in a multiple-application environment.

SSO is more complicated than reduced sign-on. In an SSO environment, each participating application must be able to communicate with a centralized authentication controller and act accordingly by requiring a new user to log in, or not.

Access Control Lists *Access control lists* (ACLs) are a common means to administer access controls. ACLs are used by many operating systems and other devices such as routers as a simple means to control access to a resource such as a server or a network.

On many devices and systems, the list of packet-filtering rules (which give a router many of the characteristics of a firewall) is known as an ACL. In the Unix operating system, for instance, ACLs can control which users are permitted to access files and directories and run tools and programs. ACLs in these and other contexts are often simple text files that can be edited with a text editor.

Access Control Processes

Sound business processes must be in place for access controls to manage user access effectively and protect critical systems and sensitive information. These business processes should be documented and detailed business records kept that document all related activities. Formal roles and responsibilities must be defined so that only authorized persons may perform various functions. Access control processes generally fall into two categories: the processing of access requests and periodic access reviews.

Access Requests   Formal access request processes should be used to control the provisioning of user access. Using the principle of separation of duties, the actions of requesting access, approving access, and providing access should be performed by three different individuals.

Each step in an access request process should be recorded. These business records permit audits of access requests processes to confirm that only properly issued and processed access requests result in the granting of user access.

 **NOTE**   The approver of an access request should be the system owner, typically an individual in a business unit or department. IT personnel, who act as stewards for information systems, should not be approvers of access requests.

Access Reviews   The rate of change in organizations creates the need for periodic reviews of access rights to ensure that all subjects that have access to systems and sensitive data still require that access. Several types of reviews ensure that provisioning and deprovisioning processes are effective, accurate, and timely.

Reviews are warranted even in organizations with automation and workflow in their identity and access management processes. Reviews are even more critical in organizations that use manual processes. The objectives of access reviews ensure that access management processes remain effective and accurate, and that access rights remain valid and justified.

Several types of access reviews are performed:

- **Access certifications**   System owners review access rights for subjects and confirm that each subject still requires access rights. Any subjects that no longer require access rights are flagged, and their accesses are removed.

- **Provisioning certifications**   Access management personnel examine subjects' access rights and confirm that there is valid evidence of properly executed requests, reviews, approvals, and execution for each.

- **Deprovisioning certifications**   Security personnel obtain lists of terminated personnel from human resources and confirm that deprovisioning was properly and timely executed for each.

- **Activity reviews**   Security personnel examine systems to determine whether subjects have logged into them recently. Inactive user accounts can be flagged for removal if subjects have not logged into them for extended periods of time, because they probably do not require access.

- **Segregation of Duties (SOD) matrix reviews**   Periodic reviews of SOD matrices help to determine whether all disallowed combinations of access are represented in SOD matrices. This is only a review of the roles themselves, not the persons who have the roles.

- **Segregation of duties reviews**   Reviews of subject accesses to detect SOD exceptions confirm whether any persons have access rights that violate the segregation of duties policy.

- **Temporary worker reviews**   In organizations lacking centralized management of temporary workers, additional reviews will be needed to ensure that no active user accounts exist for temporary workers who are no longer active in the organization.

- **Privileged account reviews**   All of the reviews listed here should be performed at a higher frequency for privileged accounts. This higher frequency is warranted because of the additional powers associated with privileged accounts and the greater damage that may result in cases of abuse and compromise.

- **Service account reviews**   These reviews determine whether service accounts are still being used, where and how they are used, and who manages them to ensure that there is no unauthorized use of service accounts.

In the absence of access governance tools, some of these reviews may be labor intensive. Because of these, organizations will perform these reviews on a risk basis: reviews of more critical systems will be performed more frequently than others.

Access Monitoring   Since many cyberattacks begin with attempts to compromise individual user and system accounts, continuous monitoring of user and system account activity is considered an essential practice in cybersecurity. This monitoring is typically achieved through the real-time transmission of user account events to a centralized log server, or better yet to a SIEM system, so that alerts on suspicious behavior can be created and such matters investigated.

The events that should be sent to a log server or SIEM include the following:

- **All successful logins**   Ideally these will include the originating IP address and geolocation of the login event.

- **All unsuccessful logins**   This also needs to include IP address and location.

- **All user account permission changes**   This should include the IP address and/or user account that performed the change.

- **All user account creations**   This should include IP address and user account performing the change.

- **Privileged account changes**   This includes permission changes and password changes.

- **Service account changes**   This includes the creation of and modification to any service account.

Organizations need to develop "use cases" in their SIEM systems to alert security personnel of events that warrant investigation and action. For instance, if a user account logs in from the United States, and then a short time later there is a login for the same user account in another country, an investigation should immediately commence to determine whether the user account has been compromised, resulting in an adversary logging into the account from the foreign location.

Chapter Review

Control objectives are statements of desired states or outcomes from business operations. When building a security program, and preferably prior to selecting a control framework, the organization must establish high-level control objectives.

It takes time (at least one full risk management cycle) for an organization to establish a framework of controls that address all identified risks, and additional time for control assessment is needed to determine whether all controls are effective.

Whether networked computers are located across the room or halfway around the world, networks facilitate all computer communications, both wired and wireless, whether telecommunications services are used or not.

The TCP/IP protocol suite employs a technique known as encapsulation, whereby messages in higher level protocols are encapsulated within messages in lower level protocols, which in turn are encapsulated in messages in the physical network medium.

Some organizations employ the technique of network segmentation, in which the network is divided into security zones, and controls between them (such as firewalls) restrict network traffic only to that which is considered necessary.

Encryption is used to hide information in plain sight. It works by scrambling the characters in a message using a method known only to the sender and receiver, making the message useless to any party that intercepts the message.

A private key cryptosystem relies on the use of an encryption key known to both parties, whereas a public key cryptosystem utilizes public/private key pairs. The selection of a private versus public key cryptosystem depends upon how it will be used, and by whom.

Hashing is the process of applying a cryptographic algorithm on a block of information that results in a compact, fixed-length "digest." A message digest can be used to verify the integrity of a large file, thus assuring that the file has not been altered.

A digital signature is a cryptographic operation where a sender "seals" a message or file using his or her identity. The purpose of a digital signature is to authenticate a message and to guarantee its integrity.

The term *key management* refers to the various processes and procedures used by an organization to generate, protect, use, and dispose of encryption keys over its lifetime.

De-identification refers to any of several techniques available that all serve a single purpose: to effectively remove from business records any traceable reference to a specific natural person.

Event monitoring is the practice of examining the events occurring on information systems, including applications, operating systems, database management systems, end-user devices, file servers, and every type and kind of network device, and being aware of what is going on throughout the entire operating environment.

Monitoring of activities related to data access can help an organization identify improper uses of personal information.

A security information and event management (SIEM) system collects and analyzes log data from many or all systems in an organization and produces alerts to inform personnel of specific events.

Organizations intent on proactively protecting sensitive information, including personal information about customers, constituents, and employees, may implement one or more types of data loss prevention (DLP) systems.

DLP systems can be used to scan static data, observe data usage, and even block unwanted data usage events.

Human- and machine-readable threat intelligence feeds help organizations understand threats occurring on the Internet and thus can help organizations better protect themselves from attack.

Orchestration refers to a scripted, automated response that is automatically or manually triggered when specific events occur.

Threat hunting is the practice of conducting searches—typically in SIEM logs and configuration management databases—to see whether traces of intrusions are present in their systems.

Identity and access management is a collection of activities in an organization that is concerned with the inventory of an organization's workers and their access rights in information systems and applications.

Access controls are used to determine whether and how subjects (usually persons, but also running programs and computers) are able to access objects (usually systems and/or data).

Because access management is central to information protection, privacy professionals need to become familiar with access management techniques as well as the vocabulary of access management, including terms such as subject, object, fail open/closed, least privilege, segregation of duties, and split custody.

System hardening techniques contribute to the security of access control systems; practices such as patching, hardening, and changing default passwords make systems more resistant to attack.

Remote access is defined as the means of providing remote connectivity to a corporate LAN through a data link. Remote access is provided by many organizations so that employees who are temporarily or permanently working offsite can access LAN-based resources from their remote location.

Identification is the act of asserting an identity without providing any proof of it.

Authentication is similar to identification in that a subject asserts an identity, but it also requires proof of identity through the use of a password or access token.

Authorization is the process through which a system determines what access rights an authenticated user will be given.

Multifactor authentication (MFA) is so called because it relies not only on "something you know" (namely, a user ID and password), but also upon "something you have" (such as a key card or smart card) and/or "something you are" (such as a fingerprint).

Reduced sign-on and single sign-on refer to authentication and authorization mechanisms intended to streamline users' access to multiple business applications.

Access request processes need to be carefully designed so that only valid access requests are fulfilled.

Access review processes ensure that access request processes are effective and confirm that users still require access to do their jobs.

Access monitoring helps an organization detect anomalous behaviors that may be signs of malicious activities that could lead to a breach.

Quick Review

- Privacy control objectives resemble ordinary control objectives but are set in the context of information security and privacy.

- A valid approach to control framework selection has more to do with the structure of controls than the controls themselves.

- The nature of network media is generally independent of the protocols used. However, newer media types typically do not support older protocols.

- Ethernet and Wi-Fi are the dominant media used in business and residential data networks, with TCP/IP as the universal network protocol.

- The term *segmentation* does not by itself imply whether network access controls exist to separate network segments.

- Encryption is considered another form of access protection. While powerful, encryption needs to be carefully considered, designed, and implemented.

- Often, a failure of cryptosystems results from poor implementation or a failure to protect encryption keys.

- A SIEM can and is often used to inform personnel of not only security events, but also of operational and privacy-related events.

- Orchestration can help an organization respond to threats more quickly and effectively.

- The potency and frequency of threats on a system are directly proportional to the perceived value of assets that the system contains or protects.

- Due to the migration of on-premises systems to the cloud, the concept and practice of remote access is giving way to VPN in order to protect network traffic for personnel working offsite.

- The terms *reduced sign-on* and *single sign-on* are often confused by persons who are not familiar with the inner workings of authentication.

- The frequency and rigor of various types of access reviews should be determined by the degree of automation as well as the prior history of such reviews.

Questions

1. What is the most significant factor that compels an organization to implement a new control?

   A. Security or privacy breach

   B. New regulation

   C. Results of a risk assessment

   D. Contents of a control framework

2. All of the following are forms of control assessment except:

   A. Document review

   B. Control self-assessment

   C. Internal audit

   D. External audit

3. Which of the following network media is used to carry broadband traffic in bulk?

   A. Twisted-pair

   B. 4G

   C. 5G

   D. Fiber-optic cable

4. Which protocol is most often transported on fiber-optic cabling by telecommunications providers?

   A. SONET

   B. DSL

   C. ISDN

   D. T-1

5. An auditor is interviewing a network engineer who describes the enterprise network as being "flat." To which of the following is the network engineer referring?

   A. The organization's internal firewalls are set to "any any."

   B. The organization's network uses private addressing.

   C. The organization's network consists of several collision domains.

   D. The organization's network contains no internal access controls.

6. In the context of cryptosystems, the term plaintext refers to which of the following?

   A. An unformatted text file

   B. An encryption key

   C. An unencrypted message

   D. An encrypted message

7. A privacy auditor has observed that PII fields in a relational database are encrypted with the DES algorithm with 64-bit keys. Keys are held in split custody between two teams of operations specialists. What should the auditor conclude from this observation?

   A. The database encryption is strong.

   B. The database encryption is weak.

   C. The key management method is weak.

   D. The encryption cipher is adequate.

8. Which of the following best describes symmetric encryption?

   A. Plaintext and ciphertext occupy the same amount of storage.

   B. Encryption and decryption use the same algorithm.

   C. All parties have a copy of the encryption key.

   D. All parties have a copy of public keys.

9. A messaging system employs hashes that accompany each message. What function can hashing perform in this context?

   A. Verify the integrity of a message.

   B. Verify the integrity and origination of a message.

   C. Guarantee the confidentiality of a message.

   D. Verify the origination of a message.

10. An auditor has noted that an organization's network routers are administered via the TELNET protocol. What should the auditor conclude from this?

    A. The organization employs a flat network.

    B. A sight-impaired administrator administers network routers.

    C. Network routers are adequately protected.

    D. A more secure protocol than TELNET should be used.

11. An organization periodically copies its customer database to a test environment. When doing so, names and other sensitive fields are substituted with made-up names and numbers. What substitution process is the organization performing?

    A. Data scrubbing

    B. Anonymization

    C. Pseudonymization

    D. Field erasure

12. Which of the following is considered a best practice with regard to event logging?

    A. Retain all event logs on the systems that create them.

    B. Transmit all event logs to a central log server.

    C. Suppress the creation of event logs on all systems.

    D. Encrypt all event logs on the systems that create them.

13. An organization wants to implement a data loss prevention (DLP) system. Which of the following is considered the best approach for such an implementation?

    A. Employ DLP in passive mode initially.

    B. Employ DLP in active mode initially.

    C. Set DLP in high-sensitivity mode.

    D. Employ DLP on e-mail systems first.

14. A privacy manager has directed that the team managing encryption keys update the password protecting encryption keys in a way that half the team members know one half of the password, and the other half of the team knows the other half of the password. What control has been implemented?

    A. Fail closed

    B. Least privilege

    C. Segregation of duties

    D. Split custody

15. An organization relying on physical access controls has migrated its on-premises applications to cloud service providers. What compensating control should be enacted for access to cloud-based applications since physical access is less of a factor?

    A. Multifactor authentication

    B. Biometrics

    C. Single sign-on

    D. Reduced sign-on

Answers

1. **C.** In an effective risk management process, risk assessments identify risks that sometimes result in an organization developing a new control to ensure that the risk is reduced in some way.

2. **A.** Document review is not considered a form of control assessment. Control self-assessment (where control owners self-attest to their controls' performance), internal audit, and external audit are all forms of control assessment.

3. **D.** Fiber-optic cable is used for virtually all bulk high-speed data traffic. Mostly used by telecommunications providers, fiber-optic cables crisscross continents and are also used in undersea cables.

4. **A.** SONET, or synchronous optical networking, is the protocol used by telecommunications providers over fiber-optic cables.

5. **D.** A "flat network" generally signifies that there is no segmentation nor internal access controls on the network; every device is free to communicate with every other device.

6. **C.** In cryptography, plaintext is an original, unencrypted message. A plaintext message can take any form, including a text file, an office artifact (document, worksheet, or presentation), an image file, or a video file. An encrypted file is referred to as ciphertext.

7. **B.** The DES encryption algorithm is considered inadequate by today's standards. The key length of 64 bits is also considered inadequate. Depending upon other factors, the AES algorithm with a 128-bit key length is considered a good starting point.

8. **C.** A cryptosystem using a symmetric encryption algorithm requires that all parties have advance knowledge of the encryption key. A difficulty with symmetric encryption is the need for all parties to exchange the encryption key without compromising its confidentiality: Until all parties can send encrypted messages to one another, how do they go about transmitting an encryption key to one another?

9. **B.** Hashing is a cryptographic function that can be used to verify the integrity and origination of a message when used as a part of a digital signature. Hashing does not provide confidentiality—encryption would be required instead.

10. **D.** The TELNET protocol is considered inadequate, primarily because all communication is transmitted in plaintext, including login credentials. Other protocols, such as SSH, should be used instead.

11. **C.** In a database containing some sort of PII, the process of substituting actual names and other fields, such as addresses and account numbers, with made-up names and numbers is known as pseudonymization.

12. **B.** The best practice of the management of event logs is the creation of a central log server, to which all logs from all systems are transmitted. A further best practice is the employment of a correlation engine that will produce alerts when security incidents are suspected.

13. **A.** The best approach to ensure long-term success with a DLP system is to configure it initially in passive mode. This means that the DLP system will not interfere with any access, use, or transmission of PII, but instead will silently log all such instances. The purpose of this approach is to help security and privacy professionals understand how PII is used, prior to having the DLP system intervene in what could be legitimate business processes.

14. **D.** Split custody, also known as split knowledge, is a practice whereby one person or group has knowledge or access to part of a control or password, and another person or group has knowledge or access to another part of the control or password. This necessitates that one person from each group be involved in accessing a protected resource.

15. **A.** Multifactor authentication is the best compensating control for restricting access to business applications once those applications have been migrated to cloud-based environments.

PART III

Data Cycle

Data Purpose

In this chapter, you will learn about
- Data governance components, including policy, controls, assessments, and reporting
- Data inventory as a first step for managing privacy data
- Data classification policy and handling standards
- System and site classification
- Data loss prevention techniques and tools
- Data quality and accuracy
- Data use limitation and data analytics

This chapter covers Certified Data Privacy Solutions Engineer job practice 3, "Data Cycle," part A, "Data Purpose." The entire Data Cycle domain represents 30 percent of the CDPSE examination.

Privacy programs require an effective data governance program to provide management visibility and control of personal information. Data governance is discussed in this chapter, including data inventory and data classification and handling, supplemented by data loss prevention (DLP) tools and techniques, measures to ensure data quality and accuracy, and measures to ensure data use limitation. The chapter concludes with a discussion on data analytics techniques and benefits.

Data Governance

Put simply, data governance is management's visibility and control over the use of information in an organization. By defining strict and tangible consequences for the failure to protect and use personal information transparently, privacy laws have ushered in the emergence of policies and practices that shine a light on data collection, usage, and protection. Organizations are now accountable for confronting data sprawl and indiscriminate use of personal information.

A typical data governance structure contains the following:

- High-level policy and related standards defining data management practices
- Defined roles and responsibilities for data management
- Key controls

- Assessments of key controls to ensure they are effective
- Reporting to management describing incidents, activities, and assessments

A key prerequisite to effective data governance is organizational change management—that is, management must have visibility into and control over changes made to business processes. Organizations lacking organizational change management will find that processes will change—including new and changed uses of personal information that may be contrary to policy—without management's awareness.

Policies and Standards

In the context of data governance, policies and standards define required behavior for personnel associated with data architecture, data management, and data usage. Data governance policies and standards will address topics including

- Approvals required for the acquisition of new data sources
- Approvals required for new or changed uses of existing data sources
- Safeguards to protect data from unauthorized access and use

Policies and standards will also define roles and responsibilities and imply the development of controls.

Roles and Responsibilities

A data governance charter or policies and standards should define roles and responsibilities concerning the management of data, including

- Decisions for access to data and databases
- Reviews of access rights to data and databases
- Decisions and reviews for uses of data and databases
- Ownership of individual controls
- Investigations into misuse and unauthorized access to data and databases

Readers versed in information security will recognize these roles and responsibilities as essential parts of a comprehensive information security program.

Control Objectives and Controls

Following the development of policies, standards, roles, and responsibilities, control objectives and controls can be developed. Control objectives and individual controls specify key desired outcomes to ensure that data governance policies will be carried out.

The functional areas where controls will be developed include

- Approvals for the acquisition of new data sources
- Approvals for new uses of data

- Monitoring of data usage
- Approvals for requests to access data
- Reviews of access to data

Organizations will develop processes and procedures that include these controls.

Assessments

The effectiveness of policies and controls cannot be fully known unless they are assessed or audited. The criticality of controls and the applicability of specific regulations will determine the approach and rigor needed to assess controls, whether they are reviewed, assessed, or audited.

Prior to recently enacted privacy laws, many organizations paid little attention to risks associated with the protection and use of personal information. Overall and focused risk assessments concerning the use of personal information are warranted, however.

Assessing controls alone addresses their effectiveness but may overlook aspects of privacy and security where no controls exist. Control assessments and risk assessments should be included in the organization's overall risk management life cycle, as discussed in Chapter 3 and in more detail in *CISM Certified Information Security Manager All-In-One Exam Guide*.

Reporting

Governance is incomplete if management is uninformed of routine business activities and incidents that occur in a program. Management needs to be periodically informed of how many incidents occur and how effective they have been at circumventing controls, and the effectiveness of incident response, corrective actions, and improvements.

Data Inventory

A nearly worn out but still highly relevant cliché in data security is this: *you cannot protect what you don't know you have*. This statement underscores the need for effective asset management at all levels, because only specifically identified data can be managed and protected.

For a privacy program to be effective, organizations must have a complete and accurate inventory of all personal information. Although an inventory of structured information (data residing in application database management systems) will remain fairly static, the transient nature of unstructured data creates additional challenges. Somehow, organizations must identify means for knowing about all structured and unstructured data, particularly when it contains personal information that is in scope of relevant privacy laws. Proactive data discovery, discussed later in this chapter in the section "Data Loss Prevention Automation," can be put in place to provide visibility into the creation and use of unstructured data.

For an organization's data inventory to remain current, two activities need to become a part of business-as-usual processes:

- Change management processes must require updates to data inventory whenever an addition or change to an information system impacts the data inventory.
- Business processes that interact with personal information must be documented.
- Periodic reviews of the data inventory should be performed to confirm its accuracy.

Periodic data inventory reviews should not only catalog existing instances of sensitive and personal information, but they should also determine in each instance whether data *should* exist where it is found. By understanding the business processes that interact with personal information, we can better identify where personal information is collected, processed, and stored. Thus, a data inventory should be thought of less as a census and more of a gap analysis. Every instance of sensitive information should be examined through the lens of the business processes and data management policy to determine whether each instance should exist and whether current protective controls are adequate.

Organizations with lower process maturity are more likely to use unstructured means for performing procedures and completing tasks. Often this will result in a greater use of e-mail for process workflow and a greater use of unstructured data stores for storing data. E-mail, file servers, and cloud storage services represent the majority of unstructured data in many organizations.

When inventorying data, you should include the following information in each catalog entry:

- Name of the file(s) or directory/directories
- Description of the contents, including personally identifiable information (PII) data fields
- Date of last update (this will aid in the removal of old data)
- Access permissions
- Data owner

This information will be useful for the development of data flow diagrams, as discussed later in the section "Data Flow and Usage Diagrams."

 **NOTE** Automation in larger organizations will ease the burden of otherwise manual and time-consuming processes to keep data inventories up-to-date.

Data Classification

Many types of information reside in an organization's information systems. Some of this information is highly sensitive because it contains personal information, intellectual property, and internal financial information; some information is important but not sensitive at all, and some is not very important. Because resources are required to protect information, it doesn't make much sense to apply equal rigor to protect both unimportant data and highly secretive or sensitive information. To this point, former US national security advisor McGeorge Bundy is attributed to have said, "If we guard our toothbrushes and diamonds with equal zeal, we will lose fewer toothbrushes and more diamonds."

Data Classification Levels

A *data classification policy* is a formal and intentional way for an organization to define levels of importance or sensitivity to information. A typical data classification policy will define two or more (but rarely more than five) data classification levels, such as the following:

- Registered
- Restricted
- Confidential
- Public

Along with defining levels of classification, a data classification policy will include examples that show the classification levels that should be assigned to various data sets. Table 7-1 provides an example of this concept.

 **NOTE** Data classification policies need to be as simple as possible so that workers will be able to understand the classification of data easily and handle it accordingly.

Classification Level	Examples of Information at this Level
Registered	Merger and acquisition proceedings, pre-announcement
Restricted	Customer PII
	Employee PII
	Program source code
	Unpublished financial records
Confidential	Internal e-mail messages
	Marketing plans
	Policies
Public	Web site content
	Released marketing brochures
	Social media postings
	Published financial reports

Table 7-1 Examples of Information at Varying Data Classification Levels

Classification Level	Data Sets at this Level
Registered	Merger and acquisition proceedings, pre-announcement
	Service account passwords
Restricted	Customer Relationship Management system database
	Human Capital Management system database
	Program source code repositories
	Unpublished Enterprise Resource Management system records
	Unpublished annual report, 10-K, 10-Q
	IT network diagrams
	IT data flow diagrams
	Contents of internal HR and legal investigations
	All legal contracts
Confidential	Internal e-mail messages
	Internal memos
	Contents of an internal intranet site, including policy and benefit information
	Marketing plans
Public	www.company.com web site content
	Released marketing brochures
	Official @Company social media postings on LinkedIn, Facebook, and Twitter
	Public 10-K and 10-Q filings, when published
	Annual report, when published

Table 7-2   Examples of Official Data Classification Level Assignments

Data classification policy can go still further and emphatically state the classification levels that are, by policy, assigned to specific data sets. This is shown in Table 7-2.

Data Handling Standards

Because so much information is handled on a daily basis by personnel, data classification policy goes still further to define acceptable handling procedures for data at various levels of classification and in numerous types of situations. Often called *data handling standards*, these procedures provide real-world guidance that workers can easily follow and use. Because information can be used and moved in so many different ways, data handling standards should clearly state what is expected of personnel when handling sensitive data.

Data handling standards usually take the form of a matrix, with various levels of classification as the columns and different data handling situations as rows. Each individual cell defines the standard for handling data at a given classification level in a certain way. Table 7-3 shows a part of such a matrix.

	Public	Confidential	Restricted	Registered
Laptop storage	Permitted	Must be encrypted	Must be encrypted	NOT permitted
USB drive storage	Permitted	Secure drive only	Secure drive only	NOT permitted
File server storage	Permitted	"Mars" server only	"Mars" server only	"Ares" server only
Cloud storage	Permitted	OneDrive only	OneDrive only	NOT permitted
E-mail, internal	Permitted	Permitted	Permitted	Must be encrypted
E-mail, external	Permitted	Must be encrypted	Must be encrypted	NOT permitted
Fax	Permitted	Attended destination only	Attended destination only	NOT permitted
Courier	Permitted	Permitted	Must be encrypted	Must be encrypted

Table 7-3   Example Data Handling Standards Matrix

To help the workforce better understand handling standards, organizations should develop training content or tutorials to explain in detail their meanings and to introduce appropriate procedures.

EXAM TIP   CDPSE candidates need to understand the typical structure of roles and responsibilities regarding the protection of data. While a data protection officer, general counsel, or CISO is responsible for establishing the organization's data classification policy, it is usually the responsibility of a document owner to classify and mark a document correctly. It is then the responsibility of any party who uses a document to handle it according to its classification level.

The Culture Shift

Privacy professionals in organizations introducing data classification, handling standards, training, and automation must understand that such an undertaking may represent a significant culture shift. It's potentially a tall order to expect a workforce that formerly took data handling for granted to become data aware and to understand and follow new procedures. Such a change does not happen overnight. Even when executives lead by example and in the presence of an internal marketing plan, many workers' responses will range from confusion to resistance and outright evasion. It is therefore important for the workforce to understand the purpose of data classification.

PART III

Data Loss Prevention Automation

Along with defining levels of classification, a data classification policy will define policies and procedures for handling information in various settings at these levels. For instance, a data handling standard will state the conditions at each level in which sensitive information may be e-mailed, faxed, stored, transmitted, and shipped. Note that some methods for handling may be forbidden—such as e-mailing a registered document over the Internet.

Relying on an organization's workers to apply data handling standards consistently is chancy at best—not because of the lack of good intentions, but because workers simply will not have safe data handling on their minds all of the time. This situation is not unlike workers who click on the occasional phishing message despite having attended effective security awareness training. People are simply not "on their guard" all of the time.

Data loss prevention (DLP) systems can greatly aid in the effort to provide visibility and even control over the use of personal and other sensitive information. Approaches to the implementation of DLP capabilities fill the remainder of this section.

Static DLP

Static DLP tools scan static data stores to identify files containing data matching specific patterns. Most often, static DLP scanning is performed on file servers—both the on-premises and cloud varieties. DLP scanning can also be performed on database management systems, either by specific tools or by scanning flat-file exports of databases.

Organizations undertaking static DLP scanning for the first time may find an abundance of files containing PII and other sensitive data. Privacy managers need to keep in mind that such data may have accumulated over a long period of time, and it may also represent current practices or former activities that are no longer practiced.

A careful analysis of the results of an initial DLP scan should be undertaken to determine the following:

- The age of files containing PII found in file stores
- The extent to which files containing PII are still being deposited in file stores
- The access rights of files containing PII
- Which users actively access the files (available in some DLP static scanning tools)
- Whether current use is following sanctioned policies, procedures, or practices

Privacy managers should not be overly hasty in the quest to "solve" any or all of the discovered instances of PII in static data stores. Some uses may be a part of key business processes along with adequately restricted access controls. Often, privacy managers will find that files containing PII in file stores are the result of one-time or ad hoc activities. For instance, a business analyst may be asked to perform a research task on the demographics of customers; the business analyst would perform a query or run a report on the customer relationship management (CRM) system, export the report to a spreadsheet, and save the spreadsheet on a file server. After completing the task, the business analyst will keep the file there in case questions are asked about it later. Soon the existence of the spreadsheet containing PII is forgotten, and there it will reside in perpetuity unless some purge, cleanup, or DLP scan is performed that discovers its existence.

 EXAM TIP CDPSE candidates need to understand the detective nature of static DLP tools. As it merely scans file stores to determine whether sensitive data is present, static DLP is an indicator of behavior but does nothing by itself to alter it.

Data Tagging

Through the process is similar to static DLP analysis, data files can be tagged, or marked in some way, if they are found to contain PII or other sensitive information. Such tagging can take on several forms, including these:

- **Metadata tagging**   The metadata of a data file can be updated to include a specially coded tag that will be recognized by DLP tooling.
- **Watermarking**   Visible or invisible watermarks can be added to data files.
- **Marking**   According to data file-marking policy, a human-readable word or phrase can be added to the header, footer, or other location in a data file (such as "XYZ Company restricted to internal use only").

Including human-readable watermarks or other markings on documents provides a visual reminder to workers using a data file about the sensitivity of the files they are working with. The main purpose of including machine-readable marks or tags in a data file is to facilitate appropriate action by dynamic DLP tools, discussed next.

Dynamic DLP

Dynamic DLP represents a variety of technologies used to detect and even intervene in the transfer of PII and other sensitive information. Dynamic DLP tools can take the form of network devices or tools running on operating systems with the ability to observe data in motion in many different circumstances. These tools can be configured to identify the specific sensitivity of data files in motion by reading their contents to determine whether they contain specific PII or other sensitive information, or they can be configured to look for previously applied tags.

Dynamic DLP Types   There are several common forms of dynamic DLP:

- **E-mail DLP**   DLP tools can examine the contents of an outgoing e-mail message to determine whether it contains specific sensitive information. E-mail-based DLP will consider whether the information is being sent to internal or external recipients.
- **USB storage control**   Host-based DLP tools restrict USB usage to company-approved (and usually encrypted) USB drives only, or they block USB storage entirely.

- **Local file storage control**   Host-based agents observe and optionally block actions violating policy, such as local storage of highly classified documents.

- **File server storage control**   Host- or server-based agents observe and optionally block actions violating policy, such as storage of highly classified documents on shares with broad access.

- **Cloud server storage control**   DLP capabilities in cloud storage services are configured to mimic local or file server DLP controls to monitor and optionally block actions violating policy.

- **Network DLP**   Network devices, or DLP modules in next-generation firewalls, observe the content of data in motion.

Dynamic DLP Actions   Most of these forms of dynamic DLP, when operating in the context of an interactive user, can present the following to the user:

- Silently note the occurrence

- Block the action and inform the user

- Warn the user that the intended action is forbidden by policy, and give the user the ability to permit the action anyway after providing a user-entered business justification to complete the action

Other tools may be used to assist in dynamic DLP efforts:

- **Firewalls**   Blocking access to/from specific networks or systems

- **IDS/IPS**   Blocking access to/from networks and systems thought to be hazardous

- **Web content filter**   Blocking browser access to sites based on policy

- **Cloud access security broker (CASB)**   Monitors and controls access to cloud-based service providers based on organization policy

- **NetFlow**   Monitors network traffic and produces alerts when anomalous traffic is seen on the network

 **EXAM TIP**   CDPSE candidates need to understand the role of Dynamic DLP as a detective control, a preventive control, or both.

Implementing Dynamic DLP   Dynamic DLP controls, in any of the forms discussed, can be highly valuable for preventing the mishandling of sensitive information. Unfortunately, dynamic DLP is also adept at interfering with legitimate business processes by blocking activities approved by management (including the privacy or security manager). Legitimate activities are blocked either because the DLP system is misconfigured or because of a "false positive" situation where the DLP system misidentifies data. Examples of such false positives include strings of numerals that are mistaken for social insurance numbers, bank account numbers, phone numbers, or credit card numbers.

Run in Learn Mode First Despite what readers may be told by experienced users of DLP systems or DLP vendors themselves, readers are *highly* recommended to run any dynamic DLP system in "learn" mode first for an extended period of time. The best way to do this is to configure the DLP system to silently log occurrences that are thought to be file handling policy violations. This enables privacy and security personnel to see whether the DLP system properly identifies the actual motion of private and other sensitive information.

Another highly useful benefit of running DLP in learn mode is similar to that of running static DLP scans: learning what data movement currently takes place in the organization to help personnel better understand existing business processes, as well as those occasional (hopefully not frequent) actions that represent actual violations of policy.

When privacy and security personnel become confident in their dynamic DLP system's ability to identify sensitive data movement correctly without false positives, they can proceed to begin activation of preventive actions to be performed by the DLP system. It is suggested that organizations proceed slowly as they build their confidence in the proper operation of the system.

Enlist Friendly Users Privacy and security personnel should identify "friendly" departments or groups when first implementing the preventive features of a DLP system, to increase confidence in the system and work out any remaining configuration problems.

Develop Response and Exception Procedures Before activating any rules in a dynamic DLP system that will block file handling actions, privacy and security personnel—preferably in cooperation with the IT service desk—should develop a playbook of response procedures when end users encounter DLP systems blocking their intended actions. Rather than be caught by surprise, IT service desk personnel (or others designated to work with end users) must have clear procedures in place for handling users who insist that the DLP system is blocking legitimate actions. Often, IT service desk personnel will need to contact privacy or security personnel who can look into these matters to determine the best course of action. The response will often require that a privacy or security manager approve exceptions and direct the configuration of the DLP system to permit specific activities that are contrary to policy. Recordkeeping of all such exceptions is important, whether as a part of an existing change control, incident management, or policy exception, or as part of another process.

System and Site Classification

Once an organization has established its data classification policy and completed its data inventory, it can proceed to classify its systems and sites. These additional classification activities can help the organization improve its security and privacy controls and use its resources more effectively.

The principle of system classification is this: a system can be classified according to the highest classification of data that the system stores, processes, or transmits. An organization can develop a system classification scheme that directs varying levels and types of controls and monitoring that are commensurate to the classification level. In other words, a system that stores or processes information at the highest level of classification can be protected and monitored with more controls than a system at a lower classification level.

Done properly, system classification should involve more than just the classification of data that systems store, process, or transmit. Other system classification criteria should include operational criticality, which can be derived from the results of a business impact analysis (BIA) that is a part of an organization's business continuity program. Business continuity planning is discussed in detail in *CISM Certified Information Security Manager All-In-One Exam Guide.*

Site classification is a further extension of the same concept. Organizations with multiple work locations can enact a site classification scheme that is based on the sensitivity of data stored or accessed there, as well as operational criticality. Sites with higher levels of classification might be equipped with more advanced physical security controls and other controls, whereas sites with lower classification levels would have fewer controls. For instance, a call center with workers who routinely access PII would be equipped with keycard access controls and extensive video surveillance, while a sales office would have fewer of these controls and features. Workplace safety is an important consideration for physical security controls, aside from operational criticality and information sensitivity.

Data Quality and Accuracy

In the context of data privacy, data quality or data accuracy is a gauge of the care that an organization places on the fidelity of its stores of personal data. Since the reality of data usage includes personal information being passed from organization to organization, the task of maintaining the accuracy of PII is an important one.

Article 5(1)(d) of the EU General Data Protection Regulation (GDPR) reads, "Personal data shall be accurate and, where necessary, kept up to date; every reasonable step must be taken to ensure that personal data that are inaccurate, having regard to the purposes for which they are processed, are erased or rectified without delay ('accuracy')." Once just a good idea, data accuracy is now required by law.

Data quality and accuracy are more than just the completeness and accuracy of data fields for data subjects; it also includes whether records for specific data subjects should even reside in an organization's database at all. Data subjects' information sometimes ends up in an organization's database simply by accident, often because of a matching error. Here are some examples:

- **Matches by name**   Some organization databases key off a subject's name only, resulting in snafus of every sort. In a real-life instance, this book's author and another person of the same name were in the Seattle job market at the same time, applying for some of the same jobs. Communications between companies and the two applicants were frequently crossed-up.

- **Matches by characteristic**   Some organizations use ancillary information to associate people. In a real-life instance decades ago, parking tickets in Reno, Nevada, were entered into a computer system; if there were no license plate on the offending vehicle, the word "none" was entered. Then a citizen ordered a vanity plate that read, "NONE," and he was soon arrested and charged with tens of thousands of dollars in unpaid parking tickets over a period of many years (all predating the existence of the issued vanity plate).

- **Data entry errors**   These are certainly the most common reason that things get fouled up. With literally billions of people using e-mail today, countless errors occur because e-mail addresses are miskeyed, resulting in messages being sent to the wrong persons. This book's author regularly receives e-mails intended for a physician located in the US Northeast, as well as e-mails intended for the owner of a private jet aircraft maintenance company in Southeast Asia. This problem goes way beyond e-mail addresses: miskeying dates of birth and other personal characteristics results in communications and records being crossed-up, as well as many intended actions not being carried out because some of the information is incomplete or incorrect.

Prior to the introduction of modern privacy laws such as GDPR, many private sector organizations had little reason to care about the accuracy of personal information, unless the accuracy had a monetary impact on them. For instance, an automobile manufacturer's database of vehicle owners is surely going to have incorrect mailing addresses for customers who move and do not inform the manufacturer. The result is twofold: marketing materials sent by mail are no longer reaching the customer and neither are safety recall notices that represent cost-per-vehicle repairs.

Data Flow and Usage Diagrams

The efforts undertaken to build and maintain data inventories do not end when it is known where all data is stored, who the owners are, and what access permissions are granted. An essential aspect of data inventory is the knowledge of two additional data characteristics: data flow and data usage.

Understanding data flow requires a deeper study of information systems where data resides, to understand how data arrives in the system and to where data is sent from the system. Interviews with business users with regard to the business process and IT personnel at the application layer as well as in system and network layers are needed to build a complete picture. The term *picture* is used deliberately: it's often useful to build a visual schematic of data flows, as this can help privacy and security professionals, as well as business leaders, better understand the usage of personal information in an organization. A *data flow diagram* (DFD), like the one shown in Figure 7-1, is a visual depiction of the flow of information between systems.

In this effort, privacy professionals need to understand that data flow implies usage, and data usage implies flow. One is often perceived with the other.

NOTE   privacy professionals mapping data flow and usage will discover both sanctioned and unsanctioned uses of data.

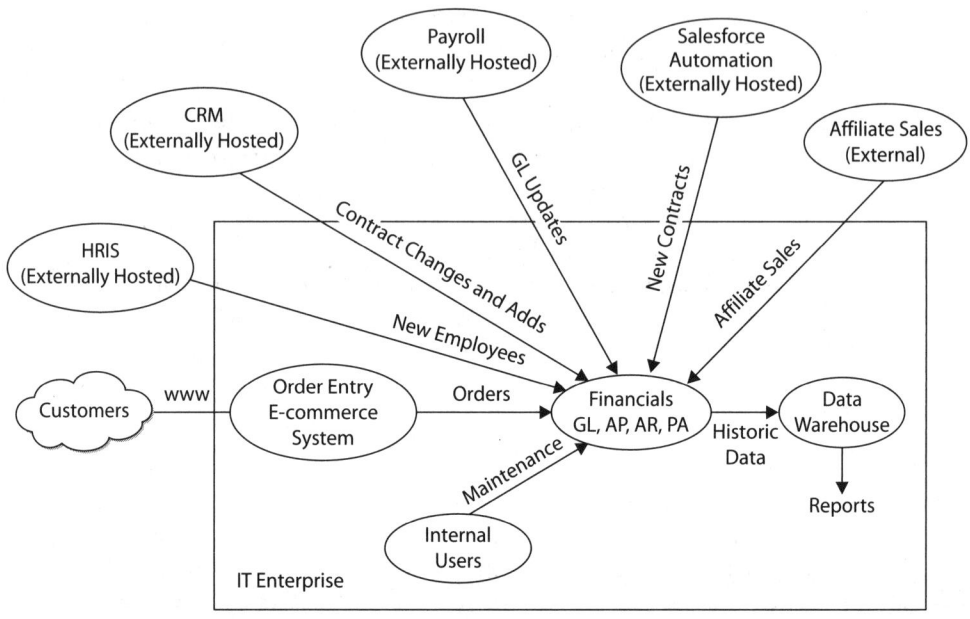

Figure 7-1 A high-level DFD depicts general data flow between IT applications.

Discovering Data Flow and Usage

Prior to the passage of privacy laws, including GDPR and California Consumer Privacy Act (CCPA), many organizations simply had no idea of the extent of the movement of their data. In many organizations, a common first step in working toward GDPR compliance was the development of a data inventory, along with the creation of data flow diagrams and the discovery of data usage. Amazing as it sounds, in many organizations, nobody was responsible for knowing these things.

The next step in most organizations is the development of data governance, providing management with visibility and control over the storage and use of personal information. It's as though there were no real rules for the management of personal information prior to GDPR.

Data Use Limitation

A key tenet of GDPR and other privacy laws is the concept of limiting the use of personal information. Following numerous abuses of PII by private organizations, privacy laws now restrict how organizations can use the personal information they collect.

Article 5(1)(b) of GDPR reads,

> Personal data shall be collected for specified, explicit and legitimate purposes and not further processed in a manner that is incompatible with those purposes; further processing for archiving purposes in the public interest, scientific or historical research purposes or statistical purposes shall, in accordance with Article 89(1), not be considered to be incompatible with the initial purposes ("purpose limitation").

Title 1.81.5 of the CCPA states,

> A business that collects a consumer's personal information shall, at or before the point of collection, inform consumers as to the categories of personal information to be collected and the purposes for which the categories of personal information shall be used. A business shall not collect additional categories of personal information or use personal information collected for additional purposes without providing the consumer with notice consistent with this section.

Data Use Governance

One could say that these and other privacy laws are "sunshine laws" that pertain to organizations' use of personal information. They have sparked lively debate in organizations that were accustomed to doing "whatever they wished" with personal information they collected from customers and others.

Organizations need to include governance structures to provide visibility and control on the topic of data usage. Such a structure would include the following:

1. Internal policies stating permitted use of personal information
2. External privacy policy accessible by relevant parties that describes all such uses
3. Establishment of controls to ensure these outcomes
4. Monitoring of these controls to verify their effectiveness
5. Corrective action to remedy all deviations
6. Metrics that measure all of the foregoing

External Privacy Policy

A privacy policy (sometimes known as the *notice of privacy practices*) published for consumption by affected parties must include descriptions of all primary and secondary uses of parties' personal information. Because a privacy policy functions as a legal contract, the organization's legal counsel should review the policy to ensure that it meets the requirements of applicable privacy laws.

> ### Market Forces and Personal Choice
> As modern privacy laws shape new norms for the collection and use of personal information, we may soon enter an era in which consumers make conscious choices regarding their perception of an organization's uses of their personal information. Consumers will choose to patronize organizations whose privacy practices best meet their preferences.

Data Analytics

Mainstream organizations have access to advanced data management and data analytics capabilities that can provide them with additional insight into their businesses. Indeed, monetization is a primary impetus for an organization mining its own data to improve how it can exploit its customers' buying preferences to increase revenues. Data analytics techniques have other purposes as well, including the discovery of new potential customers and improved insight into the use of an organization's products and services.

Some of the methods used in support of these objectives are discussed in this section.

Data Aggregation

Marketing departments in organizations frequently embark on targeted marketing and advertising, whether it's by e-mail, postal mail, online ads, or other means. To reach their target markets with the right message at the right time, organizations often purchase lists of targeted individuals from data brokers and merge that data into their marketing databases. *Data aggregation* is the practice of combining databases to enrich available data. For instance, suppose a marketing department wants to send flyers out about a new line of luxury vehicles to wealthier persons (who are more likely to buy than those with lower income), and it purchases data from a data broker that includes household income and other details. The organization merges this information into its database and then selects those wealthy persons as targets for their campaign.

This sort of activity occurs far more frequently than most people realize. There exists an entire industry of organizations with vast dossiers on virtually all adults in the United States and many other countries. This data is traded, bought, sold, merged, sorted, culled, updated, and recirculated in an endless cycle. Most of this occurs in companies most people have never heard of until a breach occurs—and, even then, personal notices are rarely sent to affected parties.

Returning to the main point of data aggregation and embellishment, another activity that frequently takes place is this: organizations seeking to aggregate customer data purchase additional data from data brokers in order to add specific data to their databases. Sometimes they receive additional data fields that are also retained, resulting in the organization having more details about its customers and prospects than it really wants or needs. This phenomenon is an example of *data sprawl*.

Aggregation works in other ways. Organizations with large customer and prospect databases can purchase data from data brokers in attempts to keep their data up-to-date.

For instance, a motor vehicle manufacturer can purchase data from data brokers in an attempt to obtain up-to-date mailing addresses for its customers so that its safety recalls will actually reach these people.

Citizens are most concerned about the potential for data aggregation among various government agencies. One can only imagine the abuses that could occur if data from one agency were accessible by malevolent persons in other agencies. To this end, citizens need to be aware of whether privacy laws apply to government agencies as well as private businesses. For example, while GDPR applies to both government agencies and private businesses, CCPA applies only to private businesses, while government agencies are exempt.

Data Lakes and Big Data

Advances in big data analysis techniques are driving organizations to develop data lakes and other large datasets to further analyze and monetize their data. A *data lake* is an aggregation of structured and unstructured data into a single data store, enabling various analytics or discovery of previously unknown relationships within the datasets. The value of a data lake is that the structure of storage and data does not need to be defined, which means you can store all types of unrelated data without needing to design relational structures carefully, or without even knowing what questions the data might help answer. This can enable valuable insights and analytics into company operations and customers, which become more accurate as more data in included. Indeed, there are many legitimate business reasons for building data lakes for big data analysis, including quality assurance and control and real-time analytics to understand customers' uses of their products and services. A data lake can also enable powerful analytics and business intelligence insight, which can require close oversight for privacy, especially if data enrichment feeds are included from third parties or some level of automated profiling or decision-making affects individuals.

Machine Learning and Artificial Intelligence

Machine learning (ML) and artificial intelligence (AI) are becoming more mainstream and are appearing in enterprise business applications to improve customer service and product and service quality, and to help organizations better understand their current and future customers and their preferences. When properly used, ML and AI can improve product and service quality and help the organization find new and better customers, and it can improve how the organization serves existing customers.

The value of ML and AI depend on the quality and quantity of data that can be analyzed. More data makes analysis more powerful and more detailed. However, this power can enable unanticipated inferences or unintended decisions if it's not properly understood and controlled. For instance, if AI or ML is tasked with making inferences about persons, an organization may find itself making misinformed marketing or operational decisions. For example, an e-commerce organization may use ML in an attempt to determine particular preferences of individual customers based upon past buying choices, and then market to them accordingly. Or analysis can enable intrusion into sensitive areas—for example, an organization may send advertisements for baby products to the family of an individual who has not yet revealed that she is pregnant.

 **NOTE** While organizations may not be prohibited from undergoing activities, including building data lakes and applying ML and AI, they do need to determine whether or not to include these activities in their privacy policies.

Chapter Review

Data governance is management's visibility and control over the use of information in an organization. Data governance takes on new meaning and relevance for organizations seeking to comply with recent privacy laws.

The governance structure concerning the use of personal information will include policies, defined roles and responsibilities, controls, processes, procedures, assessments of these, and reporting.

For a privacy program to be effective, organizations must have a complete and accurate inventory of all personal information. This inventory must include both structured as well as unstructured data. Periodic data inventory reviews should not only catalog existing instances of sensitive and personal information, but also determine in each instance whether data *should* exist in places where it is found.

A *data classification policy* is a formal and intentional way for an organization to define levels of importance or sensitivity to information. A typical data classification policy will define two or more (but rarely more than five) data classification levels.

Static DLP tools scan static data stores to identify files containing data matching specific patterns. Generally, static DLP scanning is performed on file servers, both the on-premises and cloud varieties.

Dynamic DLP can greatly improve compliance with a data classification policy; however, organizations need to tread lightly and slowly to avoid interfering with sanctioned business processes.

Data tagging can give a needed assist to DLP automation by enabling DLP to act based on the tag, without having to examine an entire data file in real time.

System and site classification schemes can supplement data classification and contribute to a holistic information and organization protection program.

Data quality or data accuracy is a gauge of the care that an organization places on the fidelity of its stores of personal data.

Data flow diagrams (DFDs) and descriptions of data usage serve to complete a data inventory effort. The visual aspect of DFDs helps organizations better understand the flow and use of personal information and guide privacy and security personnel to focus on systems and processes where personal information is used.

Data use limitation is a key tenet of privacy laws, including GDPR and CCPA. Data use limitation should be a part of overall data governance.

Data analytics is a powerful tool for improving the understanding of business operations, customer behaviors, and service quality.

Data aggregation is the practice of combining databases to enrich available data.

Advances in big data analysis techniques, including machine learning and artificial intelligence, are driving organizations to develop data lakes and other large datasets to further analyze and monetize their data.

Quick Review

- Because of the itinerant nature of unstructured data, inventories of such data will change frequently and present operational challenges.

- Data classification policies and standards need to be as simple as possible to give the workforce the greatest opportunity to understand and apply them.

- Privacy and security leaders must be especially cognizant of the potential impact of data classification on the culture of the organization.

- Data discovery with static DLP tools is a good first step to understanding the extent to which sensitive data resides on file servers.

- Organizations implementing dynamic DLP need to consider identified risks and existing controls to understand what problem they're trying to solve.

Questions

1. A new privacy leader is making recommendations for a set of activities to ensure proper management of personal and other information across the organization. What needs to be put into place?

   A. Controls

   B. Data classification

   C. Data governance

   D. Data handling

2. An organization performs annual true-ups of its data inventory and finds numerous discrepancies. What change should be undertaken to reduce or eliminate these discrepancies?

   A. Automate the data inventory with daily scans.

   B. Increase the frequency of data inventory.

   C. Incorporate data inventory updates into the change management process.

   D. Implement dynamic DLP.

3. A privacy manager is developing a data classification program. She has established a matrix that consists of a total of 12 classifications that align to privacy, as well as intellectual property and payment information. What is the most likely scenario for the adoption of this program in the organization?

   A. Orderly adoption if training takes place

   B. Workforce will refuse to adopt

   C. Clear and determined adoption

   D. Confusion as the classification scheme is too complicated

4. The purpose of a data classification and handling policy consists of all of the following *except* which one?

   A. A single method for data protection

   B. Efficient protection of information

   C. Risk-driven protection of information

   D. Direction to the workforce to apply proper handling procedures

5. An organization has begun implementation of its data classification program and wants to know the extent of storage of personal information on file servers. What is the first step that the organization should undertake?

   A. File integrity monitoring

   B. Dynamic DLP

   C. Dynamic discovery scan

   D. Static discovery scan

6. A privacy manager has developed a policy that requires that all human-readable files be marked according to their classification. What is the meaning of document marking?

   A. A metadata tag

   B. A machine-readable watermark

   C. A human-readable phrase citing the classification level

   D. A human-readable watermark

7. An organization is implementing dynamic DLP in the form of USB storage device control. The use of USB storage devices will be prohibited according to data classification and security policy. How should the organization implement this control?

   A. After announcements, activate the control after giving adequate notice.

   B. Initially implement in detective mode.

   C. Implement in active mode, one department at a time.

   D. Implement with a pilot group first.

8. All of the following tools can supplement dynamic DLP tools *except* which one?

   A. Cloud access security broker

   B. Web content filtering

   C. File integrity monitoring

   D. NetFlow

9. What is the purpose of system classification?

   A. Determine what files can be stored on a system

   B. Develop levels of protection

C. Prerequisite to network segmentation

D. Determine which systems require FIM

10. Why would an organization with operations in Europe implement controls to ensure the accuracy of its customers' PII?

 A. Required by CCPA

 B. Increases profit margins

 C. Increases revenue

 D. Required by GDPR

11. What is the best course of action for an organization to ensure that its customers' PII is always properly handled?

 A. Implement a cloud access security broker (CASB).

 B. Implement NetFlow to detect unauthorized data movement.

 C. Implement data usage governance with policies, controls, and assessments.

 D. Implement static DLP discovery scanning of databases and file servers.

12. An e-commerce organization has elected to purchase information from a data broker in order to add more details to its existing customer database. What term describes this process?

 A. Data embellishment

 B. Data aggregation

 C. Data infiltration

 D. Data exfiltration

13. An organization has developed data governance to gain visibility and control over the protection and use of personal information. What does management need to do to determine whether governance is having its intended effect?

 A. Direct control assessments to determine control effectiveness.

 B. Implement data management policies.

 C. Develop data management and handling training.

 D. Lead by example and demonstrate proper data handling.

14. What is the purpose of identifying a data owner in a data inventory?

 A. Data owners are responsible for data protection.

 B. Data owners respond to security incidents.

 C. Data owners approve DLP scanning.

 D. Data owners approve access requests.

PART III

15. An organization has performed a first-time data discovery scan on file servers and has identified numerous files that violate data handling standards. What is the best course of action to take?

A. Investigate all files to determine their legitimacy.

B. Delete all files that violate policy.

C. Encrypt all files that violate policy.

D. Contact the data owners.

Answers

1. C. The privacy manager is recommending that the organization implement data governance, which will consist of policies, definitions of roles and responsibilities, controls, processes and procedures, assessments, and reporting. Governance will ensure that management has visibility and control over business operations—in this instance, over data management.

2. A. In this situation, the population of personal information is changing frequently. Automated scans will help keep the inventory more up-to-date.

3. D. A data classification policy with 12 different classifications is going to be confusing for most any workforce to adopt. First, workers will probably have trouble correctly classifying data in every instance. Next, the handling procedures are likely going to be lengthy, since presumably, they will differ somehow for each classification. A better approach is a simpler scheme with no more than four classification levels.

4. A. Data classification programs are developed to provide for efficient, risk-driven protection of information by directing the workforce to apply proper handling procedures. Data classification provides for multiple types of protection for each respective classification, not a single method of protection.

5. D. A static discovery scan is a recommended first step. It will provide an organization with a detailed report of personal and sensitive information on its file servers. Static discovery scans can be configured to search for specific types of sensitive information.

6. C. Most commonly, document marking consists of a word or phrase affixed to a document to remind users of the classification level of the document.

7. B. The best path to implementation of a policy that prohibits the use of USB storage devices is to start in passive mode first to begin learning about all uses of USB storage devices. Starting in active mode, even after a pilot, is certain to interfere with legitimate business activities.

8. **C.** Several security-related tools can supplement a dynamic DLP system, including web content filtering (by restricting access to specific web sites), cloud access security broker (by directing users to sanctioned sites for file handling), and NetFlow (by detecting anomalous activity that may represent unauthorized data transfers).

9. **B.** The purpose of system classification is to develop levels of protection for systems, based upon the highest classification of data stored or processed by each system.

10. **D.** Article 5(1)(d) of the GDPR requires that personal data shall be accurate and kept up-to-date, and measures should be taken to ensure its continued accuracy. Secondary benefits could include increased revenue.

11. **C.** To ensure that its customers' PII is always properly handled, an organization first needs to establish data usage governance. As an organization develops policies and controls, it may determine that automated controls are needed in places to detect and even block activity that could constitute violations of data usage policy.

12. **B.** Data aggregation is the process of adding data fields to an existing database. Sometimes data embellishment is a casual term used for this process.

13. **A.** If management is unsure of whether its data management controls are effective, it should first execute control assessments to determine their effectiveness. Further steps may include risk assessments with the possibility of enacting additional controls.

14. **D.** Data owners approve access requests and perform and approve access reviews.

15. **A.** After performing a first-ever data discovery scan, the best course of action is to investigate identified files to determine whether they are a part of legitimate business processes.

Data Persistence

In this chapter, you will learn about
- Data minimization techniques, including pseudonymization and anonymization
- Privacy consideration in data migrations
- Identifying privacy requirements and issues in data storage
- Data destruction techniques and challenges

This chapter covers Certified Data Privacy Solutions Engineer job practice 3, "Data Cycle," part B, "Data Persistence." The entire Data Cycle domain represents 30 percent of the CDPSE examination.

Privacy programs rely on active and intentional management of personal information in an organization's databases and unstructured data stores. Privacy leaders need to identify and manage sometimes opposing needs through business practices, including data minimization, data retention, and data destruction. Privacy leaders need to be involved in large-scale data management and data migration operations to ensure the balance of needs from data subjects, regulators, and the organization itself.

Data Minimization

Data minimization refers to the practice of collecting and retaining only those specific data elements necessary to perform agreed-upon functions. In other words, organizations should be careful to collect or accept only the specific personally identifiable information (PII) details required to perform whatever services they provide.

In the same way that double-entry accounting describes each item on a balance sheet as both asset and liability, personal and sensitive information can provide value to an organization as an asset, but it also represents a liability. While the asset value aspect of personal and sensitive information may be clear, organizations are slower to realize that accumulating and retaining personal and sensitive information also represents a liability.

Unfortunately, the financial liability portion of retained personal information rarely shows up on an organization's financial balance sheet. And yet it is indeed a liability: the impact on an organization when cybercriminals steal that information or when the information is misused is real, in the form of breach response costs, the costs related to reducing harm inflicted on affected parties (think of credit monitoring services, a frequent remedy for stolen credit card numbers), fines from governmental regulators, and the occasional class-action lawsuit.

Data minimization has multiple dimensions:

- Collect only required fields.
- Collect only required records.
- Retain only as long as is needed.
- De-identify as soon as possible.
- Reduce accessibility.

Collecting Only Required Fields

When collecting personal information directly from data subjects, organizations should collect only data items that are required for the organization to fulfill the intended purpose for the collection. Every item collected must be rationalized and the reason for collecting it documented. Data items that cannot be justified as necessary should not be collected.

Any data item proposed to be collected that does not have a present business purpose should not be collected. Such collection would introduce liability to the organization with no corresponding benefit. For example, suppose an e-commerce company that sells books developed a customer portal where customers can select their favorite categories of books and save shipping addresses so that they do not need to enter them for each order. An analyst is proposing that the organization collect the date of birth (including year) for each customer so a birthday discount code can be sent to customers during the month of their birthday. The privacy officer successfully argues that this function does not require the year of the customer's birth, but only the month. The organization decides not to collect the birth year and day for its customers because there was no purpose identified that would require it (such as selling adults-only merchandise). Although there was a valid decision made in this case, more information may be required, such as a business record in the form of a detailed inventory of data items collected, including the collection purpose(s). This way, months or years later, a privacy professional can examine the records to discern reasons for specific data collection decisions without relying upon workers' memory.

There remains some inferred responsibility for providing personal information that falls on data subjects who are providing it. Data subjects should be aware of the information they provide to an organization or government and attempt to withhold any information they believe is unnecessary for the organization to fulfill the intended purpose. For example, most e-commerce sites should not require a data subject's date of birth to complete transactions (the sale of products prohibited for minors is one exception; applying for a credit card is another). If an e-commerce site requests a date of birth, a data subject should avoid providing it unless there is some clear purpose that the data subject agrees with.

Providing False Information

Some privacy-conscious individuals known to the author have an interesting approach to providing certain sensitive information to other organizations. They provide valid but deliberately false information to avoid revealing sensitive information about them. For instance, if an e-commerce site that sells tobacco products wants to confirm that visitors are over the age of 21, they enter a date of birth that asserts their age is over 21, without entering their actual birthdate. There are, however, a couple of complications to keep in mind if one chooses to do this. If the e-commerce site later wants to confirm a date of birth, the customer must keep track of the date of birth provided to enter the same value again (a password vault would be a handy way of storing this). If the e-commerce site later wants to see an image of a government-issued ID, the customer will have to "correct" the falsified date of birth to match their actual date of birth.

Collecting Only Required Records

If organizations acquire personal information in bulk (such as purchasing it from a data broker), only those records required to fulfill the intended business purpose should be collected. Any collection of records brings some liability to an organization, especially when the records provide no value or benefit: if an intruder breaks in and steals this information, the organization that collected the data may need to make reparations to all persons whose records were collected and stolen.

Organizations that purchase bulk PII data need to ensure that they obtain only the records they need to meet their business objectives. Sometimes, however, only large data sets are available, even if only certain records are requested or needed. In such cases, organizations should remove unneeded records as soon as it's practical to do so.

On a record-by-record basis, organizations should devise options to enable choices when collecting individual records. One example is e-commerce, where many online shopping sites permit customers to "check out as a guest." In this case, the buyer provides only enough information to complete the transaction, instead of being required to create a persistent user account that is usually accompanied by the collection of additional information such as a credit card number, billing address, shipping address, and other information.

One option for record removal is pseudonymization or anonymization of selected records, discussed later in this section.

Discarding Data When No Longer Needed

Organizations collecting and retaining personal data should understand the uses of personal data fields to determine whether long-term retention is appropriate. For instance, an e-commerce web site accepting a credit card payment may collect the CVV (card verification value) from the customer and then discard it as soon as the transaction has been approved or rejected.

In another example, a web site providing services only to adults who are 21 or older may require evidence of a customer's age before any purchase can be made. This practice may involve the uploading of a government-issued ID or some other action. Once the organization has verified that the subject is at least 21, the organization can discard the PII collected to prove the subject's age and simply indicate that their age has been verified.

Minimizing Access

Data minimization is all about risk reduction by limiting the amount of data available for various functions. In "risk-speak," data minimization reduces the impact of improper data usage or a breach of personal information. An organization can achieve effective data minimization through access controls in several ways:

- **Reducing access volume**   Organizations can limit the number of records that a worker can access, extract, or download in bulk operations. In B2C (business to customer) organizations, few persons in the organization genuinely need access to customers' entire database; most should be limited by various means.

- **Reducing personnel with data access**   Organizations can limit the number of workers with access to customer data to those whose jobs require it.

- **Reducing access to sensitive fields**   Organizations can limit the data fields accessible by its workers. For instance, most workers may not need to access customers' birthdays or full credit card numbers; hence, the ability to see these or other sensitive fields should be reduced.

Data *masking* can protect the contents of personal information from personnel who do not need to see it. A typical example is the display of a credit card number or social insurance number. Although an information system may store the full contents of these values, some of the characters can be masked so that personnel cannot see them. Most often, programs will display only the last four digits of a credit card or social insurance number.

 **EXAM TIP**   CDPSE candidates need to understand that encryption is another form of access control that may be applicable in some situations for limiting access to personal information.

Minimizing Storage

Organizations can significantly enhance the security of sensitive and personal information by limiting where and how workers can store such data. When workers use downloads or extracts from business applications, organizations can enact controls that limit where that data can be stored. Primarily, organizations should limit the storage of personal information about their customers, constituents, and employees to organization-managed systems.

Organizations can enact policies that state that all sensitive and personal information must be stored only on organization file servers, and not on laptops or desktop computers, mobile devices, removable storage devices, or personal cloud-based storage services. Further, sensitive and personal information can be blocked from being sent via company e-mail or personal e-mail. These controls are typically implemented with data loss prevention (DLP) controls, but web content filtering systems and cloud access security broker (CASB) capabilities can supplement DLP solutions.

The ultimate objective is to limit storage of personal information only to those locations permitted by policy and controlled through tooling. However, privacy and security professionals need to tread carefully to prevent disruption of sanctioned business processes. This topic is explored further in Chapter 7.

Minimizing Availability

Minimizing of the availability of information is another option that organizations can pursue to achieve data minimization. The most common approach is the migration of data on widely accessed systems to archival systems that few personnel can access. For example, suppose a regional hospital's patient care and billing systems contain medical records and billing records for the past 15 years. To reduce the risk of exposure, the hospital migrates all records more than two years old to an archival system that few hospital personnel can access. This permits the hospital to comply with minimum data retention requirements while reducing risks associated with hospital personnel having access to large volumes of medical and financial information. Further, since the data archival system is accessible only from internal networks and by few personnel, there is a correspondingly lower risk of a break-in by intruders.

Organizations can implement additional controls to further protect the archival system, including DLP, EUBA (end user behavior analytics), and NetFlow.

Minimizing Retention

Organizations are accustomed to retaining data for very long periods, often in perpetuity. For generations, the risks associated with long-term data retention have been quite low. Digital transformation has changed all of that: data sets that contain more details about data subjects (more fields with sensitive information and more records) are higher value targets sought by cybercriminal organizations. Privacy professionals often call highly sensitive information "toxic data," since its theft can have dire consequences on the organization.

Put another way, the value of information that an organization accumulates has the "credit" and "debit" characteristics of double-entry accounting. Although the accumulation of sensitive data may bring more value to the organization, it increases liability as well: the theft of a larger trove of sensitive data will incur greater costs than the theft of a smaller dataset. For example, suppose each of two similar e-commerce organizations has about 5 million customers. One of the organizations keeps only two years of transaction data and moves dormant customer data to an offline storage system. The other organization keeps all customer data online, even for customers who have not patronized the organization for years. Suppose each organization's customer databases are stolen. In that case, the organization that reduced its customer database size will incur fewer costs than the organization that kept all of its customer data online.

The approach to data retention is the development of a data retention schedule, a chart that specifies the minimum and maximum periods that specific types or sets of data will be retained by the organization. A data retention schedule is considered policy, and organizational departments are expected to comply. Security and privacy personnel may perform reviews or audits periodically to determine whether the organization complies with its data retention policy and to require corrective action when violations are found.

Enacting the purging of older records is not always as easy as it sounds. Several challenges may present themselves:

- **Database referential integrity**   The design of relational databases may make the prospect of removing records a bit tricky. Primarily, a record cannot be removed if another record elsewhere in the database refers to that record through a foreign key. The referential integrity concept refers to the restriction where a record (or "row" as it is often called) cannot be removed if another table has a row whose foreign key points to the record to be removed. The other table's row would first have to be modified or removed. This is typically a problem in older databases that were not designed with data retention in mind.

- **Comingling of data**   Some storage media cannot be modified once created. For instance, magnetic tape is an "all or nothing" medium; it is impossible to remove specific data from a magnetic tape while retaining other data. If a system was backed up to magnetic tape and that system contains specific data that must be purged after two years, but the system contains other data that must be retained for ten years, backing all of this data to magnetic tape will create a conflict, and the organization will not be able to conform to both of these retention requirements.

- **Unstructured data**   Because data can be extracted and further manipulated at users' workstations, it can become difficult to know whether individual workbooks contain information that has exceeded its retention. The date stamp on a workbook does not indicate the transaction dates of rows in the workbook, which could be recent or far in the past. Variations in the ways that data can be represented in workbooks make it infeasible to enforce transaction-level data retention effectively in unstructured file stores.

- **Third parties**   Organizations outsourcing business applications to third parties (through platform as a service [PaaS] or software as a service [SaaS] models) may find that one or more of their third parties cannot remove older records from their systems. There may be referential integrity issues in their databases (discussed earlier), or they may simply lack the tools to remove older records from selected customers' databases. This kind of a situation often arises when an organization, after selecting and using third-party applications, enacts a data retention policy only to discover that one or more third parties are unable to comply.

- **E-mail**   In some organizations, workers send sensitive and personal information to one another via e-mail. Searching for and removing specific e-mail messages that contain sensitive information may not be feasible; this problem is similar to the unstructured data problem discussed earlier. Taking the approach of removing all older e-mail messages may be a viable approach. Still, the organization needs to understand the nature of the organization's use of e-mail completely so that purging older e-mail messages does not introduce unintended consequences, such as the destruction of other records that should be retained for long periods.

Data retention does not require an "all or nothing" approach, in which an organization will simply delete older records. Instead, organizations can define storage locations for business records that can be tightly managed and can establish a generic retention schedule for other storage locations. Additionally, records that have reached their expiration dates can be pseudonymized or anonymized, thereby removing personal information from records but retaining other aspects of records for historical purposes. Pseudonymization and anonymization are discussed in the next section.

Minimization Through De-identification

Depending upon the purpose of acquiring personal information, organizations can consider de-identification as a method for minimizing the amount of PII that they retain. For instance, if an organization acquires personal information for statistical purposes, it can pseudonymize or anonymize the records so that they no longer are associated with specific persons but can still provide statistical value.

When implemented correctly and from the perspective of privacy, de-identification is as effective as the outright removal of records. When implemented correctly, an organization will continue to derive value from de-identified data through analytical value. For instance, after a database of user transactions has been de-identified, although the organization won't know precisely who performed individual transactions, the organization will still understand the details of sales trends and other big-picture insights.

Two primary techniques are used in de-identification: pseudonymization and anonymization. While these techniques differ, the results are similar.

Pseudonymization

Pseudonymization is the irreversible substitution of data in sensitive data fields with alternate values to de-identify data records with specific persons. Pseudonymization is generally an irreversible substitution technique: fields that identify actual persons are modified and replaced with pseudonym values. Here are some examples:

- *Peter Gregory* becomes *Qoem Rebnurvo*
- *118 Elm Street* becomes *539 Tlo Uepv*
- *peterhgregory@gmail.com* becomes *juwnfodpwlrmg@drep.com*

The substitution technique permits software to function correctly, while substitutions eliminate the association of the record from the actual person. However, this is different from anonymization, because pseudonymization still may enable an individual to be singled out and linkable across different data sets.

NOTE From a privacy perspective, de-identifying a record (when performed properly) is equivalent to its removal.

Anonymization

Anonymization is the process of irreversibly altering or removing sensitive data fields from records so that an individual can no longer be identified directly or indirectly. Anonymization can also be a simple removal technique: data fields that could associate a record with a specific person are removed. Here are some examples:

- *Peter Gregory* becomes (blanks)
- *118 Elm Street* becomes (blanks)
- *peterhgregory@gmail.com* (blanks)

Note that anonymization may cause software to behave in unexpected ways. Further, database management systems may resist anonymization, as this could endanger referential integrity. For anonymization by removing field data to work correctly, it may be necessary to copy records in a database to a separate database whose structure lacks the removed fields.

NOTE The challenge with de-identification is ensuring that data records can no longer be associated with natural persons while retaining the information's value for other purposes.

Data Migration

Data migration is a process of transferring data from one production system to another. In earlier days of information processing, a data migration was a time-consuming, painstaking effort to move business information from an older (and perhaps unsupported) system to a newer one. A data migration was a part of a larger effort, or conversion, which also involved the updating of business software (which, in those days, was often custom written by in-house programmers) to ensure that it would function in precisely the same way on the new system as it did on the old. Numerous rounds of testing would ensue during a migration, which in total could take a year or longer.

Modern data migrations include these concerns, although the context of migrations is more often the movement of data from one SaaS solution to another, or from an aging on-premises application to a SaaS system. The velocity by which data can be moved increases the exposure for mishandling data unless a proper data migration strategy is developed that includes accounting for security and privacy requirements. Some concerns during such a migration will include the following:

- **Business functionality** An organization needs to understand whether the new system performs its primary functions in the same way, providing the same results (or different results that the organization can deal with acceptably) as the old system.

- **Information format** When data is migrated from the old system to the new system, sometimes individual data fields need to be manipulated or transformed, as the entire set of data fields in the old and new systems rarely will match exactly. Further, the format of data stored on the old system may differ from the format of the new system, requiring that each data field be transformed. This process, called extract, transform, and load (ETL), simply means that extracted data needs to go through a series of functions in preparation for loading into the target system. For example, if an old system stores a complex date field as *YYYYMMDD* and a new system uses the format *MMDDYYYY*, an ETL process will have to transform each date field's contents in each record as it is migrated from the old system to the new one.

- **Data types** An organization migrating data from an older system to a newer one needs to understand thoroughly how individual data fields are used, including obvious and subtle differences in usage. For example, an older system may accept the inclusion of some special characters in data subjects' names, addresses, or other fields, while the newer system's rules for special characters may differ.

- **Privacy** Organizations need to understand the controls and constraints designed into the old and new systems to achieve privacy requirements and controls in the new system. In some circumstances, organizations may need to develop compensating controls to manage those differences and to ensure the sustainment of privacy expectations.

- **Security** Organizations must understand how roles and data access techniques differ between the old and new systems and adjust their practices accordingly. For instance, the structure of access roles between the old and new systems may be different and somewhat rigid. This would necessitate that the organization remap job functions to the new system's roles so that personnel have the access they require to perform their jobs, but not too much access to increase risk.

- **Training** Persons using the older system will require training to understand how to use the new system properly. This should result in fewer errors and ensure better security and privacy.

A proper migration strategy will address these requirements as well as mitigate the potential for inadvertently exposing data in temporary storage locations that can (and have) resulted in data breach events. Additionally, organizations migrating to a newer system need to understand whether it has capabilities not present in the older system. Examples include these:

- A newer system may integrate with federated identity services to be a part of a single sign-on function.

- A newer system may send security events to a security information and event management system (SIEM).

- A newer system may be able to integrate with a data loss prevention (DLP) system and thereby provide additional visibility and control over the use of personal information.

Data Storage

Organizations that process information, including sensitive and personal information, need to store that information temporarily or permanently. How sensitive and personal information is stored can make a vast difference in terms of privacy and security.

Flippantly, the decision-making process for storing information can come down to "put that data here" or "put that data there." However, organizations need to make data storage decisions with utmost seriousness and within the auspices of a formal data governance function. The stakes are high: the consequences of poor decision-making can reverse the fortunes of an organization.

An organization must keep in mind several considerations for data storage, including these:

- **Data location and sovereignty**   Some privacy laws include stipulations regarding the geographic location where data about persons is allowed to be stored. Many countries shun or outright prohibit the export of personal information. While on the surface, this concept is obvious, in practice, it is sometimes difficult to know the physical locations of data storage for many cloud-based services. Organizations are accountable for making sound decisions in the face of these uncertainties.

- **Data protection**   Security specialists need to identify the security controls in the entire stack, from physical protection to human access and usage. This can be somewhat vague in cloud-based service providers, but this is often offset through external attestations such as SSAE 18 or ISAE 3402 audits.

- **Data access**   Security and privacy specialists must identify the controls in place for data access, whether by machine or by workers. Organizations need to understand a cloud service provider's workers' ability to access the organization's data and whether the organization would be made aware if they do so.

- **Access monitoring**   The ability to monitor and log data access and data access rights changes should be a significant consideration in any data storage decision. The lack of visibility and control effectively means that an organization has given up control over data protection and usage. Generally, data access visibility requirement means that workers cannot store sensitive and personal data on removable (USB) storage devices or personally owned devices.

EXAM TIP CDPSE candidates need to understand the concept and purpose of cloud shared responsibility models.

Data Warehousing

A *data warehouse* is a set of data that has been copied from a production system and used for reporting, research, or archival purposes. Organizations create data warehouses when the size or transaction volume of their production databases makes using them infeasible for the shared purposes of performing online transactions and research and reporting. While it may sound similar to a data lake (discussed in Chapter 7), a data warehouse is a structured repository and typically supports defined business operational use cases.

Transaction processing in database management systems generally involves the insertion of records and the simultaneous updating of indexes. In contrast, research and reporting in database management systems involve searching for and retrieving records. The conflict is this: database management systems are generally tuned for optimum performance—tuning a database management system for transaction processing results in poor performance for searching, while tuning a database for searching results in poor performance for inserting new records. Generally, the best course of action is the creation of a separate data warehouse that is tuned for searching, reporting, and research, and then periodically copying transaction data to the data warehouse. This leaves the online transaction processing system ideally tuned for its purpose.

NOTE If an organization is not required to retain personal information for long periods, moving records to a data warehouse can be a good opportunity to de-identify those records as they are removed from the production environment.

Data Retention and Archiving

Organizations are accustomed to retaining data for very long periods of time, often in perpetuity. For generations, the risks associated with long-term data retention have been quite low. Digital transformation has changed all of that: with business functions implemented in a workflow and records retained online, sensitive data such as personal information can be used, misused, abused, stolen, and subsequently monetized by cyber-criminals in numerous ways, bringing harm to affected persons. Finally, the liability of excess data retention is being felt; this results in a greater emphasis on establishing data retention schedules to limit how long organizations retain sensitive information.

Industry Data Retention Laws

Privacy and security professionals need to work with business unit leaders and legal counsel to determine appropriate data retention periods for various data types. Often, there are laws on the context of said data that need to be identified and understood. For instance, nation, state, or provincial laws specify that organizations must retain employment "human resources" records throughout an employee's length of employment, plus several years afterward. Financial services and banking laws have similar requirements for minimum retention periods for financial transaction data.

Right to Be Forgotten

European Union General Data Protection Regulation (EU GDPR), California Consumer Privacy Act (CCPA), and other privacy laws give data subjects a "right to be forgotten." In other words, data subjects can request that their data be removed. Still, such requests can be granted only to the extent that an organization can remove such information without violating other laws that specify that information be retained. For example, suppose Fred was unceremoniously fired from his job at Best State Bank. Thinking he can improve his employment prospects elsewhere, Fred asks Best State Bank to remove his employment record data under the "right to be forgotten" concept in applicable privacy laws. However, Best State Bank refuses to fulfill the request, citing employment laws that require that all employment records be retained for the length of employment, plus seven years after the end of employment relationships.

Data Archival

In the early years of computing, data was often copied to magnetic tapes, which were kept in a vault for many years. Nowadays, with low data storage costs, organizations are apt to retain some data in perpetuity, particularly business transactions and records about employees and customers. However, the risks associated with long-term retention have compelled organizations to consider alternatives; one is *data archival*, the process of preparing data for long-term storage. When organizations are bound by specific laws to retain data for many years, archival provides a viable opportunity to remove data from online transaction systems to other systems or media.

The terms *online*, *nearline*, and *offline* connote related approaches to archival. Online data remains in a primary processing system along with current records. Nearline data may exist in a different system such as a data warehouse, and it can be accessed in the data warehouse or even returned to the primary transaction processing and storage system. Offline data generally resides in another form, such as backup media—whether tapes, a virtual tape library (VTL), or disk storage of some sort—but not in a form that is immediately available for returning to normal processing.

Archival may be a viable option for risk-averse organizations that want to be rid of their older records systems but are required by law to store records long-term. In this case, techniques such as pseudonymization and anonymization may not be available, because organizations are generally obligated to retain original information in business transactions, including subject names and other personal information.

The Data Retention Tug of War

Privacy professionals and business leaders are sometimes at odds with one another, particularly with regard to data retention. The core of the conflict is this: business leaders want to keep information for as long as possible, to mine every last morsel of value from collected information; on the other side, privacy professionals want data retained for the shortest possible time, if it is retained at all.

Neither side is entirely correct. Instead, business leaders and privacy professionals need to understand the facts, applicable laws, use cases, and options available to enable the organization to derive maximum value from its information, while applying techniques to reduce risks as much as possible.

Data Destruction

For all of the effort that organizations undergo to ensure data availability, an equal amount of effort is required when organizations no longer need to retain data. *Data destruction* is the purposeful act of destroying data so that it cannot be recovered. Data destruction is invoked as a part of two processes: data classification and handling, and data retention.

Data destruction policy should include directives for the safe removal of data in numerous use cases, including data that is

- Stored on a laptop computer, desktop computer, tablet computer, smartphone, or USB drive
- Stored on an application server
- Stored on a file server
- Stored on a hard disk drive (HDD) or a solid-state drive (SSD)
- Stored as a record in a database management system
- Stored as a record in a business application
- Stored on backup media
- Stored on printed paper

The rigor used to destroy data safely should depend upon the sensitivity of the affected data and a risk or threat assessment to determine the extent to which an adversary would attempt to reconstitute discarded or destroyed data.

Chapter Review

Data minimization is the practice of collecting and retaining only the specific data elements necessary to perform agreed-upon functions. Organizations should be careful to collect or accept only the specific PII details required to perform whatever services they provide.

When collecting personal information directly from data subjects, organizations should collect only data items that are required for the organization to fulfill the intended purpose for the collection. Every item collected must be rationalized and the reason for collecting it documented.

Organizations should collect only information necessary to meet specific business objectives, and they should discard records that are no longer needed. Fields and records are required to perform specific functions, but those that are no longer needed should be discarded.

Another dimension of data minimization involves limitations on the numbers of personnel who can access specific fields, records, or databases. Only personnel with specific business needs should be able to access these items.

Organizations should develop data retention schedules to limit the time that sensitive data can be retained. Periodic checks and reviews should be performed to ensure compliance.

Pseudonymization is the irreversible process of substituting data in sensitive data fields with alternate values (pseudonyms) to de-identify data records about specific persons. Pseudonymization is a substitution technique: fields that identify actual persons are modified and replaced with pseudonym values.

Anonymization is the process of irreversibly altering or removing sensitive data fields from records so that an individual can no longer be identified directly or indirectly. Anonymization can be a simple removal technique: data fields that could associate a record with a specific person are removed.

Data migration is the process of transferring data from one production system to another. Close attention is required to ensure the preservation of data accuracy and integrity during a migration.

Privacy and security professionals should pay close attention to data storage environments to ensure that the organization can exert data protection, access control, and access usage controls on them.

Data destruction is the purposeful act of destroying data so that it cannot be recovered. Data destruction is carried out on fields, records, databases, and files that have reached the end of their retention life.

Quick Review

- Organizations that understand the risks associated with retaining volumes of personal information are more apt to embrace data minimization efforts.

- Additional controls to monitor and protect personal information can be implemented to reduce risks of abuse and compromise, including DLP, EUBA, and NetFlow.

- Technical details such as database referential integrity may represent obstacles to data minimization efforts, including retention and de-identification.

- Organizations must thoroughly understand any shared responsibility model implemented by Internet service providers and other organizations to ensure that all necessary controls are in place.

- If an organization is not required to retain personal information for long periods, moving records to a data warehouse can be a good opportunity to de-identify those records as they are removed from the production environment.

- When developing data retention schedules, organizations need to identify all applicable regulations and requirements that address minimum and maximum periods that sensitive data may be retained.

- The rigor used to destroy data safely should depend upon the sensitivity of the data being destroyed and a risk or threat assessment that helps determine the extent to which an adversary would attempt to reconstitute discarded or destroyed data.

Questions

1. A privacy manager is attending a planning meeting in which marketing personnel argue for the collection of PII from customers that may be used sometime in the future. How should the privacy manager respond?

   A. Permit the collection of the additional PII fields.

   B. Require that the additional PII fields be encrypted.

   C. Forbid the collection of the additional PII fields.

   D. Update the data retention schedule to include the additional PII fields.

2. An organization's marketing department purchases PII data from a data broker to embellish and update specific data fields for its existing customers. Upon examining the purchased data contents, marketing personnel realize that additional subjects are contained in the purchased data. What should be done with this additional data?

   A. Discard the additional data.

   B. Encrypt and retain the data for future use.

   C. Declare a privacy breach and begin response proceedings.

   D. Develop a marketing campaign and target the additional subjects.

3. A privacy manager is reviewing the organization's practices of data collection from its customers. The privacy manager has observed that the organization collects PII fields that are not subsequently used. What recommendation should the privacy manager make?

   A. Discard all customer records containing the unneeded fields.

   B. Change the entry of unneeded fields from "required" to "optional."

   C. Discontinue collection of unneeded PII fields.

   D. Discontinue collection of unneeded PII fields and discard those already collected.

4. After an organization implemented dynamic DLP controls, the organization has observed numerous instances where PII is copied to USB external storage devices. What first steps should be taken?

   A. Investigate the usage of PII being copied to USB storage devices.

   B. Block the use of USB storage devices.

   C. Discipline the personnel copying PII to USB storage devices.

   D. Declare a privacy breach and begin incident response procedures.

5. Katherine recently resigned her position from a company after an investigation wrongly accused her of violating company policy. Using "the right to be forgotten" provisions in applicable privacy law, Katherine has requested the former employer remove her from employment records. How should the company respond?

   A. Update its records retention schedule to comply with the request.

   B. File a countersuit, arguing that the organization is not permitted to remove this data.

   C. Comply with applicable privacy law and discard the records as requested.

   D. Reply that applicable employment law forbids erasure of this data.

6. An organization performs its periodic data retention procedure in which specific data files are being identified for removal. Analysts have identified some data files on backup tapes that qualify for removal. How should the organization proceed?

   A. Remove expired files from backup tapes.

   B. Retain backup tapes until all files have expired.

   C. Retain backup tapes until they are rotated out.

   D. Discard the backup tapes containing expired files.

7. An organization stores unstructured data in a cloud-based storage service. In its routine data retention procedures, the organization has identified specific files stored by the storage service that need to be destroyed. How should the organization implement this control?

   A. Ask the owners of expired files to delete them.

   B. Delete expired files and remove any file recovery copies that may exist.

   C. Delete expired files.

   D. Ask the cloud storage service to shred the respective HDDs.

8. Personnel in an organization are discussing the de-identification of its older customer records. Marketing personnel are arguing that de-identification removes their ability to learn how specific customers buy services. How should the privacy manager respond?

   A. No de-identification is necessary.

   B. Records should be archived.

C. Records should be pseudonymized.

D. Records should be anonymized.

9. When reviewing the classification of data files and databases, a privacy manager has identified a set of data files containing customer PII that has been classified as Public. What should the privacy manager do about this?

A. No action is required.

B. Reclassify the data files according to the data classification policy.

C. Direct the de-identification of these files.

D. Direct the removal of these files.

10. An organization is migrating its customer database from an on-premises CRM to a cloud-based CRM. In the process of the migration, the organization created an intermediate flat-file database. How long should the intermediate flat-file database be retained?

A. In perpetuity

B. Until the migration is verified as completed

C. According to the data classification guidelines

D. According to the data retention schedule

11. Marketing analysts want to create a data lake containing all CRM records and customer purchase information to help them better understand purchasing patterns. Because this is not a production system, marketing argues that PII should remain in the data lake to fulfill their research objectives. How should the privacy manager respond to this request?

A. The data lake should be created as requested.

B. All PII should be anonymized after insertion into the data lake.

C. All PII should be pseudonymized before insertion into the data lake.

D. All PII should be anonymized before insertion into the data lake.

12. The computers used by call center personnel utilize solid-state drives (SSDs). Upon retirement, computers are removed from service and donated to a charity. What precautions should first be taken to ensure that all PII on these computers is destroyed?

A. Run an erasure program on the HDDs.

B. Shred the SSDs.

C. Delete all files and reformat the SSDs.

D. Run an erasure program on the SSDs.

PART III

13. To reduce risk, a privacy manager is advocating removing PII fields from an older database. What process is the privacy manager proposing?

    A. Anonymization

    B. Pseudonymization

    C. Hashing

    D. Masking

14. To reduce the risk of credit card fraud, an organization has modified its CRM system so that only the last four digits of customers' credit card numbers are displayed to call center personnel. What technique is being used?

    A. Data hiding

    B. Pseudonymization

    C. Anonymization

    D. Data masking

15. An organization wants to implement a control to provide the ability to detect bulk data transfers at network boundaries. What solution should be used?

    A. NetFlow

    B. Static DLP

    C. USB storage limitation

    D. Data tagging

Answers

1. C. According to data minimization principles, specific personal information fields should not be collected from data subjects unless there is a clear and present need for their use. Otherwise, such collection introduces additional risk without any benefit.

2. A. An organization purchasing PII from a data broker to embellish and update data on existing customers should immediately discard all records (and fields) that are not required for this purpose.

3. D. An organization that finds that it is not using specific PII fields should stop collecting those fields and discard those already collected. Any field that does not have a clearly defined business purpose brings no business benefit but represents only a potential liability should that data be misused or compromised.

4. A. An organization that is learning of instances where PII is being copied to USB storage devices and other destinations should first understand why these activities are taking place. Rushing to judgment may result in the disruption of sanctioned and legitimate business processes. Still, a privacy manager may be able to find a safer method for storing, processing, or transmitting PII than USB storage, which in most cases is not considered particularly secure.

5. D. Any organization that has received a request from a data subject to have their records removed must first understand applicable laws and their requirements to preserve records. If there are no requirements to retain the specific data, the organization is free to remove the data or explain why it still may need to be retained.

6. C. Individual files cannot be removed from backup tapes. Generally, backup media is handled differently from random-access media: backup tapes usually are retained according to a separate schedule.

7. B. Files stored in a cloud-based storage service that have reached their retention life need to be deleted. As many cloud-based storage services have features that permit users to recover accidentally deleted files, any files in these recovery capabilities need to be removed.

8. C. A careful pseudonymization process should result in the marketing team still understanding the buying patterns of individual customers, even though they will no longer know their specific identities.

9. B. Upon discovering that data files or databases have been misclassified, the first step is to assign the proper classification. Further action may be required, such as changing access controls or the media in which the files or databases are stored.

10. D. Intermediate files created during a migration should be retained no longer than what is specified by the data retention policy. There may be an opportunity to discard intermediate files earlier if the organization is confident that the migration was successful and other means are available for recovering data.

11. C. To fulfill customer buying patterns, the data should be pseudonymized so that individual customers' purchasing habits can still be understood, though their actual identities will have been removed.

12. B. Discarding data from SSDs is often problematic, as traditional erasure programs do not always give the desired results. Of the available options, the best course of action is to shred the SSDs.

13. A. Anonymization is the removal of PII fields that could be used to identify a data subject.

14. D. Data masking is the technique of concealing parts of sensitive information so that personnel cannot see it in its entirety.

15. A. NetFlow is a network-based traffic anomaly detection capability used to detect bulk file transfers.

PART IV

Appendix and Glossary

About the Online Content

This book comes complete with TotalTester Online customizable practice exam software with 300 practice exam questions.

System Requirements

The current and previous major versions of the following desktop browsers are recommended and supported: Chrome, Microsoft Edge, Firefox, and Safari. These browsers update frequently, and sometimes an update may cause compatibility issues with the TotalTester Online or other content hosted on the Training Hub. If you run into a problem using one of these browsers, please try using another until the problem is resolved.

Your Total Seminars Training Hub Account

To get access to the online content you will need to create an account on the Total Seminars Training Hub. Registration is free, and you will be able to track all your online content using your account. You may also opt in if you wish to receive marketing information from McGraw Hill or Total Seminars, but this is not required for you to gain access to the online content.

Privacy Notice

McGraw Hill values your privacy. Please be sure to read the Privacy Notice available during registration to see how the information you have provided will be used. You may view our Corporate Customer Privacy Policy by visiting the McGraw Hill Privacy Center. Visit the **mheducation.com** site and click **Privacy** at the bottom of the page.

Single User License Terms and Conditions

Online access to the digital content included with this book is governed by the McGraw Hill License Agreement outlined next. By using this digital content you agree to the terms of that license.

Access   To register and activate your Total Seminars Training Hub account, simply follow these easy steps.

1. Go to this URL: **hub.totalsem.com/mheclaim**

2. To register and create a new Training Hub account, enter your e-mail address, name, and password on the **Register** tab. No further personal information (such as credit card number) is required to create an account.

   If you already have a Total Seminars Training Hub account, enter your e-mail address and password on the **Log in** tab.

3. Enter your Product Key: `3p5s-ch5p-c620`

4. Click to accept the user license terms.

5. For new users, click the **Register and Claim** button to create your account. For existing users, click the **Log in and Claim** button.

   You will be taken to the Training Hub and have access to the content for this book.

Duration of License   Access to your online content through the Total Seminars Training Hub will expire one year from the date the publisher declares the book out of print. Your purchase of this McGraw Hill product, including its access code, through a retail store is subject to the refund policy of that store.

The Content is a copyrighted work of McGraw Hill, and McGraw Hill reserves all rights in and to the Content. The Work is © 2021 by McGraw Hill.

Restrictions on Transfer   The user is receiving only a limited right to use the Content for the user's own internal and personal use, dependent on purchase and continued ownership of this book. The user may not reproduce, forward, modify, create derivative works based upon, transmit, distribute, disseminate, sell, publish, or sublicense the Content or in any way commingle the Content with other third-party content without McGraw Hill's consent.

Limited Warranty   The McGraw Hill Content is provided on an "as is" basis. Neither McGraw Hill nor its licensors make any guarantees or warranties of any kind, either express or implied, including, but not limited to, implied warranties of merchantability or fitness for a particular purpose or use as to any McGraw Hill Content or the information therein or any warranties as to the accuracy, completeness, correctness, or results to be obtained from, accessing or using the McGraw Hill Content, or any material referenced in such Content or any information entered into licensee's product by users or other persons and/or any material available on or that can be accessed through the licensee's product (including via any hyperlink or otherwise) or as to non-infringement of third-party rights. Any warranties of any kind, whether express or implied, are disclaimed. Any material or data obtained through use of the McGraw Hill Content is at your own discretion and risk and user understands that it will be solely responsible for any resulting damage to its computer system or loss of data.

Neither McGraw Hill nor its licensors shall be liable to any subscriber or to any user or anyone else for any inaccuracy, delay, interruption in service, error or omission, regardless of cause, or for any damage resulting therefrom.

In no event will McGraw Hill or its licensors be liable for any indirect, special or consequential damages, including but not limited to, lost time, lost money, lost profits or good will, whether in contract, tort, strict liability or otherwise, and whether or not such damages are foreseen or unforeseen with respect to any use of the McGraw Hill Content.

TotalTester Online

TotalTester Online provides you with a simulation of the CDPSE exam. Exams can be taken in Practice Mode or Exam Mode. Practice Mode provides an assistance window with hints, references to the book, explanations of the correct and incorrect answers, and the option to check your answer as you take the test. Exam Mode provides a simulation of the actual exam. The number of questions, the types of questions, and the time allowed are intended to be an accurate representation of the exam environment. The option to customize your quiz allows you to create custom exams from selected domains or chapters, and you can further customize the number of questions and time allowed.

To take a test, follow the instructions provided in the previous section to register and activate your Total Seminars Training Hub account. When you register you will be taken to the Total Seminars Training Hub. From the Training Hub Home page, select **CDPSE All-in-One TotalTester** from the Study drop-down menu at the top of the page, or from the list of Your Topics on the Home page. You can then select the option to customize your quiz and begin testing yourself in Practice Mode or Exam Mode. All exams provide an overall grade and a grade broken down by domain.

Technical Support

For questions regarding the TotalTester or operation of the Training Hub, visit **www.totalsem.com** or e-mail **support@totalsem.com**.

For questions regarding book content, visit **www.mheducation.com/customerservice**.

acceptable use   A security policy that defines the types of activities that are acceptable and those that are not acceptable.

access control   Any means that detects or prevents unauthorized access and that permits authorized access.

access control list (ACL)   An access control method whereby a list of permitted or denied users (or systems or services, as the case may be) is used to control access.

access control log   A record of attempted accesses.

access control policy   A statement that defines the policy for granting, reviewing, and revoking access to systems and work areas.

access management   A formal business process used to control access to networks and information systems.

access point   A device that provides communication services using the 802.11 (Wi-Fi) protocol standard.

access review   A review of the users, systems, or other subjects that are permitted to access protected objects to ensure that all subjects are authorized to have access.

administrative audit   An audit of operational efficiency.

administrative controls   Controls such as policies, processes, procedures, and standards that define personnel or business practices in accordance with an organization's privacy and security goals.

advertising cookie   *See* persistent cookie.

aggregation   *See* data aggregation.

agile development   A software development process in which a large project team is divided up into smaller teams, and project deliverables are divided up into smaller pieces, each of which can be attained in a few weeks.

algorithm   In cryptography, a specific mathematical formula used to perform encryption, decryption, message digests, and digital signatures.

annualized loss expectancy (ALE)   The expected loss of asset value resulting from threat realization. ALE is defined as single loss expectancy (SLE) × annualized rate of occurrence (ARO).

annualized rate of occurrence (ARO)   An estimate of the number of times that a threat will occur every year.

anonymization   An irreversible de-identification procedure in which specific identifiers that relate personal information to a specific individual are removed. *See also* de-identification, pseudonymization.

antimalware   Software that uses various means to detect and block malware. *See also* antivirus software.

antivirus software   Software designed to detect and remove viruses and other forms of malware.

appliance   A type of computer with preinstalled software that requires little or no maintenance.

application programming interface (API)   A machine-connected computer interface through which programs can communicate with one another.

application programming language   *See* programming language.

application server   A server that runs one or more business application software programs.

application whitelisting   A tool that permits only approved programs to execute on a computer.

architecture standard   A standard that defines technology architecture at the database, system, or network level.

ARCI   *See* RACI.

artificial intelligence (AI)   The study and practice of the use of software or hardware agents that mimic human cognition and are designed for tasks such as learning and problem-solving.

asset inventory   The process of confirming the existence, location, and condition of assets. Also, the results of such a process.

asset management   The processes used to manage the inventory, classification, use, and disposal of assets.

asset value (AV)   The value of an IT asset, which is usually (but not necessarily) the asset's replacement value.

assets   The collection of property that is owned by an organization, including digitally stored information.

asymmetric encryption   A method for encryption, decryption, and digital signatures that uses pairs of encryption keys consisting of a public key and a private key.

attack surface　The people, process, and technology components present on a system or in an environment that can potentially be exploited by an attacker.

attorney–client privilege　A common legal practice that states that the communications between employees and an organization's general counsel are exempt from legal discovery.

attribute sampling　A sampling technique used to study the characteristics of a population to determine how many samples possess a specific characteristic. *See also* sampling.

audit logging　A feature in an application, operating system, or database management system that enables events to be recorded in a separate log.

audit objective　The purpose or goals of an audit. Generally, the objective of an audit is to determine whether controls exist and are effective in some specific aspect of business operations in an organization.

audit procedures　The step-by-step instructions and checklists required to perform specific audit activities. Procedures may include a list of people to interview and questions to ask them, evidence to request, audit tools to use, sampling rates, where and how evidence will be archived, and how evidence will be evaluated.

audit report　The final, written product of an audit. An audit report will include a description of the purpose, scope, and type of audit performed; persons interviewed; evidence collected; rates and methods of sampling; and findings on the existence and effectiveness of each control.

audit scope　The process, procedures, systems, and applications that are the subject of an audit.

authentication　The process of asserting one's identity and providing proof of that identity. Typically, authentication requires a user ID (the assertion) and a password (the proof). However, authentication can also require stronger means of proof, such as a digital certificate, token, smart card, or biometric.

authorization　The process whereby a system determines what rights and privileges a user has.

automatic control　A control that is enacted through some automatic mechanism that requires little or no human intervention.

availability management　Processes that ensure the sustainment of IT service availability.

back door　A section of code that permits someone to bypass access controls and access data or functions. Back doors are commonly placed in programs during development but are removed before programming is complete.

backup　The process of copying important data to another media device in the event of a hardware failure, error, or software bug that causes damage to data.

background check   The process of verifying an employment candidate's employment history, education records, professional licenses and certifications, criminal background, and financial background.

background verification   *See* background check.

beacon   *See* web beacon.

big data   The field of study into large databases for improved analysis and discovery.

biometrics   Any use of a machine-readable characteristic of a user's body that uniquely identifies the user. Biometrics can be used for strong authentication. Types of biometrics include voice recognition, fingerprint, hand scan, palm vein scan, iris scan, retina scan, facial scan, and handwriting. *See also* authentication, strong authentication.

block cipher   An encryption algorithm that operates on blocks of data.

blockchain   A distributed ledger used to record cryptographically linked transactions.

Bluetooth   A short-range air-link standard for data communications between peripherals and low-power consumption devices.

bring-your-own-device (BYOD)   The phenomenon whereby workers use personally owned devices to connect to organization networks and conduct organization business.

budget   A plan for allocating resources over a certain time period.

buffer overflow   An attack on a computer program in which input data overruns input buffers, causing arbitrary code to overwrite the program's code, thereby altering the execution of the program.

bug bounty   A business arrangement whereby an organization will solicit independent researchers to perform security tests against a target system, and the organization will then pay researchers for identifying formerly unidentified vulnerabilities.

bus   A component in a computer that provides the means for the different components of the computer to communicate with one another.

business associate agreement (BAA)   A legal agreement between two parties in which the parties agree to enact general and specific measures to protect personal information, usually protected health information (PHI).

business case   An explanation of the expected benefits to the business that will be realized as a result of a program or project.

business functional requirements   Formal statements that describe required business functions that a system must support.

business impact analysis (BIA)   A study used to identify the impacts of different disaster scenarios to ongoing business operations.

business plan   A formal statement that describes the new business activity, its contribution and impact to the organization, any resources required to operate the activity, the benefits from operating the activity, and any risks associated with the activity.

California Consumer Privacy Act (CCPA)   Privacy regulation enacted in 2020 in the state of California.

California Consumer Privacy Rights Act (CPRA)   Privacy regulation enacted through a ballot initiative in 2020 in the US state of California.

capability maturity model   A model used to measure the relative maturity of an organization or its processes.

Capability Maturity Model Integration (CMMI)   A maturity model that represents the aggregation of other maturity models.

capacity management   Activities that confirm sufficient capacity in IT systems and IT processes to meet service needs.

captcha   A challenge-response test requiring human interaction to distinguish human from machine input.

Center for Internet Security Critical Security Controls (CIS CSC)   A security controls framework developed by the Center for Internet Security (CIS).

central processing unit (CPU)   The main hardware component of a computer that executes program instructions.

certificate authority (CA)   A trusted party that stores digital certificates and public encryption keys.

certificate revocation list (CRL)   An electronic list of digital certificates that have been revoked prior to their expiration date.

chain of custody   Documentation that shows the acquisition, storage, control, and analysis of evidence. The chain of custody may be needed if the evidence is to be used in a legal proceeding.

change advisory board   The group of stakeholders from IT and business who propose, discuss, and approve changes to IT systems.

change control   *See* change management.

change control board   *See* change advisory board.

change management   The IT function that is used to control changes made to an IT environment. *See also* IT service management (ITSM).

change request   A formal request for a change to be made in an environment. *See also* change management.

change review   A formal review of a requested change. *See also* change management, change request.

charter   A document that describes a program (such as a privacy program), including its scope, mission, objectives, roles and responsibilities, authorities, and key business processes.

ciphertext   A message, file, or stream of data that has been transformed by an encryption algorithm and rendered unreadable.

citizen IT   *See* shadow IT.

clientless VPN   A virtual private network (VPN) connection that can be established without the need for a VPN client program. *See also* remote access, VPN client.

cloud   A generalization referring to the use of remote computers, storage, software, or networks, typically through a commercial service.

cloud access security broker (CASB)   A system that monitors and, optionally, controls users' access to cloud-based resources.

cloud computing   A technique of providing a dynamically scalable and usually virtualized computing resource as a service.

cloud responsibility model   *See* shared responsibility model.

cluster   A set of connected computers used to solve a common task. In a cluster, one or more servers actively perform tasks, while zero or more computers may be in a "standby" state, ready to assume active duty should the need arise.

colocation   The use of a commercial data center where customers lease space in equipment cabinets or cages.

code-division multiple access (CDMA)   An air-link standard for wireless communications between mobile devices and base stations.

code of conduct   *See* code of ethics.

code of ethics   A statement that defines acceptable and unacceptable professional conduct within an organization.

collection   In the context of data privacy, the acquisition of personal information.

colocation   The use of a commercial data center where customers lease space in equipment cabinets or cages.

Common Vulnerability Scoring System (CVSS)   A standard methodology for rating system vulnerabilities based on the ease and impact of exploitation.

compensating control   A control that is implemented because another control cannot be implemented or is ineffective.

compliance audit   An audit to determine the level and degree of compliance to a law, regulation, standard, contract provision, or internal control.

compliance risk   Risk that is associated with failure to comply with laws, regulations, and other legal obligations.

compliance testing   A type of testing that is used to determine whether control procedures have been properly designed and implemented, and are operating properly.

configuration management   The process of recording and maintaining the configuration of IT systems.

configuration standard   A standard that defines the detailed configurations that are used in servers, workstations, operating systems, database management systems, applications, network devices, and other systems.

connected devices   Devices that are connected to wired or wireless networks that have little or no human interaction features. *See also* industrial control system (ICS), Internet of Things, supervisory control and data acquisition (SCADA).

consent   Permission granted by a data subject for the collection and/or processing of his or her personal data.

contact tracing   The process of identifying persons who have been in close proximity to a person carrying an infectious disease.

containerization   A method of virtualization in which several isolated operating zones are created in a running server operation, so that programs and data can be isolated within their respective containers.

continuous integration/continuous deployment (CI/CD)   A software development and deployment methodology in which developers' source code changes are merged into a mainline codebase multiple times each day, and multiple incremental software updates are deployed through automated deployment multiple times each day.

contract   A binding legal agreement between two parties that may be enforceable in a court of law.

control   A policy, process, or procedure that is created to achieve a desired event or to avoid an unwanted event.

control objective   A foundational statement that describes desired states or outcomes from business operations.

control risk   The risk that a material error exists that will not be prevented or detected by the organization's control framework.

control self-assessment (CSA)   A methodology used by an organization to review key business objectives, risks, and controls. Control self-assessment is a self-regulation activity.

cookie   A block of data stored by a browser on a user's computer, as directed by a web site.

corroboration   An audit technique whereby an information systems (IS) auditor interviews personnel to confirm the validity of evidence obtained from others who were interviewed previously.

countermeasure   Any activity or mechanism that is designed to reduce risk.

covered entity   An organization that is obligated to comply with Health Information Portability and Accountability Act (HIPAA).

cross-border data transfer   The transfer of data, typically personal information, across a state, provincial, or national border.

cryptanalysis   An attack on a cryptosystem whereby the attacker is attempting to determine the encryption key that is used to encrypt messages.

cryptography   The practice of hiding information from unwanted persons.

cryptosystem   A set of algorithms used to generate an encryption key, perform encryption, and perform decryption.

culture   The behavioral norms in an organization.

custodian   A person or group delegated to operate or maintain an asset.

Cybersecurity Framework (CSF)   *See* NIST CSF.

data aggregation   The process of combining data sets to enrich available data.

data archival   The process of preparing or transforming data for long-term storage.

data classification   The process of assigning a sensitivity classification to a data set or information asset.

data classification policy   A policy that defines sensitivity levels and handling procedures for information.

data controller   An entity that determines the purposes and means of the processing of personal data, which can include directing third parties to process personal data on the entity's behalf.

data destruction   The purposeful act of destroying data so that it cannot be recovered.

data discovery   A process, usually automated, whereby data stores are scanned to determine the presence of specific information, typically personal and other sensitive information.

data definition language (DDL)   A procedural language used to describe the structure of data contained in a database.

data dictionary (DD)   A set of data in a database management system that describes the structure of databases stored there.

data discovery   Any manual or automated means for examining databases and data stores to identify specific types or forms of data.

data flow diagram (DFD)   A diagram that illustrates the flow of data within and between systems.

data governance   Policies and processes that result in management's visibility and control over all data management and data processing activities in an organization.

data lake   An aggregation of an organization's databases into a single store.

data loss prevention (DLP)   Any of several tools and methods used for gaining visibility and control into the presence and movement of sensitive data.

data marking   Any means for applying machine- or human-readable marks on a document to identify it as containing personal or other sensitive information. *See also* data tagging.

data migration   A process of transferring data from one production system to another.

data minimization   The practice of collecting and retaining only those data elements required to perform agreed upon processing. *See also* data retention.

data processor   An entity that processes data at the direction of a data controller. *See also* data controller.

data protection impact assessment (DPIA)   An analysis of how planned changes to a business process or information system will impact an organization's ability to protect specific types of data, such as personally identifiable information (PII). *See also* privacy impact assessment (PIA).

data protection officer (DPO)   A position tasked with ensuring that the organization's privacy policies and practices are compliant with applicable laws, regulations, and other legal obligations.

data retention   A portion of data governance. A formal approach for determining how long data should reside in an organization, and the process of developing procedures for removing data that has exceeded its retention limit.

data retention schedule   A formal statement that specifies how long various types of business records should be retained in an organization.

data sovereignty   The concept of data that is subject to laws within the jurisdiction in which it is collected.

data sprawl   The proliferation of data beyond its intended bounds.

data subject   An identifiable natural person.

data subject request (DSR)   A request or inquiry sent to an organization by a data subject for the purpose of verification, correction, or removal of personal information.

data tagging   A typically automated process of applying an identifying mark on a data file if its contents meet specific criteria.

data warehouse   A set of data that has been copied from a production system and used for reporting, research, or archival purposes.

database   A collection of structured or unstructured information.

database management system (DBMS)   A software program that facilitates the storage and retrieval of potentially large amounts of structured or unstructured information.

database server   A server that contains and facilitates access to one or more databases.

DBMS   *See* database management system.

decryption   The process of transforming ciphertext into plaintext so that a recipient can read it.

default password   A password associated with a user account or system account that retains its factory default setting.

degaussing   The application of a strong magnetic field to erase the contents of magnetic storage media.

de-identification   Any procedure through which specific identifiers about a data subject are removed or replaced. *See also* anonymization, masking, pseudonymization.

denial of service (DoS)   An attack on a computer or network with the intention of causing disruption or malfunction of the target.

detective control   A control that is used to detect events.

deterrent control   A control that is designed to deter people from performing unwanted activities.

DevOps   An agile software development and operations model that blurs the line between development and operations.

DevSecOps   An agile and secure software development and operations model that relies upon automated security testing in the development-deployment process.

Diffie-Hellman   A popular key exchange algorithm. *See also* key exchange.

digital certificate   An electronic document that contains an identity that is signed with the public key of a certificate authority (CA).

digital envelope   A method that uses two layers of encryption: a symmetric key is used to encrypt a message, and then a public or private key is used to encrypt the symmetric key.

digital signature   The result of encrypting the hash of a message with the originator's private encryption key, used to prove the authenticity and integrity of a message.

digital subscriber line (DSL)   A common carrier standard for transporting data from the Internet to homes and businesses.

directory   A structure in a file system that is used to store files and, optionally, other directories. *See also* file system.

disaster recovery and business continuity requirements   Formal statements that describe required recoverability and continuity characteristics that a system must support.

Do Not Track   A web browser feature that requests web site operators to refrain voluntarily from tracking individual visitors.

documentation   The inclusive term that describes charters, processes, procedures, standards, requirements, and other written documents.

Domain Name System (DNS)   A TCP/IP application layer protocol used to translate domain names (such as www.isecbooks.com) into IP addresses.

dynamic data loss prevention (dynamic DLP)   The use of tools to detect the movement of PII and other sensitive information. *See also* data loss prevention, static DLP.

eavesdropping   The act of secretly intercepting and recording a voice or data transmission.

elasticity   A feature of cloud environments that enables resources to be dynamically added or removed to service varying levels of work.

electronic protected health information (ePHI)   Patient-related healthcare information in electronic form, as defined by the US Healthcare Insurance Portability and Accountability Act (HIPAA).

elliptic curve   A public key cryptography algorithm.

e-mail   A network-based service used to transmit messages between individuals and groups.

encapsulation   A practice whereby a method can call on another method to help perform its work. *See also* method.

encryption   The act of hiding sensitive information in plain sight. Encryption works by scrambling the characters in a message using a method known only to the sender and receiver, making the message useless to anyone else who intercepts the message.

encryption key   A block of characters used in combination with an encryption algorithm to encrypt or decrypt a stream or block of data.

end user behavior analytics (EUBA)   *See* user behavior analytics (UBA).

endpoint   Any of several types of end user devices, including desktop computers, laptop computers, tablet computers, and smartphones.

enterprise architecture   Activities that ensure that important business needs are met by IT systems; the model that is used to map business functions into the IT environment and IT systems in increasing levels of detail.

entity-relationship diagram (ERD)   A diagram that depicts information in detail and that functions as a logical data model.

Ethernet   A standard protocol for assembling a stream of data into frames for transport over a physical medium from one station to another on a local area network. On an Ethernet network, any station is free to transmit a packet at any time, provided that another station is not already doing so.

evidence   Information gathered by the auditor that provides proof that a control exists and is being operated.

exposure factor (EF)   The financial loss that results from the realization of a threat, expressed as a percentage of the asset's total value.

extract, transfer, and load (ETL)   Software that transforms data sets, typically for migration from one system to another.

extraterritorial regulation   A government regulation that exercises jurisdiction beyond its land borders.

facial recognition   A capability, often coupled with video surveillance, to identify specific individuals automatically based upon their facial characteristics.

facility classification   A method for assigning classification or risk levels to work centers and processing centers, based on their operational criticality and other risk factors.

Factor Analysis of Information Risk (FAIR)   An analysis method that helps a risk manager understand the factors that contribute to risk, as well as the probability of threat occurrence and an estimation of potential losses.

fail closed   A situation in which all access will be denied if an access control system fails.

fail open   A situation in which all access will be permitted if an access control system fails.

false negative   An activity that is mistakenly *not* identified as an event. *See also* false positive.

false positive   An activity that is mistakenly identified as an event. *See also* false negative.

feasibility study   An activity that seeks to determine the expected benefits of a program or project.

fiber-optics   A cabling standard that uses optical fiber instead of metal conductors.

field A unit of storage in a relational database management system that consists of a single data item within a row. *See also* relational database management system (RDBMS), row, table.

file A sequence of zero or more characters that is stored as a whole in a file system. A file may be a document, a spreadsheet, an image, a sound file, a computer program, or data that is used by a program. *See also* file system.

file allocation table (FAT) A file system used by the MS-DOS operating system as well as by early versions of the Microsoft Windows operating system. *See also* file system.

file server A server that is used to store files in a central location, usually to make them available to many users.

file store *See* file server.

file system A logical structure that facilitates the storage of data on a digital storage medium such as a hard drive, CD/DVD-ROM, or flash memory device.

File Transfer Protocol (FTP) An early and still widely used TCP/IP application layer protocol that is used for the batch transfer of files or entire directories from one system to another.

File Transfer Protocol Secure (FTPS) A TCP/IP application layer protocol that is an extension of the FTP protocol whereby authentication and transport are encrypted using SSL or TLS. *See also* File Transfer Protocol (FTP), Secure Sockets Layer (SSL), Transport Layer Security (TLS).

financial audit An audit of an accounting system, accounting department processes, and accounting procedures to determine whether business controls are sufficient to ensure the integrity of financial statements.

firewall A device that controls the flow of network messages between networks. Placed at the boundary between the Internet and an organization's internal network, firewalls enforce security policy by prohibiting all inbound traffic except for the specific few types of traffic that are permitted to a select few systems.

first-party cookie A cookie whose origin matches the domain of the web server. *See also* cookie, third-party cookie.

foreign key A field in a table in a relational database management system that references a unique primary key in another table. *See also* field, relational database management system (RDBMS), row, table.

functional requirements Statements describing the required characteristics that a system must have to support business needs.

PART IV

functional testing   The portion of software testing in which functional requirements are verified.

gate process   Any business process that consists of one or more review/approval gates, which must be completed before the process may continue.

gateway   A device that acts as a protocol converter or that performs some other type of transformation of messages.

general computing controls (GCC)   *See* IT general controls (ITGC).

General Data Protection Regulation (GDPR)   A European Union regulation enacted in 2018 that defines the data privacy rights and remedies of European residents.

governance   Management's control over policy and processes.

grid computing   A large number of loosely coupled computers that are used to solve a common task.

guest   A virtual machine running under a hypervisor.

guideline   A document that provides suggestions for compliance to a policy or standard.

hacker   Someone who interferes with or accesses another's computer without authorization. Formerly, a hobbyist who strives to understand how complex mechanisms function.

hardening   The technique of configuring a system so that only its essential services and features are active and all others are deactivated, as well as improvements in security configuration settings. This helps to reduce the attack surface of a system to its essential components.

hardware   The physical machinery of computing, storage, and networks.

hardware security module (HSM)   A device used to store and protect encryption keys.

hash function   A cryptographic operation on a block of data that returns a fixed-length string of characters, used to verify the integrity of a message.

Health Insurance Portability and Accountability Act (HIPAA)   A US regulation regarding the protection of electronic protected health information (ePHI) that applies to healthcare delivery organizations, health insurance companies, and other healthcare industry organizations.

hierarchical database management system   A database management system employing a top-down hierarchy data model.

hierarchical file system (HFS)   A file system formerly used on computers running the macOS operating system. *See also* file system.

hub   An Ethernet network device used to connect devices to the network. A hub can be thought of as a multiport repeater.

hybrid cloud   A distributed computing environment consisting of cloud-based and on-premises systems.

hybrid cryptography   A cryptosystem that employs two or more iterations or types of cryptography.

Hypertext Transfer Protocol (HTTP)   A TCP/IP application layer protocol used to transmit web page contents from web servers to users who are using web browsers.

Hypertext Transfer Protocol Secure (HTTPS)   A TCP/IP application layer protocol that is similar to HTTP in its use for transporting data between web servers and browsers. HTTPS is not a separate protocol, but instead is the instance of HTTP that is encrypted with SSL or TLS. *See also* Hypertext Transfer Protocol (HTTP), Secure Sockets Layer (SSL), Transport Layer Security (TLS).

hypervisor   Virtualization software that facilitates the operation of one or more virtual machines.

identification   The process of asserting one's identity without providing proof of that identity. *See also* authentication.

identity management   The activity of managing the identity of each employee, contractor, temporary worker, and, optionally, customer for use in a single environment or multiple environments.

implementation   A step in the software development life cycle at which new or updated software is placed into the production environment and started.

incident (privacy)   A suspected or confirmed violation of privacy policy or privacy regulation.

incident (security)   A suspected or confirmed violation of security policy or security regulation.

incident response   The organized response to a privacy or security incident. *See also* privacy incident response, security incident response.

independence   The characteristic of an auditor and his or her relationship to a party being audited. An auditor should be independent of the auditee; this permits the auditor to be objective.

index   An entity in a relational database management system that facilitates rapid searching for specific rows in a table based on one of the fields other than the primary key. *See also* field, primary key, relational database management system (RDBMS), row, table.

industrial control system (ICS)   Networks and systems used to monitor and control industrial processes such as manufacturing or utilities. *See also* connected devices, Internet of Things (IoT), Supervisory control and data acquisition (SCADA).

information classification   *See* data classification.

information security management   The aggregation of policies, processes, procedures, and activities to ensure that an organization's security policy is effective.

information security management system (ISMS)   The collection of activities for managing information security, as defined by ISO/IEC 27001.

information security policy   A statement that defines how an organization will classify and protect its important assets.

infrastructure   The collection of networks, network services, devices, facilities, and system software that facilitates access to, communications with, and protection of business applications.

infrastructure as a service (IaaS)   A cloud computing model whereby a service provider makes computers and other infrastructure components available to subscribers. *See also* cloud computing.

inherent risk   The risk that there are material weaknesses in existing business processes and no compensating controls to detect or prevent them.

initialization vector (IV)   A random number that is needed by some encryption algorithms to begin the encryption process.

input authorization   Controls that ensure that all data input into an information system is authorized by management.

input controls   Administrative and technical controls that determine what data is permitted to be input into an information system. These controls exist to ensure the integrity of information in a system.

input validation   Controls that ensure that the type and values of information input into a system are appropriate and reasonable.

integrated audit   An audit that combines an operational audit and a financial audit. *See also* financial audit, operational audit.

integrated development environment (IDE)   A class of desktop software development tools that includes source code editing, source code version control, compilation, and debugging into a single tool.

intellectual property   A class of assets owned by an organization; includes an organization's designs, architectures, software source code, processes, and procedures.

Internet   The interconnection of the world's TCP/IP networks.

Internet Key Exchange (IKE)   A protocol used to establish associations (logical connections) between stations within the IPsec protocol.

Internet of Things (IoT)   The connection of physical objects other than human-interactive computers to networks and the Internet.

Internet Protocol Security (IPsec)   A suite of protocols used to secure IP-based communications by using authentication and encryption.

intrusion detection system (IDS)   A hardware or software system that detects anomalies that may be signs of an intrusion.

intrusion prevention system (IPS)   A hardware or software system that detects and blocks anomalies that may be indications of an intrusion.

IP address   An address assigned to a station on a TCP/IP network.

ISACA   An international professional association focused on IT governance.

ISACA audit guidelines   Published documents that help the IS auditor apply ISACA audit standards.

ISACA audit procedures   Published documents that provide sample procedures for performing various audit activities and for auditing various types of technologies and systems.

ISACA audit standards   The minimum standards of performance related to security, audits, and the actions that result from audits. The standards are published by ISACA and updated periodically. ISACA audit standards are considered mandatory by IS auditors worldwide.

ISAE 3402 (International Standard on Assurance Engagement)   An international standard for the external audit of a service provider. An ISAE 3402 audit is performed according to rules established by the International Auditing and Assurance Standards Board (IAASB).

ISO/IEC 20000   An ISO/IEC standard for IT service management (ITSM).

ISO/IEC 27001   An ISO/IEC standard for IT security management.

ISO/IEC 27002   An ISO/IEC standard for IT security controls.

ISO/IEC 27005   An ISO/IEC standard for cybersecurity risk assessments and risk management.

ISO/IEC 27035   An ISO/IEC standard for cybersecurity incident response.

ISO/IEC 27701   An ISO/IEC standard for privacy information management.

ISO/IEC 38500   An ISO/IEC standard for corporate governance of information technology.

ISO/IEC 9660   An ISO/IEC standard file system format used on CD-ROM and DVD-ROM media.

IT Assurance Framework (ITAF)   An end-to-end framework developed to guide organizations in developing and managing IT assurance and IT audits.

IT balanced scorecard   A balanced scorecard used to measure IT organization performance and results. Also known as the standard IT balanced scorecard.

IT general controls (ITGC)   A framework of controls established in an organization that apply across core business systems. Generally used in the context of Sarbanes-Oxley compliance.

IT governance   Management's control over IT policy and processes.

IT Infrastructure Library (ITIL)   *See* IT service management (ITSM).

IT service desk   *See* service desk.

IT service management (ITSM)   A set of business processes used to manage an IT organization.

IT steering committee   A body of senior managers or executives that discusses IT-related high-level and long-term issues in the organization.

iterative development process   A software development process that consists of one or more loops of planning, requirements, design, coding, and testing until development and implementation are considered complete.

job description   A written description of an employee's responsibilities within an organization. A job description usually contains a job title and responsibilities, experience requirements, and knowledge requirements.

judgmental sampling   A sampling technique by which items are chosen based upon the auditor's judgment, usually based on risk or materiality. *See also* sampling.

key   *See* encryption key.

key compromise   Any unauthorized disclosure or damage to an encryption key. *See also* key management.

key custody   The policies, processes, and procedures regarding the management of keys. *See also* key management.

key disposal   The process of decommissioning encryption keys. *See also* key management.

key encrypting key   An encryption key used to encrypt another encryption key.

key exchange   A technique used by two parties to establish a symmetric encryption key when no secure channel is available.

key fingerprint   A short sequence of characters used to authenticate a public key.

key generation   The initial generation of an encryption key. *See also* key management.

key goal indicator (KGI)   A measure of business activities related to the achievement of strategic goals and objectives.

key length   The size (measured in bits) of an encryption key. Longer encryption keys mean that it takes greater effort to attack a cryptosystem successfully.

key logger   A hardware device or a type of malware that records a user's keystrokes and, optionally, mouse movements and clicks and sends them to the key logger's owner.

key management   The various processes and procedures used by an organization to generate, protect, use, and dispose of encryption keys over their lifetime.

key performance indicator (KPI)   A measure of business processes' performance and quality, used to reveal trends related to efficiency and effectiveness of key processes in the organization.

key protection   All means used to protect encryption keys from unauthorized disclosure and harm. *See also* key management.

key risk indicator (KRI)   A measure of business risk, used to reveal trends related to the risk levels of various activities, processes, and systems in an organization.

key rotation   The process of issuing a new encryption key and re-encrypting data protected with the new key. *See also* key management.

laptop computer   A portable computer used by an individual user.

lean   A project management approach that emphasizes focus on value and efficiency. Lean is derived from lean manufacturing techniques developed at Toyota in Japan in the 1990s.

least privilege   The concept whereby an individual user should have the lowest privilege level possible that will still enable him or her to perform required tasks.

legitimate interest   A rationale expressed for the justification for processing personal information that presupposes benefits for the data subject and the processor.

Lightweight Directory Access Protocol (LDAP)   A TCP/IP application layer protocol used as a directory service for people and computing resources.

local area network (LAN)   An interconnection of computers within a limited area such as a single building. *See also* network.

logic bomb   A set of computer instructions designed to perform some damaging action when a specific event occurs; a popular example is a time bomb that alters or destroys data on a specified date in the future.

LTE (Long Term Evolution)   A wireless telecommunications standard for use by mobile devices, considered an upgrade of older Global System for Mobile Communications (GSM) and CDMA2000 standards.

PART IV

machine learning   Computer algorithms that improve and self-correct through the acquisition of experience.

main storage   A computer's short-term storage of information, usually implemented with electronic components such as random access memory (RAM).

mainframe   A large central computer capable of performing complex tasks for several users simultaneously.

malware   The broad class of programs that are designed to inflict harm on computers, networks, or information. Types of malware include viruses, worms, Trojan horses, spyware, and rootkits.

mandatory access control (MAC)   An access model used to control access to objects (files, directories, databases, systems, networks, and so on) by subjects (persons, programs, and so on). When a subject attempts to access an object, the operating system examines the access properties of the subject and object to determine whether the access should be allowed. The operating system then permits or denies the requested access.

manual control   A control that requires a human to operate it.

marking   The act of affixing a classification label to a document.

masking   A technique of concealing the contents of data.

maturity   The degree of formality and integrity of a business process.

Media Access Control (MAC) address   Node addressing used on an Ethernet network whereby the address is expressed as a six-byte hexadecimal value. A typical address is displayed in a notation separated by colons or dashes, such as F0:E3:67:AB:98:02.

message digest   The result of a cryptographic hash function.

methodology standard   A standard that specifies the practices used by the IT organization.

microsegmentation   A segmentation technique in which individual hosts are isolated with network access controls, typically network or host firewalls.

migration   The process of transferring data from one system to a replacement system.

mitigating control   *See* compensating control.

mobile backend as a service (MBaaS)   A class of cloud-based services consisting of backend storage and APIs, in support of mobile applications. *See also* cloud computing.

mobile device   A portable computer in the form of a smartphone, tablet computer, or wearable device.

mobile device management (MDM)   A class of enterprise tools used to manage mobile devices such as smartphones and tablet computers.

monitoring   The continuous or regular evaluation of a system or control to determine its operation or effectiveness.

multifactor authentication   Any means used to authenticate a user that is stronger than the use of a user ID and password. Multifactor authentication includes a user ID and password, plus any one of the following: digital certificate, token, smart card, or biometric.

near-field communications (NFC)   A standard for extremely short-distance radio frequency data communications.

NetFlow   A network diagnostic tool that collects all network metadata and can be used for network diagnostic or security purposes.

network   An interconnection of computers for the purpose of exchanging information.

network access control (NAC)   An approach for network authentication and access control for devices designed to attach to a LAN or wireless LAN.

network architecture   The overall design of an organization's network.

network segmentation   The design process that results in the creation of network security zones that are defined and controlled by firewalls or other stateful ACLs that limit access between zones.

Network Time Protocol (NTP)   A TCP/IP application layer protocol used to synchronize the time-of-day clocks on systems with time reference standards.

NIST CSF (National Institute for Standards and Technology Cybersecurity Framework)   A controls and controls management framework developed by the US National Institute for Standards and Technology.

NIST SP 800-30   A NIST special publication regarding a standard methodology for conducting a risk assessment.

nonrepudiation   The property of digital signatures and encryption that can make it difficult or impossible for a party to deny having sent a digitally signed message—unless they admit to having lost control of their private encryption key.

notebook computer   *See* laptop computer.

NoSQL   An inclusive term referring to several nonrelational database management system designs.

NT File System (NTFS)   A file system used by newer versions of the Microsoft Windows operating system. *See also* file system.

object   A resource, such as a computer, application, database, file, or record. *See also* subject.

object database   *See* object database management system (ODBMS).

object database management system (ODBMS)   A type of database management system in which information is represented as objects that are used in object-oriented programming languages.

objectivity   The characteristic of a person that relates to his or her ability to develop an opinion that is not influenced by external pressures.

operating system   A large, general-purpose program that is used to control computer hardware and facilitate the use of software applications.

operational audit   An audit of IS controls, security controls, or business controls to determine control existence and effectiveness.

Operationally Critical Threat, Asset, and Vulnerability Evaluation (OCTAVE)   A risk analysis approach developed by Carnegie Mellon University and used to assess privacy and security risks.

orchestration   A scripted, automated response that is triggered when specific events occur.

organization chart   A diagram that depicts the manager–subordinate relationships in an organization or in a part of an organization.

owner   A person or group responsible for the operation of an asset.

passphrase   A longer password that is constructed from a string of words.

password   An identifier that is created by a system manager or a user; this secret combination of letters, numbers, and other symbols is known (or should be known) only to the user who uses it.

password complexity   The characteristics required of user account passwords. For example, a password may not contain dictionary words and must contain uppercase letters, lowercase letters, numbers, and symbols.

password length   The minimum and maximum number of characters permitted for a password that is associated with a computer account.

password reset   The process of changing a user account password and unlocking the user account so that the user's use of the account may resume.

password reuse   The act of reusing a prior password for a user account. Some information systems can prevent password reuse in case of password compromise with or without the user's knowledge.

password vaulting   The process of storing a password in a secure location for later use.

patch management   The process of identifying, analyzing, and applying patches (including security patches) to systems.

Payment Card Industry Data Security Standard (PCI-DSS) A security standard whose objective is the protection of credit card numbers while in storage, while processed, and while transmitted. The standard was developed by the Payment Card Industry, a consortium of credit card companies, including VISA, MasterCard, American Express, Discover, and JCB.

penetration test A simulation of an attack on a system or network to identify the presence of exploitable vulnerabilities.

persistent cookie A cookie used to identify a user and store user preferences. *See also* cookie.

Personal Information Protection and Electronic Documents Act (PIPEDA) A Canadian data privacy law that went into effect in 2000 that seeks to ensure consumer data privacy in the context of e-commerce.

personally identifiable information (PII) Any information relating to an identifiable natural person.

phishing A social engineering attack on unsuspecting individuals whereby e-mail messages that resemble official communications entice victims to visit imposter web sites that contain malware or request credentials to sensitive or valuable assets.

physical control Controls that employ physical means.

plaintext An original message, file, or stream of data that can be read by anyone who has access to it.

platform as a service (PaaS) A cloud computing delivery model whereby the service provider supplies the platform on which an organization can build and run software. *See also* cloud computing.

playbook A detailed procedure, typically the instructions to be followed in response to an event or incident.

policy A statement that specifies what must be done (or not done) in an organization. A policy usually defines who is responsible for monitoring and enforcing it.

population A complete set of subjects, entities, transactions, or events that are the subject of an audit.

preventive action An action that is initiated to prevent an undesired event or condition.

preventive control A control that is used to prevent unwanted events from occurring.

primary key One of the fields in a table in a relational database management system whose values are unique for each record (row). *See also* field, relational database management system (RDBMS), row, table.

privacy The protection of personal information from unauthorized disclosure, use, and distribution.

privacy awareness   A formal program used to educate employees, users, customers, or constituents on required, acceptable, and unacceptable privacy-related behaviors. *See also* security awareness.

privacy governance   Management's control over an organization's information privacy program.

privacy impact assessment (PIA)   An analysis of how personally identifiable information is collected, used, shared, and maintained as part of planned changes to a business process or information system to identify any changes in privacy risk. *See also* data protection impact assessment (DPIA).

privacy incident   An event in which personal information has been misused, accessed by unauthorized persons, or affected by a security incident. *See also* security incident.

Privacy Information Management System (PIMS)   The collection of activities for managing information privacy, as defined by ISO/IEC 27701.

privacy office   A corporate oversight function that ensures that the organization complies with applicable privacy laws and other related requirements.

privacy policy   A policy statement that defines how an organization will protect, manage, and handle private information.

privacy requirements   Formal statements that describe required privacy safeguards that a system must support.

privacy steering committee   A body of senior managers or executives that establishes priorities and provides oversight for activities and issues related to information privacy in the organization.

problem   As defined in IT service management, a situation characterized by several similar incidents. *See also* incident, IT service management.

procedure   A written sequence of instructions used to complete a task.

process   A collection of one or more procedures used to perform a business function. *See also* procedure. Also, a logical container in an operating system in which a program executes.

process isolation   A basic feature of an operating system that prevents one process from accessing the resources used by another process.

program   An organization of many large, complex activities; it can be thought of as a set of projects that work to fulfill one or more key business objectives or goals.

program charter   A formal definition of the objectives of a program, its main timelines, sources of funding, the names of its principal leaders and managers, and the business executive(s) who are sponsoring the program.

program management The management of a group of projects that exist to fulfill a business goal or objective.

programming language A vocabulary and set of rules used to construct a human-readable computer program.

project A coordinated and managed sequence of tasks that result in the realization of an objective or goal.

protected health information (PHI) Patient-related healthcare information, as defined by the US Healthcare Insurance Portability and Accountability Act (HIPAA). *See also* electronic protected health information (ePHI).

protocol standard A standard that specifies the protocols used by the IT organization.

provided by client (PBC) list A list of evidence requested of an auditee at the onset of an audit.

provisioning The creation of a user account and the issuance of credentials to the user.

pseudonymization An irreversible de-identification procedure whereby a specific identifier is replaced by other values to make it less identifiable to the original data subject. *See also* anonymization, de-identification.

public cloud A commercial computing, storage, or processing service accessible over the Internet.

public key cryptography *See* asymmetric encryption.

public key infrastructure (PKI) A centralized function that is used to store and publish public keys and other information.

qualitative risk analysis A risk analysis methodology whereby risks are classified on a nonquantified scale, such as "High, Medium, Low," or on a simple numeric scale, such as 1 through 5.

quality assurance testing (QAT) The portion of software testing in which system specifications and technologies are formally tested.

quantitative risk analysis A risk analysis methodology whereby risks are estimated in the form of actual cost amounts.

RACI (Responsible-Accountable-Consulted-Informed) The responsibility model used to track individual responsibilities in a business process or a project.

random access memory (RAM) A type of semiconductor memory usually used for a computer's main storage.

ransomware Malware that performs some malicious action, requiring payment from the victim to reverse the action. Malicious actions include data erasure, data encryption, and system damage.

records   Documents describing business events such as meeting minutes, contracts, financial transactions, decisions, purchase orders, logs, and reports.

reduced sign-on   The use of a centralized directory service (such as LDAP or Microsoft Active Directory) for authentication into systems and applications. Users will need to log in to each system and application, using one set of login credentials. *See also* Lightweight Directory Access Protocol (LDAP), single sign-on.

Redundant Array of Independent Disks (RAID)   A family of technologies used to improve the reliability, performance, or size of disk-based storage systems.

referential integrity   The characteristic of relational database management systems that requires the DBMS maintain the parent–child relationships between records in different tables and prohibits activities such as deleting parent records and transforming child records into orphans. *See also* relational database management system (RDBMS).

registration authority (RA)   An entity that works within or alongside a certificate authority (CA) to accept requests for new digital certificates.

regulatory requirements   Formal statements, derived from laws and regulations, that describe the required characteristics a system must support.

relational database management system (RDBMS)   A database management system that permits the design of a database consisting of one or more tables that can contain fields that refer to rows in other tables. This is currently the most popular type of database management system.

remediation   The correction of a defect.

remote access   A capability that permits a user to establish a network connection from a remote location so that the user can access internal network resources from a remote location.

reperformance   An audit technique whereby an IS auditor repeats actual tasks performed by auditees to confirm that they were performed properly.

replication   An activity in which data that is written to a storage system is also copied over a network to another storage system and written. The result is the presence of up-to-date data that exists on two or more storage systems, each of which could be located in a different geographic region.

request for information (RFI)   A formal solicitation to outside organizations for detailed information about specific types of products or services.

request for proposal (RFP)   A formal solicitation to outside organizations for business proposals for specific types of products or services.

requirements   Formal statements that describe required (and desired) characteristics of a system that is to be changed, developed, or acquired.

residual risk   The risk that remains after being reduced through other risk treatment options.

responsibility   A stated expectation of activities and performance.

return on investment (ROI)   The ratio of money gained or lost as compared to an original investment.

right to audit   A clause in a contract that indicates that one party has the right to conduct an audit of the other party's operations.

risk   Generally, the fact that undesired events can happen that may damage property or disrupt operations; specifically, an event scenario that can result in property damage or disruption.

risk acceptance   The risk treatment option by which management chooses to accept the risk as-is.

risk analysis   The process of identifying and studying risks in an organization.

risk appetite   The level of risk that an organization is willing to accept while in pursuit of its mission, strategy, and objectives, and before action is needed to treat or manage the risk. *See also* risk capacity.

risk assessment   A process by which risks, in the form of threats and vulnerabilities, are identified for each asset.

risk avoidance   The risk treatment option involving a cessation of the activity that introduces identified risk.

risk capacity   The objective amount of loss that an organization can tolerate without its continued existence being called into question. *See also* risk appetite.

Risk IT Framework   A risk management model that approaches risk from the enterprise perspective.

risk management   The management activities used to identify, analyze, and treat risks.

risk mitigation   The risk treatment option involving implementation of a solution that will reduce the impact of an identified risk.

risk tolerance   *See* risk appetite.

risk transfer   The risk treatment option involving the act of transferring risk to another party, such as an insurance company.

risk treatment   The decision to manage an identified risk. The available choices are mitigate the risk, avoid the risk, transfer the risk, or accept the risk.

role   A set of privileges in an application. Also, a formally defined set of work tasks assigned to an individual.

PART IV

rollback   A step in the system development life cycle where system changes need to be reversed, returning the system to its previous state.

router   A device that is used to interconnect two or more networks.

row   A unit of storage in a relational database management system that consists of a single record in a table. *See also* relational database management system (RDBMS), table.

sample   A portion of a population of records selected for auditing.

sampling   A technique used to select a portion of a population when it is not feasible to test an entire population.

Sarbanes-Oxley Act   A US law requiring public corporations to enact business and technical controls, perform internal audits of those controls, and undergo external audits.

scanning   *See* static DLP, vulnerability scanning.

secondary storage   A computer's long-term storage of information, usually implemented with hard disk drives or static random access memory (SRAM).

Secure File Transfer Protocol (SFTP)   A TCP/IP application layer protocol that is an extension of FTP, whereby authentication and file transfer are encrypted using Secure Shell (SSH). Sometimes referred to as SSH File Transfer Protocol. *See also* File Transfer Protocol (FTP), Secure Shell (SSH).

Secure Hypertext Transfer Protocol (SHTTP)   A protocol used to encrypt web pages between web servers and web browsers. Often confused with Hypertext Transfer Protocol Secure (HTTPS).

Secure Multipurpose Internet Mail Extensions (S/MIME)   An e-mail security protocol that provides sender and recipient authentication and encryption of message content and attachments.

Secure Shell (SSH)   A TCP/IP application layer protocol that provides a secure channel between two computers whereby all communications between them are encrypted. SSH can also be used as a tunnel to encapsulate and thereby protect other protocols.

Secure Sockets Layer (SSL)   An encryption protocol used to encrypt web pages requested with the HTTPS URL. Deprecated by Transport Layer Security (TLS). *See also* Hypertext Transfer Protocol Secure (HTTPS), Transport Layer Security (TLS).

security awareness   A formal program used to educate employees, users, customers, or constituents on required, acceptable, and unacceptable security-related behaviors.

security governance   Management's control over an organization's security program.

security incident   An event in which the confidentiality, integrity, or availability of information (or an information system) has been compromised.

security incident response   The formal, planned response that is enacted when a security incident has occurred. *See also* security incident.

security information and event management system (SIEM)   An information system that collects event logs and generates alerts to inform personnel of events occurring that warrant attention and potential action.

security policy   *See* information security policy.

security requirements   Formal statements that describe the required security characteristics that a system must support.

segmentation   The practice of dividing a network into two or more security zones, with network access controls controlling access between those zones.

segregation of duties   The concept that ensures that single individuals do not possess excess privileges that could result in unauthorized activities such as fraud or the manipulation or exposure of sensitive data.

separation of duties   *See* segregation of duties.

server   A centralized computer used to perform a specific task.

serverless computing   An application operations cloud service model in which server operating systems are provided and managed by the cloud service provider.

service desk   The IT function that handles incidents and service requests on behalf of customers by acting as a single point of contact. *See also* IT service management (ITSM).

service level agreement (SLA)   A formal commitment by an individual or group to provide services at stated levels of quantity and quality.

service provider audit   An audit of a third-party organization that provides services to other organizations.

session cookie   A cookie used by a web server that uniquely identifies a logged-in user from other logged-in users. *See also* cookie.

shadow IT   The phenomenon whereby organization departments procure IT services directly, bypassing corporate IT.

shared responsibility model   A logical model that depicts and describes operational responsibilities in a cloud services environment. The model indicates the responsibilities belonging to the cloud service provider and those belonging to the customer.

Simple Mail Transfer Protocol (SMTP)   A TCP/IP application layer protocol that is used to transport e-mail messages.

single loss expectancy (SLE)   The financial loss when a threat is realized one time. SLE is defined as AV × EF. *See also* asset value (AV), exposure factor (EF).

single sign-on (SSO)   An interconnected environment in which applications are logically connected to a centralized authentication server that is aware of the logged-in and/or logged-out status of each user. A user can log in once to the environment; each application and system is aware of a user's login status and will not require the user to log in to each one separately. *See also* reduced sign-on.

site classification policy   A policy that defines sensitivity levels, security controls, and security procedures for information processing sites and work centers.

smart card   A small, credit card–sized device that contains electronic memory and is accessed with a smart card reader and used in multifactor authentication.

smartphone   A mobile phone equipped with an operating system and software applications.

snapshot   A continuous auditing technique that involves the use of special audit modules embedded in online applications that sample specific transactions. The module copies key database records that can be examined later on.

SOC   *See* system and organization controls (SOC) audit

SOC 1   A system and organization controls audit of a financial services provider using a bespoke set of controls.

SOC 2   A system and organization controls audit of a service provider using a standard set of controls.

social engineering   The act of using deception to trick an individual into revealing secrets.

software as a service (SaaS)   A software delivery model whereby an organization obtains a software application for use by its employees and the software application is hosted by the software provider, as opposed to the customer organization. *See also* cloud computing.

software development life cycle (SDLC)   *See* systems development life cycle (SDLC).

software licensing   The process of maintaining accurate records regarding the permitted use of software programs.

software maintenance   An activity in the software development life cycle whereby modifications are made to the software code.

source code management   The techniques and tools used to manage application source code.

SOX   *See* Sarbanes-Oxley Act.

spam   Unsolicited and unwanted e-mail.

spam filter   A central program or device that examines incoming e-mail and removes all messages identified as spam.

spiral model   A software development life cycle process whereby the activities of requirements definition and software design go through several cycles until the project is complete. *See also* systems development life cycle (SDLC).

split custody   The concept of splitting knowledge of a specific object or task between two persons.

split tunnel   A characteristic of a VPN connection in which network traffic destined to the organization's internal network will traverse the VPN connection, whereas network traffic destined to the Internet will proceed directly to the Internet, bypassing the VPN connection.

sprint   A portion of an agile development project in which an individual or a team will accomplish a set of objectives.

spyware   A type of malware in which software performs one or more surveillance-type actions on a computer, reporting back to the spyware owner.

SSAE 16 (Statements on Standards for Attestation Engagements No. 16)   An external audit of a service provider. SSAE 16 has been superseded by SSAE 18.

SSAE 18 (Statements on Standards for Attestation Engagements No. 18)   An external audit of a service provider. An SSAE 18 audit is performed according to rules established by the American Institute of Certified Public Accountants (AICPA).

SSL decryption   The practice of decrypting encrypted network traffic to examine the contents of the traffic, as a part of an overall data loss prevention program designed to prevent security and privacy breaches.

standard   A statement that defines the technologies, protocols, suppliers, and methods used by an IT organization.

standard IT balanced scorecard   *See* IT balanced scorecard.

static DLP   The use of scanning tools to identify files containing PII or other sensitive information on file servers and other file stores. *See also* data loss prevention, dynamic DLP.

statistical sampling   A sampling technique whereby items are chosen at random; each item has a statistically equal probability of being chosen. *See also* sampling.

stop-or-go sampling   A sampling technique used to permit sampling to stop at the earliest possible time. This technique is used when the auditor believes that there is low risk or a low rate of exceptions in the population. *See also* sampling.

storage area network (SAN)   A stand-alone storage system that can be configured to contain several virtual volumes and can be connected to many servers through fiber-optic cables.

strategic planning   Activities used to develop and refine and organization's long-term plans and objectives.

stratified sampling   A sampling technique whereby a population is divided into classes or strata, based upon the value of one of the attributes. Samples are then selected from each class. *See also* sampling.

stream cipher   A type of encryption algorithm that operates on a continuous stream of data, such as a video or audio feed.

strong authentication   *See* multifactor authentication.

structured data   Data that resides in database management systems and in other forms, as part of information systems and business applications. *See also* unstructured data.

Structured Query Language (SQL)   A computer language used to query or update data in a relational database management system (RDBMS).

subject   In access controls, a person or a system. *See also* object. In information privacy, a natural person.

subject data request   *See* data subject request (DSR).

supervisory authority   An organization that has been delegated to enforce laws, perform investigations, and/or resolve disputes.

supervisory control and data acquisition (SCADA)   Systems in which industrial processes are monitored and controlled through specialized devices and communications networks. *See also* connected devices, industrial control system (ICS), Internet of Things (IoT).

switch   A device used to connect computers and other devices to a network. Unlike a hub, which sends all network packets to all stations on the network, a switch sends packets only to intended destination stations on the network.

symmetric encryption   A method for encryption and decryption in which both parties must possess a common encryption key.

system and organization controls (SOC) audit   An audit of a service provider's controls performed by a public accounting firm. *See also* SOC 1, SOC 2.

systems development life cycle (SDLC)   The life cycle process used to develop or acquire and maintain information systems. Also known as software development life cycle.

system hardening   *See* hardening.

system testing   The portion of software testing in which an entire system is tested.

T-1   A common carrier standard protocol for transporting voice and data. T-1 can support up to 24 separate voice channels of 64 kbit/sec each and is used primarily in North America.

T-3   A common carrier standard protocol for transporting voice and data. T-3 can support up to 672 separate voice channels of 64 kbit/sec each and is used primarily in North America.

table   A unit of storage in a relational database management system that can be thought of as a list of records. *See also* relational database management system (RDBMS).

tablet   A mobile device with a touchscreen interface. *See also* mobile device.

tabletop   An exercise, usually of privacy incident response, security incident response, and business continuity plans, that consists of a scripted simulation of an actual incident or event.

T-carrier   A class of multiplexed carrier network technologies developed to transport voice and data communications over long distances using copper cabling.

TCP/IP network model   The four-layer network model that incorporates encapsulation of messages. The TCP/IP suite of protocols is built on the TCP/IP network model.

technical control   A control that is implemented in IT systems and applications.

technical requirements   Formal statements that describe the required technical characteristics that a system must support.

technology standard   A standard that specifies the software and hardware technologies that are used by the IT organization.

termination   The process of discontinuing the employment of an employee or contractor.

test plan   The list of tests that are to be carried out during a unit test or system test. *See also* system testing, unit testing.

third-party cookie   A cookie whose domain is different from the web server that creates the cookie. Often used for advertising tracking. *See also* cookie, first-party cookie.

third-party risk management (TPRM)   A business process used to assess and treat risks related to third-party service providers.

threat   An event that, if realized, would bring harm to an asset.

threat hunting   The proactive search for intrusions, intruders, and indicators of compromise.

threat intelligence   A human- or machine-readable feed of threat-related information that can help organizations better protect themselves from emerging threats.

time bomb   *See* logic bomb.

time synchronization   A network-based service used to synchronize the time clocks on computers to a standard time source.

token   A small electronic device used in two-factor authentication. A token may display a number that the user types into a login field, or it may be plugged into a workstation to complete authentication. *See also* multifactor authentication.

training   The process of educating personnel; to impart information or provide an environment where personnel can practice a new skill.

transfer   The process of changing an employee's job title, department, and/or responsibilities within an organization.

Transport Layer Security (TLS)   An encryption protocol used to encrypt web pages requested with the HTTPS URL. Replacement for Secure Sockets Layer (SSL). *See also* Hypertext Transfer Protocol Secure (HTTPS), Secure Sockets Layer (SSL).

tunneling   The practice of encapsulating messages within another protocol.

twisted-pair cable   A type of network cabling that consists of a thick cable containing four pairs of insulated copper conductors, all surrounded by a protective jacket.

two-factor authentication   *See* multifactor authentication.

unit testing   The portion of software testing in which individual modules are tested.

Universal Disk Format (UDF)   An optical media file system considered a replacement for ISO/IEC 9660. *See also* file system, ISO/IEC 9660.

universal serial bus (USB)   An external bus technology used to connect a computer to a peripheral such as a mouse, keyboard, storage device, printer, scanner, camera, and network adaptor. However, the USB specification indeed contains full networking capabilities, facilitated through the use of a USB hub.

unstructured data   Data that resides on end user workstations and network file shares, usually as a result of the creation of reports and extracts. *See also* structured data.

user   A worker or customer who uses an information system.

user acceptance testing (UAT)   The portion of software testing in which end users test software programs for correct functional operation and usability.

user behavior analytics (UBA)   Tools and techniques that learn end user behavior and generate alerts when end user behavior exceeds norms.

user ID   An identifier that is created by a system manager and issued to a user for the purpose of identification or authentication.

variable sampling   A sampling technique used to study the characteristics of a population to determine the numeric total of a specific attribute from the entire population. *See also* sampling.

vendor standard   A standard that specifies which suppliers and vendors are used for various types of products and services.

version control   The techniques and tools used to manage different versions of source code files.

view A virtual table in a relational database management system (RDBMS) created by a stored query.

virtual desktop infrastructure (VDI) A technology in which user workstations use operating systems that are stored and run on central servers.

virtual machine A software implementation of a computer, usually an operating system or other program running within a hypervisor. *See also* guest, hypervisor.

virtual private cloud A portion of a public cloud environment that is logically separated between customers.

virtual private network (VPN) Any network encapsulation protocol that utilizes authentication and encryption; used primarily for protecting remote access traffic and for protecting traffic between two networks. *See also* encapsulation, tunneling.

virtual server An active instantiation of a server operating system running on a system that is designed to house two or more such virtual servers. Each virtual server is logically partitioned from every other so that each runs as though it were on its own physically separate machine.

virus A type of malware in which fragments of code attach themselves to executable programs and are activated when the program they are attached to is run.

Voice over IP (VoIP) Several technologies that permit telephony that is transported over IP networks.

VPN client A software program used to establish a virtual private network connection. *See also* virtual private network (VPN).

vulnerability A weakness that may be present in a system that increases the probability of one or more threats occurring.

vulnerability management A formal business process used to identify and mitigate vulnerabilities in an IT environment.

vulnerability scanning The use of a tool that automatically identifies exploitable vulnerabilities on systems connected to a network.

waterfall model A software development life cycle process whereby activities are sequential and are executed one time in a software project. *See also* systems development life cycle (SDLC).

watering hole attack An attack on an organization that consists of the compromise of a web site that personnel in the organization are known to frequent. This compromise then injects malware into web site visitors' computers.

web beacon Use of a tiny or invisible image on a web page or e-mail message for the purposes of tracking individual views of the image, along with other properties such as the IP address of the device viewing the web page or image.

web content filter   A central program or device that monitors and, optionally, filters web communications. A web content filter is often used to control the sites (or categories of sites) that users are permitted to access from the workplace. Some web content filters can also protect an organization from malware.

web server   A server that runs specialized software that makes static and dynamic HTML pages available to users.

web services   A means for system-to-system communications using HTTP. *See also* application programming interface (API).

Web Services Description Language (WSDL)   An XML-based language used to describe web services. *See also* web services.

web tracking   The use of any of several technologies and techniques for tracking the use of web page views and web site viewing.

web-based application   An application design whereby the database and all business logic are stored on central servers and user workstations use only web browsers to access the application.

web-based application development   A software development effort in which the application's user interface is based on the HTTP (Hypertext Transport Protocol) and HTML (Hypertext Markup Language) standards.

Wi-Fi   The common name for a wireless LAN protocol.

works council   A body, similar to a labor union, that represents the rights of workers in an organization, and with whom an organization is required to negotiate on matters of the collection and use of workers' personal information.

XaaS   A term that collectively represents infrastructure as a service (IaaS), platform as a service (PaaS), software as a service (SaaS), and other as-a-service models.

zero trust (ZT)   An architecture model in which a portion of an environment is considered to be untrusted.

INDEX